T0360561

Reliability Modeling
with Applications

Reliability Modeling with Applications

Essays in Honor of
Professor Toshio Nakagawa
on His 70th Birthday

Editors

Syouji Nakamura
Kinjo Gakuin University, Japan

Cun Hua Qian
Nanjing University of Technology, China

Mingchih Chen
Fe Jen Catholic University, Taiwan

 **World Scientific**

NEW JERSEY · LONDON · SINGAPORE · BEIJING · SHANGHAI · HONG KONG · TAIPEI · CHENNAI

Published by

World Scientific Publishing Co. Pte. Ltd.
5 Toh Tuck Link, Singapore 596224
USA office: 27 Warren Street, Suite 401-402, Hackensack, NJ 07601
UK office: 57 Shelton Street, Covent Garden, London WC2H 9HE

British Library Cataloguing-in-Publication Data
A catalogue record for this book is available from the British Library.

RELIABILITY MODELING WITH APPLICATIONS
Essays in Honor of Professor Toshio Nakagawa on His 70th Birthday

ISBN 978-981-4571-93-7

Printed in Singapore

Preface

Professor Toshio Nakagawa has greatly contributed to the growth of a variety of Maintenance Policies in Reliability Theory during four decades. As the results, he has published over 200 papers in research journals, and summarizing them, has already published 4 books and will publish soon one more book from Springer series in Reliability Engineering, as shown in an impressive Publication List of Chapter 19. Furthermore, he has organized a research group named Nagoya Computer and Reliability Research (NCRR) with the prime objective of presenting and writing research papers studied by each member. The NCRR has continued for 25 years unexpectedly since 1989, and each member has presented actively many papers under his leadership at several international workshops organized mainly by Professor Shunji Osaki, Professor Hoang Pham, Professor Shigeru Yamada, Professor Tadashi Dohi and other distinguished researchers. In memory of 20th anniversary, some research results of the main members were published on the book form on Stochastic Reliability Modeling, Optimization and Applications from World Scientific Publishing edited with Professor Syouji Nakamura in 2010.

After graduating Nagoya Institute of Technology in 1967, he joined Department of Mathematics, Meijo University in Nagoya, got a Doctor of Engineering Degree from Kyoto University in 1977, and became a Professor of Department of Industrial Engineering, Aichi Institute of Technology in 1988. He retired his teaching job for 46 years in 2013 and is now a Guest Professor of Aichi Institute of Technology. However, he hopes to continue to study reliability theory from this time forth. Recently, recognizing his academic achievement, he received Service Award from International Society of Science and Applied Technologies in 2009, President Award from Aichi Institute of Technology in 2012, and Reliability Engineering Award from the IEEE Reliability Society Japan Chapter in 2012.

In celebration of his 70th birthday, we have made a plan of publishing the book titled on Reliability Modeling with Applications from World Scientific Publishing, and called out to the researchers familiar with him at international conferences and workshops. Fortunately, more than 30 excellent authors in the world have supported willingly to this plan, and the book can be published by their positive cooperation. The book is composed of 4 parts: Maintenance Policies with 5 papers, Reliability Analysis with 4 papers, Computer System with 5 papers and Reliability Applications with 4 papers. Some papers present new insights or results for future studies, give good overviews of current researches, and propose useful techniques for practical applications in reliability areas. Finally, Professor Nakagawa makes his publication list divided into 8 parts, and also, gives his future plan for further studies in Chapter 19. We believe that this book can serve as a good textbook and guidebook for students, engineers and researchers in reliability.

We would like to express our sincere appreciate to all authors to this book, and also, Dr. Xufeng Zhao for his kind help in writing and typing this book. Finally, we would like to thank for the support and Editor Chelsea Chin, World Scientific Publishing for providing the opportunity to publish this book.

Syouji Nakamura, Kinjo Gakuin University, Japan
Cun Hua Qian, Nanjing University of Technology, China
Mingchih Chen, Fe Jen Catholic University, Taiwan

Contents

Reliability Analysis 99

Reliability Applications   269

    Tetsushi Yuge and Shigeru Yanagi

    Yoshinobu Tamura and Shigeru Yamada

Studies on Reliability and Maintenance   347

    Toshio Nakagawa

PART 1
Maintenance Policies

Chapter 1

A Dynamic Programming Approach for Sequential Preventive Maintenance Policies with Two Failure Modes

Hiroyuki Okamura[1], Tadashi Dohi[1] and Shunji Osaki[2]

[1]*Department of Information Engineering, Graduate School of Engineering, Hiroshima University, 1-4-1 Kagamiyama, Higashi-Hiroshima 739-8527, Japan*
[2]*Faculty of Information Sciences and Engineering, Nanzan University, Seto 489-0863, Japan*

1 Introduction

Stochastic preventive maintenance problem consists of formalism (modeling) and algorithm (optimization) for real applications. Since the seminal contribution by Barlow and Hunter [Barlow and Hunter (1965)], a huge number of stochastic preventive maintenance problems have been discussed from both viewpoints of modeling and optimization. Sequential preventive maintenance policies to determine aperiodic maintenance schedules may be regarded as the most critical but complex ones because they are reduced to nonlinear optimization problems with multiple decision variables. First Nguyen and Murthy [Nguyen and Murthy (1981)] consider two aperiodic preventive maintenance models with/without minimal repairs, and extend the Barlow and Hunter model [Barlow and Hunter (1965)]. Nakagawa [Nakagawa (1986)] independently considers the similar model and extends it in the subsequent paper [Nakagawa (1988)]. Since the above papers by Nakagawa [Nakagawa (1986, 1988)], the stochastic preventive maintenance models with aperiodic maintenance schedules are called the *sequential preventive maintenance models*, and have attracted much attentions from many authors.

Lin *et al.* [Lin *et al.* (2000)] extend Nakagawa models [Nakagawa (1986, 1988)] in terms of age reduction effect and hazard rate adjustment. The same authors [Lin *et al.* (2001)] also take account of two failure modes and extend their models [Lin *et al.* (2000)]. El-Ferik and Ben-Daya [El-Ferik and Ben-Daya (2005)] also extend Nguyen and Murthy model [Nguyen and Murthy (1981)] with a different policy from Nakagawa [Nakagawa (1988)] and Lin *et al.* [Lin *et al.* (2001)]. Kim *et al.* [Kim *et al.* (2007)] consider a problem to determine both the optimal number of preventive maintenance and its associated time sequence. Sheu and Liou [Sheu and Liou (1995)] and Sheu and Chang [Sheu and Chang (2002)] formulate the different sequential preventive maintenance problems with the same line as Nguyen and Murthy [Nguyen and Murthy (1981)] and Nakagawa [Nakagawa (1986, 1988)]. Recently, Nakagawa and Mizutani [Nakagawa and Mizutani (2009)] give a modification of Nakagawa model [Nakagawa (1986, 1988)] with a finite planning time horizon. It is worth mentioning that these preventive maintenance policies are not impractical models. A good illustrative example of the sequential preventive maintenance policy is given by Jayabalan and Chaudhuri [Jayabalan and Chaudhuri (1992)], where they apply it to the maintenance problem of bus engines in a large transport network with more than 2500 buses.

In this way, considerable attentions have been paid to the sequential preventive maintenance models. However, the main concern devoted in the related work was just the modeling, but not the computation algorithm. More specifically, the underlying optimization problems for the sequential preventive maintenance policies can be reduced to nonlinear optimization problems with multiple decision variables. In almost all papers, it is shown that the optimal time sequence of preventive maintenance must satisfy the first-order condition of optimality, but no effective computation algorithms are not developed. It should be surprised to see that the above related papers to the sequential preventive maintenance policies give very small toy exercise and never solve a realistic level of problem with more than 100 time points. In this paper, we take an example model with two different failure modes corresponding to the minimal repair and replacement, and formulate this generalized problem by means of the dynamic programming (DP). The resulting algorithm gives an effective algorithm to compute the aperiodic optimal maintenance schedule which minimizes the relevant expected cost rate, and is independent of the kind of model. So, the proposed computation scheme provides a unified framework to determine the sequential preventive maintenance policies.

2 Sequential Imperfect PM

Nguyen and Murthy [Nguyen and Murthy (1981)] and Nakagawa [Nakagawa (1986, 1988)] propose the sequential preventive maintenance (PM) policies. Unlike the periodic PM policies with equidistant PM time period, the sequential PM policies allow the aperiodic PM time sequence at which the system should undergo preventive maintenance optimally.

Consider a system under minimal repairs and replacement. The system has two failure modes: Type I and Type II failures, where Type I failure is an error of any system component and can be fixed by a minimal repair, and Type II failure is a fatal error and is repaired by only corrective replacement. In this paper, we assume that two types of failure occur independently. More precisely, the system undergoes preventive maintenance at each time points, $t_1 < t_2 < \cdots < t_N$, where the k-th preventive maintenance is performed at t_k. From the assumption of imperfect preventive maintenance [Lin *et al.* (2001); Nakagawa (1988)], it is supposed that the failure rates for Type I and Type II failures are changed at each preventive maintenance point. Define the failure rate functions for Type I and Type II failures in the k-th period of preventive maintenance by

- Type I failure:

$$h_k^I(t|t_{k-1}), \quad 0 \le t < t_k - t_{k-1}, \tag{1}$$

- Type II failure:

$$h_k^{II}(t|t_{k-1}), \quad 0 \le t < t_k - t_{k-1}. \tag{2}$$

Suppose that the failure rate functions in the k-th period depend on the time point of the last preventive maintenance, t_{k-1}. For the sake of convenience, we set $t_0 = 0$.

Define the following cost parameters:

- c_1: the cost of a minimal repair,
- c_2: the cost of preventive maintenance,
- c_3: the cost of replacement,
- c_4: the cost of corrective maintenance.

Let $S_k(t_k|t_{k-1})$ and $T_k(t_k|t_{k-1})$ be the expected total cost incurred in the k-th period and the expected time length of the k-th period, provided that the $k-1$-st and k-th preventive maintenances are preformed at t_{k-1} and t_k, respectively. Then we have

$$
\begin{aligned}
S_k(t_k|t_{k-1}) = & c_1 \int_0^{t_k-t_{k-1}} h_k^I(t|t_{k-1}) R_k^{II}(t|t_{k-1}) dt \\
& + c_2 R_k^{II}(t_k - t_{k-1}|t_{k-1}) \\
& + c_4 \left(1 - R_k^{II}(t_k - t_{k-1}|t_{k-1})\right),
\end{aligned}
\tag{3}
$$

$$
k = 1, \ldots N-1,
$$

$$
\begin{aligned}
S_N(t_N|t_{N-1}) = & c_1 \int_0^{t_N-t_{N-1}} h_N^I(t|t_{N-1}) R_N^{II}(t|t_{N-1}) dt \\
& + c_3 R_N^{II}(t_N - t_{N-1}|t_{N-1}) \\
& + c_4 \left(1 - R_N^{II}(t_N - t_{N-1}|t_{N-1})\right),
\end{aligned}
\tag{4}
$$

$$
T_k(t_k|t_{k-1}) = \int_0^{t_k-t_{k-1}} R_k^{II}(t|t_{k-1}) dt,
\tag{5}
$$

$$
k = 1, \ldots N,
$$

where $R_k^{II}(t|t_{k-1})$ is the reliability function for Type II failure in the k-th period:

$$
R_k^{II}(t|t_{k-1}) = \exp\left(-\int_0^t h_k^{II}(s|t_{k-1}) ds\right), \quad 0 \le t < t_k - t_{k-1}.
\tag{6}
$$

Using $S_k(\cdot|\cdot)$ and $T_k(\cdot|\cdot)$, the expected cost rate under the sequential preventive maintenance policy $\boldsymbol{\pi}_N = \{t_1, \ldots, t_N\}$, provided that the number of total maintenances is N, is given by

$$
C(\boldsymbol{\pi}_N, N) = \frac{\sum_{k=1}^N \prod_{l=1}^{k-1} R_l^{II}(t_l - t_{l-1}|t_{l-1}) S_k(t_k|t_{k-1})}{\sum_{k=1}^N \prod_{l=1}^{k-1} R_l^{II}(t_l - t_{l-1}|t_{l-1}) T_k(t_k|t_{k-1})}.
\tag{7}
$$

The problem is to find the optimal N^* and $\boldsymbol{\pi}_N^*$ which minimize the expected cost rate.

The above formula comprehends the existing sequential PM models. For example, when we set $R_k^{II}(t|t_{k-1}) = 1$, the model can be reduced to the original sequential PM model in [Nakagawa (1986)]. Also, if the failure rate is defined by the base failure rate $h_0^I(t)$;

$$
h_k^I(t|t_{k-1}) = \beta_k h_0^I(t + \alpha_k t_{k-1}),
\tag{8}
$$

the model represents the hazard rate PM and age reduction PM models [Lin *et al.* (2000)]. In addition, when the failure rate $h_k^I(t|t_{k-1})$ is replaced with

$$h_k^{I+III}(t|t_{k-1}) = h_k^I(t|t_{k-1}) + h^{III}(t + t_{k-1}), \qquad (9)$$

the model corresponds to the sequential PM with unmaintainable failure mode (Type III failure) [Lin *et al.* (2001)].

3   DP Algorithm

Since the expected cost rate for each policy is given as a function of N and π_N, the optimization problem is reduced to a non-linear programming to obtain $\min_{N, \pi_N} C(\pi_N, N)$, given the number of total maintenances N. It is worth noting that there is no effective algorithm to find the optimal pair (N^*, π_N^*) simultaneously. Hence the total number of preventive maintenances must be carefully adjusted according to any heuristic manner. Instead, we focus on finding the optimal maintenance schedule π_N^* under a fixed N. In the case of a fixed N, the most popular method to find the optimal maintenance schedule might be Newton's method or its iterative variants. However, since Newton's method is a general-purpose non-linear optimization algorithm, it may not often work well to solve the minimization problem with many parameters. In our minimization problem, the decision variables $\pi_N = \{t_1, \ldots, t_N\}$ have the constraint $t_1 < \cdots < t_N$. For such sequential optimization problems, it is well known that the dynamic programming (DP) can be used effectively.

In this section, we develop a DP algorithm for finding the optimal maintenance schedule π_N^*. The idea behind our DP algorithm is the iterative algorithm based on the optimality equations which are typical functional equations. It is straightforward to give the optimality equations which the optimal maintenance schedule π_N^* must satisfy. Suppose that there exists the unique minimum expected cost rate ρ. From the principle of optimality, we obtain the following optimality equations for the minimization problem of the expected cost rate:

$$J_k = \min_{t_k} W_k(t_k | t_{k-1}^*, J_1, J_{k+1}, \rho), \quad k = 1, \ldots, N-1, \qquad (10)$$

$$J_N = \min_{t_N} W_N(t_N | t_{N-1}^*, J_1, \rho), \qquad (11)$$

where functions $W_k(t_k | t_{k-1}, J_1, J_{k+1}, \rho)$ and $W_N(t_N | t_{N-1}, J_1, \rho)$ are given by

$$W_k(t_k|t_{k-1}, J_1, J_{k+1}, \rho) = S_k(t_k|t_{k-1}) - \rho T_k(t_k|t_{k-1})$$
$$+ J_{k+1} R_k^{II}(t_k - t_{k-1}|t_{k-1}) + J_1 \left(1 - R_k^{II}(t_k - t_{k-1}|t_{k-1})\right), \qquad (12)$$
$$W_N(t_N|t_{N-1}, J_1, \rho) = S_N(t_N|t_{N-1}) - \rho T_N(t_N|t_{N-1}) + J_1. \qquad (13)$$

In the above equations, J_k, $k = 1, \ldots, N$, are called the *relative value functions*.

Equations (10) and (11) are necessary and sufficient conditions of the optimal maintenance schedule. That is, the problem can be reduced into finding the maintenance schedule which satisfies the optimality equations. In the long history of the DP research, there are a couple of algorithms to solve the optimality equations. In this paper, we apply the *policy iteration* scheme to derive the optimal maintenance schedule. Our algorithm is twofold: the policy improvement under given relative value functions and the computation of relative value functions under a maintenance schedule. These two steps are repeatedly executed until the maintenance schedule converges. In the policy improvement, we find a new maintenance schedule based on the following functions under given relative value functions $J_1, \ldots, J_N$:

$$W_k(t_k|t_{k-1}, J_1, J_{k+1}, \rho), \quad k = 1, \ldots, N-1 \qquad (14)$$

and

$$W_N(t_N|t_{N-1}, J_1, \rho). \qquad (15)$$

However, when $J_1, \ldots, J_N$ are constants, the above functions are not always convex with respect to decision variables t_k. Thus our policy improvement algorithm is based on the following composite functions for two successive periods, instead of $W_k(\cdot|\cdot)$:

$$\tilde{W}_k(t_k|t_{k-1}, t_{k+1}, J_1, J_{k+2}, \rho)$$
$$= W_k(t_k|t_{k-1}, J_1, W_{k+1}(t_{k+1}|t_k, J_1, J_{k+2}, \rho), \rho), \qquad (16)$$
$$t_{k-1} \le t_k \le t_{k+1}, \quad k = 1, \ldots, N-2,$$
$$\tilde{W}_{N-1}(t_{N-1}|t_{N-2}, t_N, J_1, \rho)$$
$$= W_{N-1}(t_{N-1}|t_{N-2}, J_1, W_N(t_N|t_{N-1}, J_1, \rho), \rho), \qquad (17)$$
$$t_{N-2} \le t_{N-1} \le t_N.$$

The above composite functions are convex in the respective ranges $t_{k-1} \le t_k \le t_{k+1}$, $k = 1, \ldots, N-1$. In addition, the function $W_N(t_N|t_{N-1}, J_1, \rho)$ is also convex in the range $t_{N-1} \le t_N < \infty$. Thus it is possible to find the improved maintenance schedule by performing the one-dimensional optimization for each period of preventive maintenance.

Under a given maintenance schedule $t_1, \ldots, t_N$, the computation step gives corresponding relative value functions and ρ by solving the following linear system:

$$Mx = b, \tag{18}$$

where

$$[M]_{i,j} = \begin{cases} -R_i^{II}(t_i - t_{i-1}|t_{i-1}) & \text{if } i = j \text{ and } j \neq N, \\ 1 & \text{if } i = j + 1, \\ T_i(t_i|t_{i-1}) & \text{if } j = N, \\ 0 & \text{otherwise}, \end{cases} \tag{19}$$

$$x = (J_2, \ldots, J_N, \rho)', \tag{20}$$

$$b = (S_1(t_1|t_0), \ldots, S_N(t_N|t_{N-1}))'. \tag{21}$$

In Eq. (19), $[\cdot]_{i,j}$ denotes the (i,j)-element of matrix, and the prime $(\prime)$ represents transpose of vector. The above linear system comes from the optimality equations (10) and (11) directly. Note that $J_1 = 0$, since we are here interested in the relative value function J_i and ρ. Finally, we derive the DP algorithm to derive the optimal maintenance schedule as follows.

- **Step 1:** Give initial values

$$k := 0,$$

$$t_0 := 0,$$

$$\pi_N^{(0)} := \{t_1^{(0)}, \ldots, t_N^{(0)}\}.$$

- **Step 2:** Compute $J_1^{(k)}, \ldots, J_N^{(k)}, \rho^{(k)}$ for the linear system (18) under the maintenance schedule $\pi_N^{(k)}$.

- **Step 3:** Solve the following optimization problems:

$$t_i^{(k+1)} := \underset{t_{i-1}^{(k)} \leq t \leq t_{i+1}^{(k)}}{\operatorname{argmax}} \tilde{W}_i(t|t_{i-1}^{(k)}, t_{i+1}^{(k)}, J_1^{(k)}, J_{i+2}^{(k)}, \rho^{(k)}),$$

$$i = 1, \ldots, N - 2,$$

$$t_{N-1}^{(k+1)} := \underset{t_{N-2}^{(k)} \leq t \leq t_N^{(k)}}{\operatorname{argmax}} \tilde{W}_{N-1}(t|t_{N-2}^{(k)}, t_N^{(k)}, J_1^{(k)}, \rho^{(k)})$$

$$t_N^{(k+1)} := \underset{t_{N-1}^{(k)} \leq t < \infty}{\operatorname{argmax}} W_N(t|t_{N-1}^{(k)}, J_1^{(k)}, \rho^{(k)}).$$

- **Step 4:** For all $i = 1, \ldots, N$, if $|t_i^{(k+1)} - t_i^{(k)}| < \delta$, stop the algorithm, where δ is an error tolerance. Otherwise, let $k := k+1$ and go to Step 2.

In Step 3, an arbitrary optimization technique can be applied. Since the composite functions are convex functions having a unique solution in the ranges $[t_{i-1}, t_{i+1})$, $i = 1, \ldots, N - 1$, it is not so difficult to calculate the optimal preventive maintenance time. In fact, the golden section method is effective to find the solution.

4   Numerical Examples

We first consider the case where the failure rate for Type I failure is given by the following Weibull-type failure rate:

$$h_k^I(t|t_{k-1}) = \alpha_k \beta_k t^{\beta_k - 1}, \tag{22}$$

where $1/\alpha_k = 100 \times 0.81^{k-1}$ and $\beta_k = 2.0$. The other parameters are $c_1 = 1.0$, $c_2 = 3.0$, $c_3 = 100.0$ and $c_4 = 1000.0$. These parameters are cited from [Nakagawa (1986)]. We investigate the sensitivity of the failure rate for Type II failure on the optimal PM sequence and the minimum cost. Table 1 presents the optimal PM timing when the total number of PMs is fixed under $h_k^{II}(t|t_{k-1}) = 0$, i.e., $R_k^{II}(t|t_{k-1}) = 1$. As mentioned before, our model is reduced to Nakagawa's model when $R_k^{II}(t|t_{k-1}) = 1$. From the table, the optimal number of PMs is $N = 11$, which minimizes the total cost rate. In [Nakagawa (1986)], the optimal PM sequence in the case of $N = 11$ was presented, and it is exactly same as our result. That is, our DP algorithm can solve the sequential PM policy stably. Figure 1 illustrates the optimal PM sequences for $N = 2$ through 16. In the figure, x-axis represents PM timing and the optimal PM sequence was plotted as points on the horizontal line. From the figure, it can be founded that the time interval of PMs becomes monotonically decreasing sequence for all the cases. Moreover, when we focus on the first PM timing, it decreases as the number of total PMs increases from $N = 2$ to $N = 11$, but it decreases from $N = 11$ to $N = 16$. In this case, the optimal number of total PMs is $N = 11$. The tendency of PM sequences are changed at the optimal number of PMs.

Next we investigate the case where the type II failure rate is given by a constant;

$$h_k^{II}(t|t_{k-1}) = \lambda. \tag{23}$$

Table 1 Optimal sequential PM timing for $N = 8, 9, 10, 11, 12, 13, 14$ without type II failure.

N	8	9	10	11	12	13	14
t_1	53.1	52.6	52.4	52.3	52.4	52.5	52.8
t_2	96.1	95.3	94.8	94.7	94.8	95.0	95.5
t_3	131.0	129.8	129.2	129.1	129.2	129.5	130.1
t_4	159.2	157.8	157.0	156.9	157.0	157.3	158.2
t_5	182.1	180.4	179.6	179.4	179.6	179.9	180.9
t_6	200.6	198.8	197.9	197.6	197.8	198.3	199.3
t_7	215.6	213.7	212.7	212.4	212.6	213.1	214.2
t_8	227.8	225.7	224.6	224.4	224.6	225.1	226.2
t_9		235.5	234.4	234.1	234.3	234.8	236.0
t_{10}			242.2	241.9	242.2	242.7	243.9
t_{11}				248.3	248.6	249.1	250.3
t_{12}					253.7	254.3	255.5
t_{13}						258.5	259.8
t_{14}							263.2
$C(\pi_N, N)$	1.062	1.053	1.048	1.047	1.048	1.051	1.056

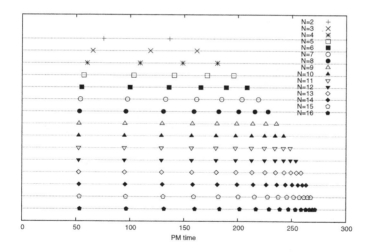

Fig. 1 Optimal PM sequences for fixed N without Type II failure.

Figures 2 through 7 depict the optimal PM sequences for respective type II failure rates $\lambda = 1/100, 1/200, 1/300, 1/400, 1/500, 1/1000$ with a fixed N. Also Table 2 presents the minimum cost rates for each optimal PM sequence with a fixed N. The last column indicates the minimum cost rates in the case where the Type II failure does not occur, i.e., $h_k^{II}(t|t_{k-1}) = 0$. The

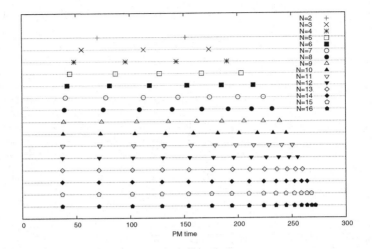

Fig. 2   Optimal PM sequences for fixed N ($\lambda = 1/100$).

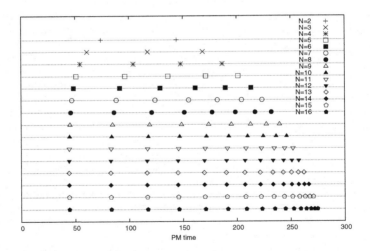

Fig. 3   Optimal PM sequences for fixed N ($\lambda = 1/200$).

asterisk means the optimal number of PMs minimizing the cost rates. For every case, we apply the DP algorithm to obtain the optimal PM sequences. Even if the number of PMs is large, e.g., $N = 16$, we can get the optimal PM sequences stably.

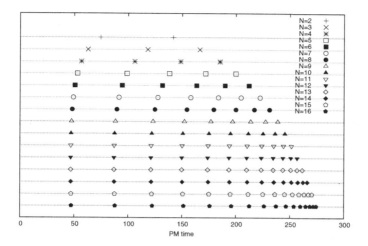

Fig. 4   Optimal PM sequences for fixed N ($\lambda = 1/300$).

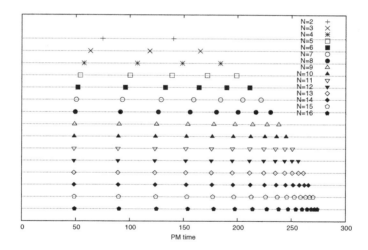

Fig. 5   Optimal PM sequences for fixed N ($\lambda = 1/400$).

From the figures, even when the Type II failure occurs, the optimal PM sequence has the similar tendency as the case where the Type II failure does not occur. Also, the PM timing tends to be earlier in the case where Type II failure rate is high; $\lambda = 1/100$. Moreover, it can be seen that the

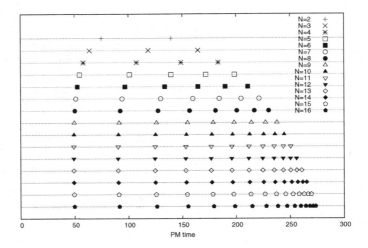

Fig. 6   Optimal PM sequences for fixed N ($\lambda = 1/500$).

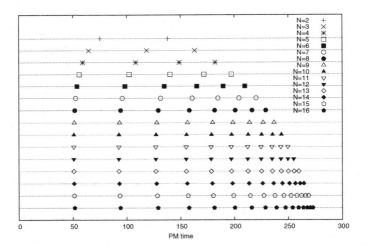

Fig. 7   Optimal PM sequences for fixed N ($\lambda = 1/1000$).

minimum cost rate strongly depend on Type II failure rate from the table. In addition, in the case of the high Type II failure rate, the optimal number of PMs is bigger than the other cases.

Table 2   Minimum cost rates.

N	1/100	1/200	1/300	1/400	1/500	1/1000	w/o Type II
2	11.009	6.236	4.656	3.867	3.394	2.451	1.509
3	10.851	6.055	4.468	3.676	3.202	2.256	1.311
4	10.766	5.957	4.367	3.574	3.100	2.152	1.206
5	10.716	5.899	4.306	3.513	3.038	2.089	1.143
6	10.683	5.861	4.268	3.474	2.998	2.050	1.104
7	10.662	5.836	4.242	3.448	2.973	2.024	1.078
8	10.647	5.819	4.225	3.431	2.956	2.008	1.062
9	10.637	5.808	4.214	3.421	2.946	1.998	1.053
10	10.630	5.801	4.208	3.414	2.939	1.992	1.048
11	10.625	5.797	4.204	3.411	2.937	1.990	1.047*
12	10.622	5.795	4.203*	3.410*	2.936*	1.990*	1.048
13	10.620	5.794*	4.203	3.411	2.937	1.992	1.051
14	10.620	5.795	4.205	3.413	2.940	1.996	1.056
15	10.619*	5.796	4.207	3.417	2.944	2.001	1.062
16	10.619	5.798	4.210	3.421	2.948	2.006	1.068

5 Concluding Remarks

In this paper, we have considered a sequential preventive maintenance model with minimal repair and replacement. The model under consideration is an extension of Nakagawa [Nakagawa (1986, 1988)] by introducing two different failure modes. To derive the optimal preventive maintenance schedule which minimizes the expected cost rate, we have developed a DP-based algorithm which is twofold; policy improvement and computation of relative value functions. In particular, we have applied composite functions for two successive periods of preventive maintenance to derive the improved maintenance schedule.

Acknowledgment

The authors are grateful to Prof. Toshio Nakagawa who stimulated their research interests in preventive maintenance modeling through his many papers. This research was partially supported by the Ministry of Education, Science, Sports and Culture, Grant-in-Aid for Scientific Research (C), Grant No. 23500047 (2011–2013) and Grant No. 23510171 (2011–2013).

References

Barlow, R. E. and Hunter, L. C. (1965). Optimum preventive maintenance policies, *Operations Research* **8**, pp. 90–100.

El-Ferik, S. and Ben-Daya, M. (2005). Age-based hybrid model for imperfect preventive maintenance, *IIE Transactions* **38**, pp. 365–375.

Jayabalan, V. and Chaudhuri, D. (1992). Sequential imperfect preventive maintenance policies: a case study, *Microelectronics and Reliability* **32**, pp. 1223–1229.

Kim, H. S., Kwon, Y. S. and Park, D. H. (2007). Adaptive sequential preventive maintenance policy and Bayesian consideration, *Communications in Statistics – Theory and Methods* **36**, pp. 1251–1269.

Lin, D., Zuo, M. J. and Yam, R. C. M. (2000). General sequential imperfect preventive maintenance models, *International Journal of Reliability, Quality and Safety Engineering* **7**, pp. 253–266.

Lin, D., Zuo, M. J. and Yam, R. C. M. (2001). Sequential imperfect preventive maintenance models with two categories of failure modes, *Naval Research Logistics* **48**, pp. 172–183.

Nakagawa, T. (1986). Periodic and sequential preventive maintenance policies, *Journal of Applied Probability* **23**, 536–542.

Nakagawa, T. (1988). Sequential imperfect preventive maintenance policies, *IEEE Transactions on Reliability* **37**, 295–298.

Nakagawa, T. and Mizutani, S. (2009). A summary of maintenance policies for a finite interval, *Reliability Engineering and System Safety* **94**, pp. 89–96.

Nguyen, D. G. and Murthy, D. N. P. (1981). Optimal preventive maintenance policies for repairable systems, *Operations Research* **29**, pp. 1181–1194.

Sheu, S.-H. and Chang, T.-H. (2002). Generalized sequential preventive maintenance policy of a system subject to shocks, *International Journal of Systems Science* **33**, pp. 267–276.

Sheu, S.-H. and Liou, C.-T. (1995). A generalized sequential preventive maintenance policy for repairable systems with general random minimal repair costs, *International Journal of Systems Science* **26**, pp. 681–690.

Chapter 2

Selective Maintenance for Complex Systems Considering Imperfect Maintenance Efficiency

Mayank Pandey[1], Yu Liu[1,2], Ming J. Zuo[1,2]

[1]*Department of Mechanical Engineering, University of Alberta, Edmonton, Alberta, T6G 2G8, Canada*
[2]*School of Mechanical, Electronic, and Industrial Engineering, University of Electronic Science and Technology of China, Chengdu, Sichuan 611731, China*

1 Introduction

All equipment and systems deteriorate with age and usage. Maintenance is required to be performed on a repairable system to improve the overall system reliability and availability. If timely maintenance is not performed, the system may fail, leading to huge costs associated with the failure and corrective actions afterward. Corrective maintenance is performed after the failure realization and aims to make the system perform the desired functions after maintenance. Corrective maintenance is expensive, hence it is important that a failure is prevented. Preventive maintenance (PM) is performed at pre-specified intervals or as per some criteria such that the system reliability is increased and failure is avoided. Undesired failure and corrective maintenance costs are saved using PM. The maintenance can be categorized as follows:

1) Perfect maintenance/repair or replacement: Replacement of a system whether it is working or failed, restores it to as good as new (AGAN) condition. Upon perfect maintenance, the failure intensity function of a component is the same as a new component.

2) Minimal repair: If failure intensity of a component after repair is the same as it had when it failed, the repair action is called minimal repair. The system operating state is as bad as old (ABAO) after minimal repair. Changing the headlight of a truck could be an example of minimal repair because it does not change the overall failure intensity of the truck.

3) Imperfect maintenance/repair: Traditionally, it is assumed that maintenance brings a system back to as good as new (AGAN) or as bad as old (ABAO) condition. However, maintenance can restore a system to a state somewhere between AGAN and ABAO conditions. Such maintenance/repair actions are called imperfect maintenance/repair. Replacing only a few parts of a system can be one example.

In practice, PM lengthens the useful lifetime of a system by reducing the occurrence of failure. One of the key characteristics of a maintenance model is the effect of different kinds of maintenance on the effective age and/or hazard rate of the system. Modeling the effectiveness of PM is widely studied and reviewed in the literatures [Barlow and Hunter (1960); Pham and Wang (1996); Dekker *et al.* (1997); Dekker and Scarf (1998); Wang (2002); Doyen and Gaudoin (2004); Desai and Mital (2006)]. With the introduction of imperfect maintenance along with the traditional replacement and minimal repair models, different approaches are proposed for imperfect maintenance modeling. One common approach is to assume that PM is equivalent to minimal repair with probability p and replacement with probability $1-p$ [Chan and Downs (1978); Nakagawa (1979); Murthy and Nguyen (1981); Brown and Proschan (1983); Sheu and Liou (1995)]. Another popular approach is to specify the effect of maintenance on the effective age and/or hazard rate of the system [Lie and Chun (1986); Nakagawa (1986, 1988); Lin *et al.* (2001); Pandey *et al.* (2013a)]. A recent comprehensive review of imperfect maintenance models is documented in [Pham and Wang (1996); Wu and Zuo (2010)]. It is noteworthy that for any specific engineered system, model validation and selection should be conducted to choose the best imperfect maintenance model among all possible candidates [Liu *et al.* (2012)]. In the ensuing paragraphs, we briefly review the imperfect maintenance models that will be used in this chapter.

[Lie and Chun (1986)] and [Nakagawa (1986)] introduced adjustment/improvements in the hazard rate and effective age after PM. [Nakagawa (1988)] used adjustment/improvement factors for the hazard rate and effective age to solve sequential PM problems and proposed that:

(i) the hazard rate in the next PM interval becomes '$ah(x)$ where $h(x)$ is the hazard rate in the previous PM interval. The adjustment factor is '$a \geq 1$' and $x \geq 0$ represents time elapsed from the previous PM time. (ii) if the effective age of a component is t right before the PM, then it reduces to 'bt' right after the PM, where $0 \leq b \leq 1$ is the improvement factor in effective age. The first model is called the hazard adjustment model, while the second model is called the age reduction model. The hazard rate adjustment model assumes that the hazard rate right after PM reduces to zero but increases more quickly in the next PM interval as compared to the previous PM interval. The age reduction model assumes that PM reduces the effective age and right after PM it may be greater than zero. The hazard rate is a function of effective age before and after maintenance.

[Lin *et al.* (2001)] proposed that in more general cases, PM cannot only reduce the effective age but may also increase the hazard rate. They combined the age reduction and hazard rate adjustment model and called it a hybrid imperfect maintenance model. If the hazard rate function for time $t \in (0, t_1)$ is $g(t)$ and PM at maintenance break, (t_1, t_2) changes the hazard rate to $h(t)$ for $t \in (t_2, t_3)$ (see Fig. 1). If the effective age of the system before and after maintenance are B and A, respectively, then the combined hybrid model, which includes the effect of the hazard adjustment and the age reduction, can be written as:

$$h(t_2 + x) = ag(bB + x) \qquad (1)$$

where, $a \geq 1$ and $0 \leq b \leq 1$, and $x \in \{0, t_3 - t_2\}$. When $a=1$, the above model is the same as the age reduction model and for $b=0$, it is the same as the hazard adjustment model. Hence, the hybrid imperfect maintenance model can be used to characterize the effect of PM in a general manner.

PM consumes time, human resources, and has associated costs. However, inadequate maintenance schedules or nonessential services may waste limited maintenance resources. The decision to perform PM becomes more complicated when a system is composed of several components. Optimal allocation of maintenance resources and selection of a subset of maintenance activities that fulfill the system requirements after maintenance are very important. The number of PM options available for a system depends on the PM options available for each component within the system. The maintenance decision of any of the components within a system will affect the system performance. It is a major challenge to consider the effect of the stochastic processes for components along with selecting maintenance options for each component in the multi-component system. However, it

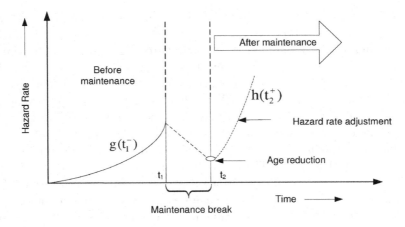

Fig. 1   Hybrid imperfect PM model

may not be possible to perform all desirable maintenance activities during a maintenance break due to limited maintenance resources like time, budget, and repairman availability. In this case, a subset of maintenance actions is chosen to make sure that the desired operation period after the maintenance break is successfully completed. This maintenance policy is called 'selective maintenance'.

For many types of equipment or systems, breaks between successive operating periods (missions) offer the best opportunity for maintenance. Some examples may include manufacturing equipment, military vehicles, power generation units, etc. Manufacturing equipment may work during the week and be maintained during weekends, and similarly, power generation units may work for the whole week and maintenance can be performed early Sunday morning. Aircraft may be maintained between flights and military equipment may be maintained between operations. In all of the above cases, maintenance is required to be performed between missions such that the system performs satisfactorily during the next mission. After inspection, there could be some possible maintenance actions to be performed during maintenance breaks, e.g., minimal repair, different preventive maintenance actions, or system replacement. After maintenance, the system should work fulfilling the desired objective during the next mission until the next scheduled maintenance break. Since each of the available maintenance options consumes some maintenance resources, e.g. time and cost, the optimal allocation of resources is required.

Our primary focus is on explaining the different selective maintenance models considering imperfect maintenance. This chapter focuses on modeling rather than statistical inference. We have tried our best to review the selective maintenance for a system fairly completely; however, some papers, which are not included, were either considered not related directly or were overlooked unknowingly. We have to discuss all relevant papers with a focus on the most recent works in the domain of selective maintenance that discusses imperfect repair/maintenance. We have discussed the results related to the effect of imperfect maintenance/repair only. Other results irrelevant to imperfect maintenance/repair are not included.

This chapter is divided as follows: Sec. 2 briefly explains previous literature related to selective maintenance for systems with either working or failed states, i.e. binary states. More explanation is provided for the most recent works on selective maintenance for binary systems discussing imperfect maintenance. For some systems, more than two performance states are possible, these systems are called multistate systems (MSS). In Sec. 3, selective maintenance strategies for the MSS are explained. Sec. 4 summarizes the conclusion.

2 Selective Maintenance for Binary Systems

Traditionally a system is assumed to be in two states; working or failed. Such systems are called binary systems. For binary systems, a selective maintenance problem was introduced by [Rice *et al.* (1998)], where a system with series-parallel configuration, constant component failure rates, i.e. exponential distribution and only one type of maintenance action (replacement of failed component) was considered. They only considered time as a resource constraint. They studied a multi-component series-parallel system as an example and maximized system reliability during the next mission such that maintenance is performed within the available time. A series-parallel system is shown in Fig. 2. In Fig. 2, there are S subsystems ($j = 1, 2, ..., S$) connected in a series and each of the jth subsystem has n_j components connected in parallel. [Rice *et al.* (1998)] assumed that all components within a subsystem were identical. They also proposed a heuristic that was good for identical components within a subsystem.

[Cassady *et al.* (2001a)] extended the model presented by [Rice *et al.* (1998)] and included cost as one more resource constraint in addition to time. By selecting reliability, cost, or time as the objective and the other two as constraints, they developed three different selective main-

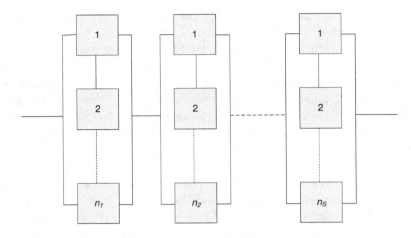

Fig. 2   A series-parallel system

tenance models. In the first model, they maximized reliability under the cost and time constraint, in the second model maintenance cost was minimized with mission reliability and maintenance time as constraints, and in the third model they minimized maintenance time under mission reliability and maintenance cost constraints. The first model is useful where system reliability is important. They mentioned that for safety-critical applications, e.g. maintaining aircraft or space shuttles, it was likely that the first model would be used due to the severity of a system failure. In a profit oriented application, like manufacturing systems, the second model is more likely to be selected. In service oriented applications, e.g. maintaining computer servers, maintenance time may correspond to the loss of service. In this case, the third model could be used.

Further, [Cassady *et al.* (2001b)] assumed that the components' lifetimes follow the Weibull distribution. Hence, rather than constant failure rates used in previous studies, an increasing failure rate was considered in [Cassady *et al.* (2001b)]. They also increased the possible maintenance actions for a component. Earlier, only the replacement of failed components was a possible maintenance option; however, [Cassady *et al.* (2001b)] included minimal repair and replacement of working components as additional maintenance options. They limited their study to only time as a resource constraint. Later, [Schneider and Cassady (2004)] considered multiple series-parallel systems simultaneously and called it a fleet. They used the model of [Rice *et al.* (1998)] and solved selective maintenance problems for a fleet (consisting of multiple systems together) performance.

There are other works that address the problem of improving the computational efficiency of the selective maintenance optimization. [Rajagopalan and Cassady (2006)] proposed four types of enumeration methods to solve selective maintenance problems for a series-parallel system. Their goal was to reduce CPU time to solve selective maintenance problems. However, they assumed that all components in a subsystem are similar and only the replacement of failed components was possible. For non-identical components, non-identical maintenance time from one component to another in a subsystem, or increases in the number of maintenance options, their heuristic becomes inefficient. Thus, it was found that the enumeration method had limited application. [Lust *et al.* (2009)] also mentioned that for a system with a large number of components, the selective maintenance problem becomes combinatorial in nature and the enumeration method was no longer useful. They proposed a heuristic to generate an initial solution and used it as input to the branch and bound procedure and the Tabu search algorithm (an evolutionary algorithm). They found that the Tabu search provided optimal or very close to optimal solutions promptly as compared to the branch and bound method. It was the first time in [Lust *et al.* (2009)] that an evolutionary approach was used to solve the selective maintenance problem. Thus, an evolutionary algorithm was found to be useful in solving a selective maintenance optimization problem. Later [Zhu *et al.* (2011)] also used the Tabu search algorithm with heuristic to solve the selective maintenance problem for a series-parallel manufacturing system.

[Iyoob *et al.* (2006)] focused on resource allocation for subsequent missions under selective maintenance. They also combined redundancy allocation problems with the selective maintenance decision making. [Maillart *et al.* (2009)] considered selective maintenance for a multi-mission problem with maintenance possible only at the beginning of the first mission, which makes its application limited. [Pandey *et al.* (2012)] used an age based imperfect repair model similar to [Nakagawa (1988)], but additionally, they defined an age reduction factor based on the cost and effective age. They mentioned that when a component is relatively newer, age reduction can be achieved using some maintenance cost; however, when the component becomes old, more cost is required to achieve the same amount of age reduction. They further extended their work in [Pandey *et al.* (2013a)] and proposed a hybrid imperfect maintenance model where the improvement factors are based on the effective age and the maintenance cost. In the following sections we explain the model used in [Pandey *et al.* (2013a)]. They used the following assumptions:

1) The system consists of multiple, repairable components.
2) The components, as well as the system, is in a binary state, i.e., it is either working or failed.
3) After replacement, the component is 'as good as new' and if minimal repair is performed, it is 'as bad as old'. Maintenance is also possible such that the component health may lie between as good as new and as bad as old, i.e., maintenance can be modeled by imperfect repair.
4) Limited resources (budget, repairman, and time) are available and the amount of resources required for maintenance activities are known and fixed.

Whenever the system comes in for maintenance after a mission, a decision is to be made for each component regarding maintenance. The component can be in either working or failed state after a mission. If a component i in subsystem j is in working state before maintenance, it is defined as $Y_{i,j} = 1$, otherwise $Y_{i,j} = 0$. Similarly, after maintenance a working state of component (i, j) is defined as $X_{i,j} = 1$, otherwise $X_{i,j} = 0$. Depending on the available resources and the component's age, a maintenance decision is determined for a system. In general, the maintenance quality improves with the amount of the budget invested during maintenance. As given in [Lie and Chun (1986)], the maintenance cost used and the age of the component are two important factors for determining the age reduction factor (b) for a component. Based on the definition given in [Lie and Chun (1986)], [Pandey *et al.* (2013a)] formulated an age reduction factor as:

$$
b\left(B_{i,j}, l_{i,j}\right) = \begin{cases} 1 - \left(\frac{\left(C_{i,j}(l_{i,j}) - C_{i,j}^{MR}\right)}{C_{i,j}^{R}}\right)^{m(B_{i,j})} & , \text{for } Y_{i,j} = 0, \quad 2 \leq l_{i,j} < N_{i,j} \\ 1 - \left(\frac{C_{i,j}(l_{i,j})}{C_{i,j}^{R}}\right)^{m(B_{i,j})} & , otherwise \end{cases}
$$

(2)

where $1 \leq l_{i,j} < N_{i,j}$ are possible maintenance actions for a component (i, j). Here, $l_{i,j} = N_{i,j}$ shows component replacement. When $Y_{i,j} = 0$, $l_{i,j} = 2$ shows minimal repair and $3 \leq l_{i,j} < N_{i,j}$ shows imperfect repair options. When $Y_{i,j} = 1$, $2 \leq l_{i,j} < N_{i,j}$ defines imperfect maintenance options for component (i, j). For a component (i, j), $l_{i,j} = 1$ defines that no maintenance is performed. In equation (2), $C_{i,j}(l_{i,j})$ is the cost related to maintenance option $l_{i,j}$, $C_{i,j}^{MR}$ is the minimal repair cost and $C_{i,j}^{R}$ is the replacement cost of the component. In the equation (2), $B_{i,j}$ is the effective age of (i,j) before maintenance and $m(B_{i,j})$ is called the characteristic constant. [Pandey *et al.* (2013a)] used $m(B_{i,j})$ to incorporate the effect of effective age of the component. They defined $m(B_{i,j})$ as:

$$m(B_{i,j}) = \frac{B_{i,j}}{\mathrm{MRL}} = \frac{B_{i,j}}{\left(\frac{\int_{B_{i,j}}^{\infty} R(x)dx}{R(B_{i,j})}\right)} = \frac{B_{i,j} \times R(B_{i,j})}{\int_{B_{i,j}}^{\infty} R(x)\,dx} \qquad (3)$$

If a component's effective age is less than its mean residual life (MRL), it is assumed to be relatively younger. However, when the mean residual life of the component is less than the effective age of the component, it is said that the component is relatively old. The younger a component is, the better it responds to the maintenance budget invested. In equation (2) use of $m(B_{i,j})$ defines how the relative age of the component will affect the amount of budget used.

2.1 *Hazard Adjustment Factor*

After maintenance, the hazard rate may also change with the effective age of a component. [Nakagawa (1988); Lin *et al.* (2001); Liao *et al.* (2010)] assumed that the hazard rate changes by a constant factor for a component during each maintenance. In addition to the time of particular maintenance break, hazard rate after maintenance may also be affected by the budget used for maintenance. If the budget is small, low improvement in component health is expected. Its hazard rate increment after maintenance will be higher compared to the case when a large budget is involved in the component's maintenance. Based on the above argument, [Pandey *et al.* (2013a)] provided the following formulation for the hazard adjustment factor,

$$a(B_{i,j}, l_{i,j}) = \begin{cases} \cfrac{p}{\left((p-1)+\left(\frac{(C_{i,j}(l_{i,j})-C_{i,j}^{MR}}{C_{i,j}^{R}}\right)^{\frac{1}{m(B_{i,j})}}\right)} \\ \quad for \quad Y_{i,j}=0, \quad 3 \le l_{i,j} < N_{i,j}, \\ 1, \\ \quad for \quad l_{i,j}=1, \quad and\ for \quad Y_{i,j}=0, \quad Y_{i,j}=2, \\ \cfrac{p}{\left((p-1)+\left(\frac{(C_{i,j}(l_{i,j})}{C_{i,j}^{R}}\right)^{\frac{1}{m(B_{i,j})}}\right)} \\ \quad otherwise, \end{cases}$$

$$(4)$$

where p determines the maximum allowable hazard increment for a component, i.e., it defines the upper limit of the hazard adjustment factor that a

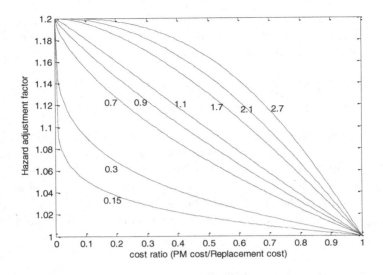

Fig. 3   Hazard adjustment factor versus cost ratio for different values of m (for $p = 6$)

component can achieve after a maintenance break. The smaller is the value of p, the larger is the maximum allowable hazard adjustment and vice-versa. The value of p is determined from the component's history. Variations of the hazard adjustment factor for different values of $m(B_{i,j})$ are shown in Fig. 3. This figure shows that for a smaller cost ratio, i.e., when a smaller budget is used for maintenance of a component, the hazard rate will increase faster after maintenance (higher value of hazard adjustment factor) and vice-versa. It also shows that for a fixed value hazard adjustment factor, the amount of budget needed increases as the component ages (i.e. $m(B_{i,j})$ increases). Once the age reduction and hazard adjustment factors are determined for maintenance decision for each component in the system, it is required to determine the system reliability.

2.2   *System Reliability Evaluation*

Let us assume that $p_{i,j,l_{i,j}}$ is the probability that a component (i, j) after undergoing maintenance option $l_{i,j}$, finishes the next mission successfully. This probability shows the reliability of the component for a given mission duration. If the length of the next mission is L, $t_{i,j}$ is the beginning of the next mission, the component hazard rate during the next mission

$\left(h_{i,j,l_{i,j}}\left(t_{i,j}+x\right)\right)$ can be obtained from the equation (1):

$$h_{i,j,l_{i,j}}\left(t_{i,j}+x\right) = a\left(B_{i,j}.l_{i,j}\right) \times g\left(b\left(B_{i,j},l_{i,j}\right) \times B_{i,j}+x\right), 0 \leq x \leq L \quad (5)$$

The probability of this component successfully completing the next mission is:

$$p_{i,j,l_{i,j}} = \exp\left(-\int_0^L h_{i,j,l_{i,j}}\left(t_{i,j}+x\right)\,dx\right) \quad (6)$$

Thus, the reliability of the component (i,j) can be defined as:

$$R_{i,j,l_{i,j}} = p_{i,j,l_{i,j}} \times X_{i,j} \quad (7)$$

Hence, system reliability for the next mission can be given as:

$$R\left(l\right) = \prod_{i=1}^{S} R_i\left(l\right) = \prod_{i=1}^{S}\left(1 - \prod_{j=1}^{n_i}\left(1 - R_{i,j,l_{i,j}}\right)\right) \quad (8)$$

where $l = [l_{1,1},...,l_{i,j},...,l_{s,n_s}]$ is a vector comprising the maintenance decision variable $l_{i,j}$ for all components in the system.

2.3 *Selective Maintenance Modeling*

If a system comes for maintenance after a mission with a known state $Y_{i,j}$, effective age $(B_{i,j})$, and the lifetime distribution parameters for all components, only a subset of maintenance actions can be performed due to limited resources. If the budget constraint during the maintenance break is given by C_0 and the available maintenance duration is limited to T_0, the selective maintenance optimization problem to maximize the probability of successfully completing the next mission is developed as:

Objective:

$$\max \quad R\left(l\right) = \prod_{i=1}^{S}\left(1 - \prod_{j=1}^{n_i}\left(1 - R_{i,j,l_{i,j}}\right)\right) \quad (9)$$

Subject to:

$$C \leq C_0 \quad (10)$$

$$T \leq T_0 \quad (11)$$

$$V_{i,j} = \begin{cases} 1, & \text{if } l_{i,j} > 1 \\ 0, & \text{otherwise} \end{cases} \tag{12}$$

$$X_{i,j} = \begin{cases} V_{i,j}, & \text{if } Y_{i,j} = 0 \\ Y_{i,j}, & \text{otherwise} \end{cases} \tag{13}$$

$$1 \leq l_{i,j} \leq N_{i,j} \tag{14}$$

In this formulation, C is the total cost of maintenance and T is the time to perform maintenance for the whole system. Further details to calculate maintenance cost and time can be found in [Pandey *et al.* (2013a)]. Constraints (10), and (11) show that the limited resources are available to perform maintenance, constraints (12), and (13), set the component state at the beginning of the next mission, depending on the state at the end of the previous mission and the maintenance action performed.

2.4   *Results*

Based on the above model, [Pandey *et al.* (2013a)] solved the selective maintenance problem of multi-component series-parallel system as shown in Fig. 4. They used Differential Evolution (DE) [Brest *et al.* (2006)] to solve the problem.

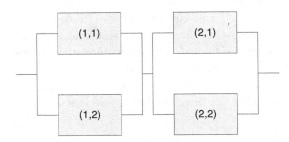

Fig. 4   A series parallel system

They assumed here that for intermediate maintenance actions, associated time and cost varied linearly as: $t_{i,j,l_{i,j}} = (l_{i,j} - 1) \times \Delta t_{i,j}^{W}$ and $c_{i,j,l_{i,j}} = (l_{i,j} - 1) \times \Delta c_{i,j}^{W}$ for $Y_{i,j} = 1$ and $t_{i,j,l_{i,j}} = T_{i,j}^{MR} + (l_{i,j} - 2) \times \Delta t_{i,j}^{F}$ and $c_{i,j,l_{i,j}} = C_{i,j}^{MR} + (l_{i,j} - 2) \times \Delta c_{i,j}^{F}$ for $Y_{i,j} = 0$. Here $\Delta t_{i,j}^{W}$, $\Delta c_{i,j}^{W}$, $\Delta t_{i,j}^{F}$ and $\Delta t_{i,j}^{F}$ indicate the time and cost required to increase the intermediate

Table 1 System parameters, maintenance time and cost

(i,j)	$\alpha_{i,j}$	$\beta_{i,j}$	$Y_{i,j}$	$B_{i,j}$	$T^{MR}_{i,j}$	$T^{WR}_{i,j}$	$\Delta t^{W}_{i,j}$	$T^{FR}_{i,j}$	$\Delta t^{F}_{i,j}$	$C^{MR}_{i,j}$	$C^{WR}_{i,j}$	$\Delta c^{W}_{i,j}$	$C^{FR}_{i,j}$	$\Delta c^{F}_{i,j}$
(1,1)	15	1.5	1	15	3	5	0.25	1	0.25	6	12	2	12	1
(1,2)	15	1.5	1	20	3	5	0.25	1	0.25	5	12	1.75	12	1
(2,1)	20	3	0	8	2	4	0.2	2	0.2	5	14	1.5	14	2
(2,2)	20	3	1	15	2	4	0.2	2	0.2	6	15	1.6	15	1.5

Table 2   Comparison when only replacement/minimal repair is used and when imperfect repair/maintenance is included ($T_o = 9$ units, $C_o = 25$ units) [Pandey *et al.* (2013a)]

Comp #(i,j)	With imperfect repair (Proposed model)					Replacement and minimal repair only				
	$l_{i,j}$	$T_{i,j}^*$	$C_{i,j}^*$	$X_{i,j}$	$A_{i,j}^*$	$l_{i,j}$	$T_{i,j}^*$	$C_{i,j}^*$	$X_{i,j}$	$A_{i,j}^*$
(1,1)	DN	0	0	1	15	DN	0	0	1	15
(1,2)	WR	5	12	1	0	WR	5	12	1	0
(2,1)	IM	2.8	13	1	2.7466	MR	2	5	1	20
(2,2)	DN	0	0	1	15	DN	0	0	1	15
		$\sum = 7.8$	$\sum = 25$				$\sum = 7$	$\sum = 17$		
$R(l)$	0.7293					0.6140				

*$T_{i,j}$ = Time spent on (i, j), $C_{i,j}$ = Cost spent on (i, j), $X_{i,j}$ = State of (i, j) after maintenance, $1 \leq l_{i,j} \leq 6$ for $Y_{i,j} = 1$, $1 \leq l_{i,j} \leq 7$ for $Y_{i,j} = 0$, $A_{i,j}$ = Effective age after maintenance, DN = Do nothing, WR = Replacement of a working component, IM = Imperfect maintenance, MR = Minimal repair.

maintenance level by unity for working and failed components, respectively. They used the parameters shown in Table 1.

In Table 1, $\alpha_{i,j}$ is the scale parameter and $\beta_{i,j}$ is the shape parameter for the Weibull distribution, $Y_{i,j}$ is the state before maintenance, $B_{i,j}$ is the effective age before maintenance, $T_{i,j}^{MR}$ and $C_{i,j}^{MR}$ is the time and cost to perform minimal repair, $T_{i,j}^{WR}$ and $C_{i,j}^{WR}$ is the time to replace the working component (i,j), $T_{i,j}^{FR}$ and $C_{i,j}^{FR}$ is the time and cost to replace the failed component (i,j). They assumed that $p=8$ for each component in the system (equation (4)). For the mission duration of $L = 8$ time units, budget limit of $C_0=25$ units, and time constraint of $T_0=9$ units, [Pandey *et al.* (2013a)] compared two scenarios. In the first scenario, only replacement and minimal repair are possible maintenance options, while in the second scenario imperfect maintenance/repair is also possible, along with replacement and minimal repair. The results are shown in Table 2.

From Table 2, it can be seen that inclusion of imperfect maintenance/repair as maintenance options increases the system reliability by more than 11%. Hence, it is important to include the imperfect maintenance/repair as a maintenance action for selective maintenance. It provides flexibility to use the available resources in an optimal manner such that system reliability is maximized. [Pandey *et al.* (2013a)] also compared the age reduction model, hazard adjustment model, and hybrid imperfect model. It was found that for the above problem, the age reduction model gives the next mission system reliability of 0.7324 and the hazard adjustment model gives a mission system reliability of 0.88, respectively. The system reliability for the hybrid model is 0.7293. The reason is that for the age reduction model, there is no change in the hazard rate and for the hazard rate model, effective age becomes zero at the beginning of the next mission, hence both individual models give higher system reliability. However, when combined, i.e. hazard rate increment as well as the effective age change is considered, lower system reliability is achieved for the hybrid model. Their results were similar to what was observed by [Lin *et al.* (2000)] for scheduling PM for a system.

3 Selective Maintenance for Multi-state Systems

For some systems or components, the binary assumption does not reflect the possible states that each of them may experience. They can perform their tasks with various discrete levels of efficiency known as 'performance rates' [Lisnianski and Levitin (2003)]. A system that has a finite number of performance rates varying from perfect operation to complete failure is defined as a multi-state system (MSS). For example, a power generation system may have four states denoted as 0, 1, 2, and 3. The performance rates for these states may correspond to the outputs of 0MW, 30 MW, 50 MW, and 80 MW, respectively [Massim *et al.* (2005)]. There are only a few works for selective maintenance of an MSS. First, [Chen *et al.* (1999)] worked on the selective maintenance optimization for a multi-state series-parallel system. They considered all components and system in multiple states and optimized cost associated with transition from one state to another for each component and the overall system. In their problem, transition probabilities and associated costs were known. However, they did not show the maintenance action required for a component or components state after maintenance. Their heuristic was good for a system with a small number of components only.

Another work on selective maintenance for an MSS was done by [Liu and Huang (2010)]. They assumed that components within a system were in a binary state, i.e., either working or failed; however, the system itself could have multiple states. The imperfect maintenance efficiency was considered in their selective maintenance, and a cost-maintenance efficiency relation that considered the age reduction factor of the Kijima imperfect maintenance model as a function of assigned maintenance cost was established. Recently, [Pandey *et al.* (2013b)] relaxed binary component assumption and solved the selective maintenance optimization problem for an MSS where components could also exhibit multiple performance levels. In this chapter, we will discuss both of the recent works of [Liu and Huang (2010)] and [Pandey *et al.* (2013b)].

3.1 *Selective Maintenance for MSS with Binary Components*

In [Liu and Huang (2010)], the basic assumptions for the studied MSS are as follows:

1) The MSS consists of M binary states components, and performance rates for each component $i(i \in \{1, ..., M\})$ are denoted by the set $g_i = \{g_{i,1}, g_{i,2}\}$, where $g_{i,2}(\neq 0)$ is a nominal performance rate, and $g_{i,1} = 0$ represents failure.
2) The MSS can be constructed by components in arbitrary configuration, such as series-parallel, bridge, complex network, etc.

3) At the beginning of the kth mission, the status of component i is represented by binary variable $X_i(k)$, where

$$X_i(k) = \begin{cases} 1 & \text{if the component } i \text{ is functioning} \\ 0 & \text{if the component } i \text{ is in failure state} \end{cases}.$$

4) $A_i(k)$ represents the effective age of component i at the beginning of the kth mission, and $B_i(k)$ is the effective age of component i when the kth mission is over.

5) The duration of the kth mission is denoted by $T(k)$.

6) Any maintenance action can only be executed during the break between two successive missions. In the break, there exists limited resources (e.g. cost, time, repairmen, etc.) to perform maintenance. Decision-makers need to determine how to allocate the maintenance resources to individual components with the aim of restoring the entire MSS to the state that maximizes the probability of successfully completing the subsequent mission.

7) The probability of the system successfully completing a mission is defined as the probability that the performance rate of the MSS is not less than the demand level during the whole mission.

8) Multiple maintenance actions can be chosen for both failed, and functioning components, including minimal repair, corrective/preventive replacement (or perfect maintenance), and imperfect maintenance. The maintenance efficiency relates with assigned maintenance cost (or resources), and their relation can be measured.

The efficiency of imperfect maintenance is characterized by the Kijima type II age reduction model [Kijima (1989)]. The effective age of any component i after the maintenance subsequent to the kth mission is given by

$$A_i(k+1) = b_i(k) B_i(k). \tag{15}$$

If the binary state component i fails during the kth mission, its effective age will immediately stop increasing with chronological time as shown in Case 3 in Fig. 5. In this figure, failures happen, respectively, at the 2nd, and 7th days of chronological time; and the effective age of the component is steady at the remaining mission time. $b_i(k)(0 \le b_i(k) \le 1)$ is the age reduction factor representing maintenance efficiency, and a smaller $b_i(k)$ means a greater improvement as plotted in Fig. 5 (see Cases 1 and 2 where no failure happens).

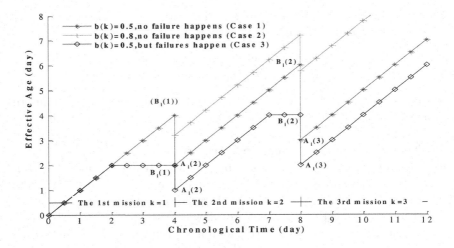

Fig. 5   Chronological time versus effective age in the Kijima type II model

Based on the effective age model, if the random quantity Y indicates the lifetime of a component and t' is the elapsed time after maintenance, the conditional survival probability of a component after a maintenance activity is given by:

$$R(t') = 1 - \Pr\{Y - t \le t' | X > t\} = 1 - \frac{\Pr\{t < Y \le t' + t\}}{\Pr\{Y > t\}} = \frac{\Pr\{Y > t' + t\}}{\Pr\{Y > t\}},$$
$$\tag{16}$$

where the component is functioning at the beginning of the mission with the effective age equal to t.

The maintenance cost of any component i after the kth mission is defined as:

$$C_i(k) = c_i(k) + c_i^0. \tag{17}$$

Let c_i^{rf} denotes the corrective repair cost for replacement of failed component i. The age reduction factor as a function of the corrective repair cost is then defined as

$$b_i(k) = 1 - \left(\frac{c_i(k)}{c_i^{rf}}\right)^{\frac{1}{m_i^f}}, \tag{18}$$

where $m_i^f \left(m_i^f > 0\right)$ is a characteristic constant that determines the exact relation between corrective repair cost and age reduction factor through

(18). In the same fashion, the age reduction factor as a function of the preventive repair cost is expressed as

$$b_i(k) = 1 - \left(\frac{c_i(k)}{c_i^{rp}}\right)^{\frac{1}{m_i^p}},$$ (19)

where $m_i^p (m_i^p > 0)$ is a characteristic constant that determines the exact relation between preventive repair cost and the corresponding age reduction factor through (19).

The probability of successfully completing a mission is defined as the probability that the MSS performance rate is not less than the mission demand level during the whole single mission period. The formula of the probability of successfully completing a single consecutive mission for an MSS is presented in Sec. 3.2.1. Hence, the selective maintenance modeling for an MSS with binary-state components is expressed as:

Objective:

$$\max \quad R_S\left(W_{k+1}, T(k+1), X(k+1), A(k+1)\right) =$$
$$\sum_{G_J(t) \geq W_{k+1}} P_J\left(T(k+1), X(k+1), A(k+1)\right) \quad (20)$$

Subject to:

$$C \leq C_0$$ (21)

$$A_i(k+1) = b_i(k) \cdot B_i(k)$$ (22)

$$X_i \in \{0, 1\}$$ (23)

where $A(k+1) = \{A_1(k+1), ..., A_M(k+1)\}$ is a vector representing the effective ages of components at the beginning of mission k. $X(k+1) = \{X_1(k+1), ..., X_M(k+1)\}$ is a binary-valued vector containing the states, either functioning or failed, of components at the beginning of mission k. $A(k+1)$ and $X(k+1)$ are completely determined by the type of maintenances (i.e. corrective or preventive) and the amount of maintenance cost performed on or assigned to each component. C is the total maintenance cost for the MSS after the last mission; whereas C_0 is the maintenance budget during the maintenance break. It should be noted that selective maintenance optimization for an MSS with imperfect maintenance is a complex, non-linear, continuous programming problem as shown in (20)-(23). An exhaustive examination of all possible solutions is not realistic due to

the computational time limitation. Meta-heuristic algorithms, such as genetic algorithm (GA), differential evolution (DE), Tabu search, simulated
annealing algorithm, and ant colony optimization (ACO), can be used to
solve the resulting optimization in a computational efficient manner.

A coal transportation system supplying coal to a boiler in a power station is presented here as an example to demonstrate the effectiveness of
our proposed method. The studied system includes five basic subsystems
as shown in Fig. 6. Feeder 1 (subsystem 1) transfers coal from the bin to
conveyor 1 (subsystem 2). Conveyor 1 transports the coal from feeder 1
to the stacker reclaimer (subsystem 3) that lifts the coal up to the burner
level. Feeder 2 (subsystem 4) then loads conveyer 2 (subsystem 5) that
transfers the coal to the burner feeding system of the boiler.

Every subsystem consists of binary state components. The parameter
settings for each component, e.g. nominal performance rate (ton/hour),
parameters of the Weibull life distribution, maintenance cost, effective age,
and status after last mission (the kth mission), are tabulated in Table 3. The
units of time, and cost are days, and \$1,000, respectively. The uncertain
demand for the $(k+1)^{th}$ mission is distributed as shown in Table 4, with
the required demand levels, and their corresponding probabilities.

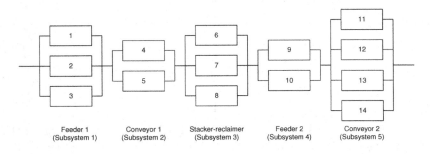

Fig. 6   Block diagram of a coal transportation system

Suppose the duration of the $(k+1)^{th}$ mission is $T(k+1) = 10\ days$. Although the system is still functioning at the end of the last mission (the kth
mission) without any maintenance action, the probability of it successfully
completing the next mission (the $(k+1)^{th}$ mission) is only 0.006. Given
the maintenance budget $C_0(k)=\$200,000$ in the break after the kth mission,
one has to optimally allocate the maintenance cost to each component to
maximize the probability of successfully completing the $(k+1)^{th}$ mission.

Table 3 Parameters of components: where performance rate is in tons/hour, time is in days, and costs are in $1,000 units

Comp. ID	$g_{i,2}$	η_i	β_i	m_i^p	c_i^{rp}	m_i^f	c_i^{rf}	c_i^0	$B_i(k)$	$Y_i(k)$
1	55	1.5	25	2.5	15	2.5	25	3	35	1
2	80	2.4	38	2.2	20	2.0	32	4	24	0
3	120	1.6	28	2.6	25	3.0	35	3	45	0
4	90	2.6	40	2.2	20	3.2	35	5	35	0
5	145	1.8	28	1.8	25	4.0	34	2	28	1
6	70	2.4	34	2.4	15	3.2	20	3	36	1
7	95	2.5	26	2.8	24	3.0	30	6	44	0
8	80	2.0	28	2.3	20	2.8	35	5	28	0
9	95	1.2	26	2.0	18	2.5	28	3	38	1
10	130	1.4	35	2.5	20	2.8	35	6	15	0
11	50	2.8	40	3.2	22	3.0	32	7	30	0
12	75	1.5	35	2.6	25	2.2	35	4	22	1
13	85	2.4	30	2.8	18	2.8	36	6	38	1
14	95	2.2	45	2.2	15	2.6	38	3	35	0

Table 4 Mission demands

Demand (ton/hour)	120	90	60	30	10
Probability	0.1	0.25	0.35	0.2	0.1

The genetic algorithm is used to search the global optimal solution, and the best maintenance strategy is presented as Scenario 1 in Table 5. The optimal allocations of repair costs are listed in the column "Cost" with the related fixed maintenance cost in parentheses. From Table 5 (Scenario 1), one can see that all the components are functioning at the start of the $(k + 1)^{th}$ mission, only components 3 and 7 are subjected to corrective replacement, and component 5 is subjected to preventive replacement. All of the other failed components are imperfectly repaired before the next mission is executed, and some functioning components (6 and 13) are subjected to imperfect PM. The probability of successfully completing the $(k + 1)^{th}$ mission is 0.77342, and total maintenance cost is $199,880. The optimal solution for the case where only minimal repair, preventive, and corrective replacement (Scenario 2) are considered is also tabulated in Table 5. The probability of successfully completing the mission is 0.7336, and corresponding total maintenance cost is $199,000. As shown in Table 5, although the

Table 5 Optimal solutions and comparison [Liu and Huang (2010)]

Comp ID	With imperfect maintenance (Scenario1)				Without imperfect maintenance (Scenario2)			
	Action	Cost	$X_i(k+1)$	$A_i(k+1)$	Action	Cost	$X_i(k+1)$	$A_i(k+1)$
1	DN	0	1	35	PR	15(3)	1	0
2	IC	5.33(4)	1	14.2	CR	32(4)	1	0
3	CR	35(3)	1	0	MC	0(3)	1	45
4	IC	17.5(5)	1	6.82	CR	35(5)	1	0
5	PR	25.0(2)	1	0	DN	0	1	28
6	IP	8.57(3)	1	7.49	DN	0	1	36
7	CR	30(6)	1	0	CR	30(6)	1	0
8	IC	5.84(5)	1	13.23	MC	0(5)	1	28
9	DN	0	1	38	PR	18(3)	1	0
10	IC	5.83(6)	1	7.89	MC	0(6)	1	15
11	IC	5.34(7)	1	13.49	MC	0(7)	1	30
12	DN	0	1	22	DN	0	1	22
13	IP	5.14(6)	1	13.71	PR	18(6)	1	0
14	IC	6.33(3)	1	17.43	MC	0(3)	1	35
$C(k+1)$	$199,880				$199,000			
$R(k+1,w)$	0.7734				0.7336			

*The value in "Cost" column is the allocated repair cost with the fixed maintenance cost in parentheses, where costs are in $1,000 units. Symbols denotation: "DN"-Do Nothing; IC-Imperfect Corrective repair; CR-Corrective Replacement; IP-Imperfect Preventive repair; PR-Preventive Replacement; MC-Minimal Corrective repair.

maintenance cost in Scenario 1 is slightly higher than Scenario 2 (\$880 or 0.44%), the probability of successfully completing the mission in Scenario 1 remarkably increases by nearly 5.43%. If maintenance resources are unlimited, and either corrective or preventive replacement is performed on each component before the next mission (Scenario 3), the probability of mission success is 0.8947 with a total maintenance cost of \$448,000. It indicates the maintenance cost in Scenario 1 decreases by 55.38% as compared with Scenario 3, while the probability of successfully completing the mission in Scenario 1 decreases by only 13.56.

3.2 *Selective Maintenance for MSS with Multistate Components*

To solve the selective maintenance optimization problem for an MSS with multi-state components, the following assumptions are considered in [Pandey *et al.* (2013b)]:

1) The system consists of multiple, repairable components.
2) The components as well as the system may be in multiple states, i.e. components and system has several discrete performance levels.
3) Replacement brings the component back to the best possible state.
4) Maintenance is possible only during maintenance breaks, no repair/maintenance can be performed during missions i.e. system and components only degrade during operation. Maintenance may bring a component to a better state.
5) At the end of a mission, the current component/system states are observable.
6) System degradation is modeled using the homogenous Markov model i.e. transition time between components state follow exponential distribution.
7) Limited resources (budget, repairman and time) are available and the amount of resources required for maintenance activities are known and fixed.

In place of binary components, [Pandey *et al.* (2013b)] considered multistate components for study. Based on the state of the components before maintenance, system demand during the next mission and available resources, the maintenance decision for each component in the system was determined. In an MSS, each component in the system can have discrete performance rates (also called states) as shown in Fig. 7.

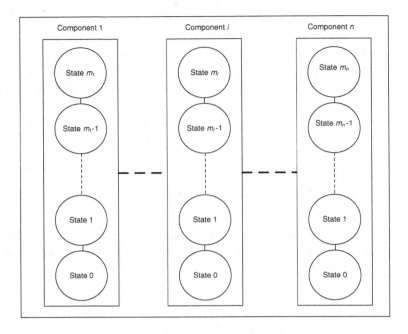

Fig. 7   An MSS with n components

In the MSS shown in Fig. 7, there are n multistate components and any component 'i' ($i=1, 2\ldots, n$) has '$m_i + 1$' states ($0 \leq k \leq m_i$). The sojourn time of a component in a state is independent of the sojourn time of another component, which is in its own state. Here, state '$j=0$' refers to the complete failure state of a component. This is the worst state for a component, whereas $'j = m_i'$ is the best possible state for component i [Pandey *et al.* (2013b)]. Any component i can be in $(m_i + 1)$ possible states ($0 \leq k \leq m_i$) and corresponding to each state there are associated performance rates given as:

$$g_i\left(t\right) = \{g_{i,0}, g_{i,1}, ..., g_{i,m_i}\}, \qquad i = 1, 2, ..., n \tag{24}$$

The performance rate $g_i\left(t\right)$ of a component 'i' at any instant ($t \geq 0$) is a discrete random variable and can have any value from $g_{i,0}$ to g_{i,m_i}. The performance rate of an MSS is also a random variable that depends on the performance rates of the components:

$$G\left(t\right) = \Psi\left(g_1\left(t\right), ..., g_n\left(t\right)\right) \tag{25}$$

where Ψ is the structure function.

The system performance rate can also have any discrete performance value from the set $G\left(t\right) = \{G_0, G_1, ..., G_S\}$, where $G_0, G_1, ..., G_S$ are the performance rates that the system can have at any given time depending on the components performance rate at that instant. In the case of a group of components connected in parallel, the performance rate of the subsystem is the sum of the performances of the components. If n_s components in a subsystem s are connected in parallel and at any time instant t, the performances of these components are given by $g_i\left(t\right)$ where $1 \leq i \leq n_s$, then the performance of this subsystem is:

$$g^{subsystem}\left(t\right) = \sum_{i=1}^{n_S} g_i\left(t\right) \tag{26}$$

In the case of a group of components connected in a series, the performance of the subsystem is:

$$g^{subsystem}\left(t\right) = \min_{1 \leq i \leq n_S}\left(g_i\left(t\right)\right) \tag{27}$$

Hence, for a series-parallel system at any instant t, , the subsystem (the group of components connected in parallel) performance is first determined, and then the system performance is calculated. Hence, the performance level $G\left(t\right)$ of the system can be evaluated as:

$$
\begin{aligned}
g_l^{subsystem}\left(t\right) &= \sum_{i=1}^{n_l} g_i\left(t\right), \qquad l = 1, 2, ..., s \\
G\left(t\right) &= \min_{1 \leq l \leq s}\left(g_l^{subsystem}\left(t\right)\right)
\end{aligned}
\tag{28}
$$

It is obvious from the above discussion that at any instant t, the system performance can be described completely if components performance levels are known. If the component i has state y_i before maintenance, it is required to find the state x_i of each component after maintenance break such that system performance during the mission is satisfied. Hence, x_i is the decision variable here. The following maintenance options are possible for the multistate components of an MSS:

1) **Do nothing:** No action is performed. Leave the component as is. In this condition the decision variable $x_i = y_i$, i.e. the component state before and after the maintenance break are the same.
2) **Replacement:** The old component is replaced and a new component is installed. After replacement, the component state becomes $x_i = m_i$.
3) **Imperfect repair/maintenance:** For imperfect repair/maintenance, the state of a component from complete failure $(y_i = 0)$/non-failure $(0 < y_i < m_i - 1)$ is improved to a higher state (other than m_i). In this

case $y_i < x_i < m_i$. The term imperfect repair is used when $y_i = 0$, i.e., the component was in a failed state before maintenance and is improved to a higher state during maintenance. The term imperfect maintenance is used when the component performance level before maintenance was higher than '0', i.e., $y_i > 0$ and during maintenance its performance level is further improved.

Once the system arrives at the maintenance depot, component states are known and once a maintenance decision is taken we also know the component state x_i at the beginning of the mission after maintenance. With known x_i, we can find the probability of a component being in any lower state during the mission. Since no maintenance is performed during the mission, a component can only degrade and go lower than x_i state. There is no chance that it will go to a higher state than x_i when it starts the mission with state x_i. Component state probability is denoted by $p_{i,k}(t, x_i)$, which shows the probability of being in state k at time t during the mission given that the component was in state x_i at the beginning of the mission. For all states $k > x_i$, $p_{i,k}(t, x_i) = 0$. For states $k \le x_i$, the Chapman-Kolmogorov equation can be used to find $p_{i,k}(t, x_i)$ [Pandey *et al.* (2013b)]. Based on the component state probabilities, the system state probabilities and the probability of successfully completing a mission by the system are calculated using the 'Universal Generating Function' UGF [Levitin (2005)].

3.2.1 *System reliability evaluation*

An MSS degrades with time over the mission duration; hence it is important that the performance of the MSS does not fall below the demand level (W). It is then required to find the system performance $G(t)$ at the end of the next mission when the system performance level is at a minimum in the given mission (see Fig. 8). Figure 8 shows that a demand level 'W' is required during a mission of duration 'T'. Even though the system performance may degrade to lower performance levels, at any time 't' during the mission, its performance rate is not allowed to fall below the demand level, i.e. $G(t) \ge W$.

The system reliability can be estimated by determining the probability that the system is in an acceptable state$(G(T) \ge W)$ at the end of the next mission $(t = T)$. Hence, the reliability of the MSS can be calculated as:

$$Prob(G(T) \ge W) = \varphi\left(\sum_{J=0}^{S} P_J(T, x).z^{G_J}\right) \qquad (29)$$

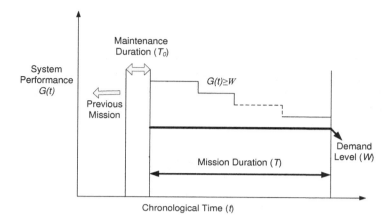

Fig. 8 System performance during a mission and demand level

where φ is the distributive operator defined by the following equations:

$$\varphi\left(P\left(T,x\right).z^{\sigma-W}\right) = \begin{cases} P\left(T,x\right) & \text{if } \sigma \geq W \\ 0 & \text{if } \sigma < W \end{cases} \tag{30}$$

Here $P_J\left(T,x\right)$ is the probability that the system is in state J at the end of the mission for given states of the components at the beginning of the mission defined by x. In equation (29) and (30), z^{G_J} is an exponent used in UGF (see [Pandey *et al.* (2013b)] for further details). Hence, depending on the mission demand W and maintenance decision variables, the system reliability at the end of mission duration T can be expressed as:

$$R_S\left(W,T,x\right) = \varphi\left(\sum_{J=0}^{S} P_J\left(T,x\right).z^{G_J-W}\right) = \sum_{J=0}^{S}\left(\varphi\left(P_J\left(T,x\right).z^{G_J-W}\right)\right) \tag{31}$$

With the use of distributive operator φ, equation (31) can also be written as:

$$R_S\left(W,T,x\right) = \sum_{G_J(t)\geq W} P_J\left(T,x\right) \tag{32}$$

Equation (32) shows that the system reliability is the sum of state probabilities of system states that are higher than the demand level at the end of the mission. Hence, using equation (32) the system reliability can be estimated. The objective is to perform selective maintenance in order to maximize the system performance during the next mission.

3.2.2 *Selective maintenance modeling*

Upon the arrival of the system for maintenance, the components' performance states $y = [y_1, y_2, ..., y_n]$ are known. The budget constraint on the total maintenance cost during the maintenance break is given by C_0 and the available maintenance time duration is limited to T_0. The non-linear formulation to maximize the probability of successfully completing the next mission is expressed as:

Objective:

$$\max \quad R_S(W, T, x) = \sum_{G_J(t) \geq W} P_J(T, x) \tag{33}$$

Subject to:

$$C \leq C_0 \tag{34}$$

$$T \leq T_0 \tag{35}$$

$$y_i \leq x_i \leq m_i \tag{36}$$

In the above formulation, C is the total cost of maintenance and T is the time to perform maintenance for the whole system (see [Pandey *et al.* (2013b)] for further details). This is a typical constrained nonlinear optimization problem involving integer variables only. An evolutionary algorithm is used to solve this optimization problem. Differential evolution (DE) is used in this case. To solve the problem of selective maintenance for multistate systems with multistate components, the example of coal transportation system presented in Sec. 3.1 is used here. However, it assumes that every component may possess more than two states, rather than only two states as in Sec. 3.1.

The capacity of each component, number of component states, transition probabilities, and other parameters are not included here due to space limitation. All these values can be read from [Pandey *et al.* (2013b)]. We aim to highlight the advantages of imperfect repair/maintenance policy; hence, we will limit ourselves to discussing the results related to imperfect maintenance repair only. Readers are encouraged to read [Pandey *et al.* (2013b)] for further details and an understanding of the problem and see other results related to the problem.

Table 6 Selective Maintenance decision with both replacement and imperfect maintenance/repair as options (costs in thousands of $ and time in hrs) [Pandey et al. (2013b)]

Comp.(i)	1	2	3	4	5	6	7	8	9	10	11	12	13	14
m_i	3	3	3	2	2	3	3	3	3	3	4	4	4	4
y_i	0	0	0	0	1	1	2	1	1	2	2	1	0	1
I* x_i	3	3	0	2	1	1	2	1	3	2	2	1	0	4
Actions	CR	CR	DN	CR	DN	DN	DN	DN	CR	DN	DN	DN	DN	CR
C_{i,x_i}	21.20	16	0	15.10	0	0	0	0	21.40	0	0	0	0	12.60
T_{i,x_i}	2.25	1.75	0	1.55	0	0	0	0	2.40	0	0	0	0	1.30

$R_s = 0.91774$, $C = 86.3$, $T = 9.25$

	1	2	3	4	5	6	7	8	9	10	11	12	13	14
II* x_i	2	2	2	2	1	2	2	2	2	2	2	1	2	1
Actions	IR	IR	IR	CR	DN	IM	DN	IM	IM	DN	DN	DN	IR	DN
C_{i,x_i}	16.20	13	16.10	15.10	0	4.75	0	6.13	6.40	0	0	0	9.83	0
T_{i,x_i}	1.75	1.45	1.75	1.55	0	0.55	0	0.63	0.90	0	0	0	1.18	0

$R_s = 0.9613$, $C = 87.5096$, $T = 9.7623$

*I = Only replacement as option, II = Imperfect maintenance included with replacement as option. C_{i,x_i} = The budget used in the maintenance of component i, T'_{i,x_i} = The time used in the maintenance of component i; DN = Do nothing, IM = Imperfect maintenance, IR = Imperfect repair, CR = Component replacement.

3.2.3 *Results*

The comparison of results when only replacement is possible and when imperfect repair/maintenance is also included are presented in Table 6. Table 6 depicts when only replacement is considered as an option, the maximum achievable system reliability is 0.91774. The total cost and time consumed in the maintenance are 86.30 units and 9.25 hrs., respectively. In this example, unit costs=$1000. While in the case of imperfect maintenance, the maximum achievable reliability increased to 0.9613, an increase of about 5%. In the imperfect maintenance/repair case, the cost and time used during maintenance are 87.50 units and 9.76 hrs, respectively.

It is clear that better utilization of resources is possible if imperfect maintenance is included as an option for selective maintenance rather than using replacement as the only maintenance option. If only replacement is opted for as a maintenance option then components # 1, 2, 4, 9, and 14 are selected. If imperfect repair/maintenance is also included as a maintenance option then only component # 4 is replaced. Components # 1, 2, 3, and 13 are selected for imperfect repair while components # 6, 8, and 9 have imperfect maintenance. No maintenance is performed on the other components.

4 Conclusion

An effective age and maintenance cost based age reduction and hazard adjustment factor is defined in [Pandey *et al.* (2013a)]. It is found that for the same investment in relatively younger components as compared to older components, better improvement can be achieved. A more generalized hybrid imperfect maintenance model is used to formulate the component improvement after repair. This assumption is more realistic and more general. To define the relative age of the component a characteristic constant 'm' is also proposed, which determines whether a component is relatively younger or older. It is found that whether a binary system is considered or an MSS, imperfect maintenance/repair is important in optimizing selective maintenance. Under imperfect maintenance/repair selective maintenance optimization provides better reliability than maintenance decisions with replacement only as an option.

All of the works focused on single objective only, however, trade-offs between maintenance budgets, time, and mission success probability, as well as other resources (e.g. multiple repairmen) need to be addressed. Multi-objective optimization approaches may be used to solve this problem. For multistate systems with multistate components a constant failure rate was used. It was assumed that sojourn time follows the Exponential distribution, however, other distributions, e.g. the Weibull distribution, can also be used.

Acknowledgment

This research was supported by the Natural Sciences and Engineering Research Council of Canada (NSERC).

References

Barlow, R. E. and Hunter, L. E. (1960). Optimum preventive maintenance policies, *Operational Research* **8**, pp. 90–100.

Brest, J., Greiner, S., Bošković, B., Mernik, M. and Zumer, V. (2006). Self-adapting control parameters in differential evolution: A comparative study on numerical benchmark problems, *IEEE Transactions on Evolutionary Computation* **10**, 6, pp. 646–657.

Brown, M. and Proschan, F. (1983). Imperfect repair. *Journal of Applied Probability* **20**, 4, pp. 851–859.

Cassady, C., Pohl, E. and Murdock, W. P. (2001a). Selective maintenance modeling for industrial systems, *Journal of Quality in Maintenance Engineering* **7**, 2, pp. 104–117.

Cassady, C. R., Murdock, W. P. and Pohl, E. A. (2001b). Selective maintenance for support equipment involving multiple maintenance actions, *European Journal of Operational Research* **129**, 2, pp. 252–258.

Chan, P. and Downs, T. (1978). Two criteria for preventive maintenance. *IEEE Trans Reliab* **R-27**, 4, pp. 272–273.

Chen, C., Meng, M. Q. H. and Zuo, M. J. (1999). Selective maintenance optimization for multistate systems, in *IEEE Canadian Conference on Electrical and Computer Engineering*, pp. 1477–1482.

Dekker, R. and Scarf, P. (1998). On the impact of optimisation models in maintenance decision making: The state of the art, *Reliability Engineering and System Safety* **60**, 2, pp. 111–119.

Dekker, R., Wildeman, R. and Van Der Duyn Schouten, F. (1997). A review of multi-component maintenance models with economic dependence, *Mathematical Methods of Operations Research* **45**, 3, pp. 411–435.

Desai, A. and Mital, A. (2006). Design for maintenance: Basic concepts and review of literature, *International Journal of Product Development* **3**, 1, pp. 77–121.

Doyen, L. and Gaudoin, O. (2004). Classes of imperfect repair models based on reduction of failure intensity or virtual age, *Reliability Engineering and System Safety* **84**, 1, pp. 45–56.

Iyoob, I., Cassady, C. and Pohl, E. (2006). Establishing maintenance resource levels using selective maintenance, *Engineering Economist* **51**, 2, pp. 99–114.

Kijima, M. (1989). Some results for repairable systems with general repair, *Journal of Application Probability* **26**, 1, pp. 89–102.

Levitin, G. (2005). *The universal generating function in reliability analysis and optimization* (Springer-Verlag, New York, USA).

Liao, W., Pan, E. and Xi, L. (2010). Preventive maintenance scheduling for repairable system with deterioration, *Journal of Intelligent Manufacturing* **21**, 6, pp. 875–884.

Lie, C. H. and Chun, Y. H. (1986). Algorithm for preventive maintenance policy. *IEEE Transactions on Reliability* **R-35**, 1, pp. 71–75.

Lin, D., Zuo, M. and Yam, R. (2000). General sequential imperfect preventive maintenance models, *International Journal of Reliability, Quality and Safety Engineering* **7**, 3, pp. 253–266.

Lin, D., Zuo, M. and Yam, R. (2001). Sequential imperfect preventive maintenance models with two categories of failure modes, *Naval Research Logistics* **48**, 2, pp. 172–183.

Lisnianski, A. and Levitin, G. (2003). *Multi-state system reliability: assessment, optimization and applications* (World Scientific Publishing Limited, MA, USA).

Liu, Y. and Huang, H.-Z. (2010). Optimal selective maintenance strategy for multi-state systems under imperfect maintenance, *IEEE Transactions on Reliability* **59**, 2, pp. 356–367.

Liu, Y., Huang, H.-Z. and Zhang, X. (2012). A data-driven approach to selecting imperfect maintenance models, *IEEE Transactions on Reliability* **61**, 1, pp. 101–112.

Lust, T., Roux, O. and Riane, F. (2009). Exact and heuristic methods for the selective maintenance problem, *European Journal of Operational Research* **197**, 3, pp. 1166–1177.

Maillart, L., Cassady, C., Rainwater, C. and Schneider, K. (2009). Selective maintenance decision-making over extended planning horizons, *IEEE Transactions on Reliability* **58**, 3, pp. 462–469.

Massim, Y., Zeblah, A., Meziane, R., Benguediab, M. and Ghouraf, A. (2005). Optimal design and reliability evaluation of multi-state series-parallel power systems, *Nonlinear Dynamics* **40**, 4, pp. 309–321.

Murthy, D. and Nguyen, D. (1981). Optimal age-policy with imperfect preventive maintenance. *IEEE Transactions on Reliability* **R-30**, 1, pp. 80–81.

Nakagawa, T. (1979). Optimum policies when preventive maintenance is imperfect. *IEEE Transactions on Reliability* **R-28**, 4, pp. 331–332.

Nakagawa, T. (1986). Periodic and sequential preventive maintenance policies. *Journal of Applied Probability* **23**, 2, pp. 536–542.

Nakagawa, T. (1988). Sequential imperfect preventive maintenance policies. *IEEE Transactions on Reliability* **37**, 3, pp. 295–298.

Pandey, M., Zuo, M. and Moghaddass, R. (2012). Selective maintenance for binary systems using age-based imperfect repair model, in *Proceedings of*

International Conference on Quality, Reliability, Risk, Maintenance, and Safety Engineering (QR2MSE), pp. 385–389.

Pandey, M., Zuo, M., Moghaddass, R. and Tiwari, M. (2013a). Selective maintenance for binary systems under imperfect repair, *Reliability Engineering and System Safety* **113**, 1, pp. 42–51.

Pandey, M., Zuo, M. J. and Moghaddass, R. (2013b). Selective maintenance modeling for a multistate system with multistate components under imperfect maintenance, *IIE Transactions* doi:10.1080/0740817X.2012.761371.

Pham, H. and Wang, H. (1996). Imperfect maintenance, *European Journal of Operational Research* **94**, 3, pp. 425–438, cited By (since 1996)297.

Rajagopalan, R. and Cassady, C. (2006). An improved selective maintenance solution approach, *Journal of Quality in Maintenance Engineering* **12**, 2, pp. 172–185.

Rice, W. F., Cassady, C. and Nachlas, J. A. (1998). Optimal maintenance plans under limited maintenance time, in *Proceedings of the Seventh Industrial Engineering Research Conference, Banff, Canada*.

Schneider, K. and Cassady, C. (2004). Fleet performance under selective maintenance, in *The Annual Reliability and Maintainability Symposium, St. Louis, USA*, pp. 571–576.

Sheu, S. H. and Liou, C. T. (1995). A generalized sequential preventive maintenance policy for repairable systems with general random minimal repair costs, *International Journal of Systems Science* **26**, 3, pp. 681–690.

Wang, H. (2002). A survey of maintenance policies of deteriorating systems, *European Journal of Operational Research* **139**, 3, pp. 469–489.

Wu, S. and Zuo, M. (2010). Linear and nonlinear preventive maintenance models, *IEEE Transactions on Reliability* **59**, 1, pp. 242–249.

Zhu, H., Liu, F., Shao, X., Liu, Q. and Deng, Y. (2011). A cost-based selective maintenance decision-making method for machining line, *Quality and Reliability Engineering International* **27**, 2, pp. 191–201.

Chapter 3

Random Replacement Policies

Won Young Yun and Li Liu

Department of Industrial Engineering, Pusan National University,
30-san, Jangjeon-dong, Geumjong-gu, Busan 609-735, South Korea

1 Introduction

In most literature related to preventive maintenance and replacement policies, it is assumed that we can maintain the system preventively at any time. But it is sometimes impractical to replace or maintain a system in the strictly pre-planned time because it may have a random work cycle and it is impossible or impractical to replace it in mid-cycle. The random age replacement model is studied analytically in [Barlow and Proschan (1965)]. [Nakagawa (2005)] summarizes random maintenance policies in which three replacement cases (age, periodic and block replacement) are studied and a random inspection model is also analyzed. For example, a unit is replaced at time T, Y, or at failure, whichever occurs first, where T is constant (pre-planned) and Y is a random variable (a random age replacement model). Secondly, a system is replaced at planned time T or at random time Y, whichever occurs first, and undergoes only minimal repair at failures between replacements (random periodic replacement model). [Nakagawa *et al.* (2011)] proposes two random replacement models with shortage and excess costs. [Chen *et al.* (2010)] studies random replacement policies in which a system is replaced preventively at time T and number N. For example, a unit is replaced at time T, N, Y, or at failure, whichever occurs first, where T and N are given and Y is a random variable (random age replacement model). [Zhao and Nakagawa (2012)] considers optimization problems of replacement first or last in which the concept of replacement last is new one and the system is replaced preventively at time T or Y, whichever occurs last. [Zhao *et al.* (2012)] studies an age replacement with overtime policy. [Chen (2013)] proposes modified models with random age and periodic replacement.

Random replacement models have been also studied as opportunistic replacement policies in which opportunistic time points of preventive replacement occur according to some stochastic processes [Dekker and Dijkstra (1992)].

In this chapter, we summarize random age and number-based replacement models and consider random replacement models with different cost terms. Two random replacement models are studied: the planned PR (Preventive Replacement) is assumed to be the mean of the actual PR time in the first model and the opportunity times of PR are assumed to occur according to stochastic processes in the second model. Corrective and preventive replacement costs are also included in the model and additionally we consider a benefit term from random preventive replacement times. Large variance of the preventive replacement time means that we have more flexibility in maintaining preventively the system. In numerical examples, we assume that the system failure distribution is a Weibull distribution and the preventive maintenance time follows Exponential, uniform, and triangular distributions. We determine the optimal mean PR time minimizing the expected cost rate numerically and compare the optimal PR times and the expected cost rates in the existing models.

This chapter is organized as follows. In Section 2, the standard age replacement and random replacement models are summarized and random age replacement models in which the planned PR time is the mean of the actual PR time with different cost terms are studied. Time-based and number-based replacement models under random opportunistic times are reviewed in Section 3. Some extended models are discussed in Section 4. Finally, Section 5 concludes the chapter. Table 1 shows notation in this chapter.

2   Random Age Replacement Policies

In this chapter, we study preventive replacement policies of a system with a stochastic lifetime under the following basic assumptions.

1. The failed system is replaced by new one correctively.
2. The system can be replaced preventively.
3. The actual preventive replacement times are random variables.

Table 1   Notation for random replacement models

Notation	
c_1	Corrective replacement cost
c_2	Preventive replacement cost
c	Profit rate
$f(t), F(t), \overline{F}(t), \lambda(t)$	Density, distribution, reliability, failure rate functions of system failure
μ	Mean time to system failure
T_p	Planned PR time
T_a	Actual PR time, a random variable
$g(\cdot), G(\cdot), \overline{G}(\cdot), r(\cdot)$	Density, distribution, reliability, failure rate functions of actual preventive maintenance time

4. The corrective and preventive replacement times are negligible.
5. The corrective and preventive replacement costs are c_1 and c_2 respectively $(c_1 > c_2)$.
6. The time horizon is infinite and the long run average cost is used as the optimization criterion.

Thus, we want to find the optimal preventive replacement policies to minimize the expected cost rate.

2.1   *Standard Age Replacement Policy*

In this subsection, we consider the age replacement policy in which the system is planned to be replaced preventively at age T_p. Thus, if the system is failed before the planned T_p, the system is replaced by new one correctively. Otherwise, the system is replaced preventively.

Firstly, if the actual preventive times can be equal to the planned preventive times, the expected cost rate is given [Nakagawa (2005)]

$$C_1(T_p) = \frac{(c_1 - c_2)F(T_p) + c_2}{\int_0^{T_p} \overline{F}(t)\mathrm{d}t}. \tag{1}$$

From $\mathrm{d}C_1(T_p)/\mathrm{d}T_p = 0$, the optimal T_p we get as solution of the equation

$$\lambda(T_p) \int_0^{T_p} \overline{F}(t)\mathrm{d}t - F(T_p) = \frac{c_2}{c_1 - c_2}. \tag{2}$$

Suppose that the failure rate is continuous and strictly increasing. If $\lambda(\infty) > c_1/[\mu(c_1 - c_2)]$, then there exists a finite and unique T_p^* that satisfies Equation (2). Otherwise, the optimal $T_p^* = \infty$.

2.2 *Random Age Replacement Policy*

In this subsection, we assume that the actual preventive replacement times are random variables dependent on the planned PR time. If the planned preventive replacement time is T_p, the actual PR times, T_a are random variables of which the distribution is dependent on T_p. In this section, we assume that the mean value of the actual PR times is the planned PR time, $E[T_a] = T_p$. Thus, the system is replaced at time T_a, or at failure, whichever occurs first. But the actual PM time T_a is a random variable and the expected cost rate is given [Nagakawa (2005)],

$$C_2(T_p) = \frac{(c_1 - c_2) \int_0^\infty \overline{G}(t; T_p) dF(t) + c_2}{\int_0^\infty \overline{G}(t; T_p)\overline{F}(t) dt}, \tag{3}$$

where $\overline{G}(t; T_p)$ is the reliability function of the actual PR time in which the mean value is T_p.

Theoretically, the random age replacement policy is worse than the standard age replacement policy (deterministic PM) [Barlow and Poschan (1965)]. Thus,

$$\min_{T_p} C_1(T_p) \le \min_{T_p} C_2(T_p).$$

Therefore, if we can maintain preventively on planned PM time, we should keep the planned PR times. But in practice, it is sometimes impossible or not economical to replace an item in the preplanned time. Sometimes, if we maintain early or delay the PR times, we can save much the production cost induced by interrupting continuous works. In random replacement models, the mean of the actual PR time is the planned PR time and its variance is related to flexibility of PR. We can get some profits from changing PR time and it is assumed that the profit is proportional to the standard deviation of the actual PR time. Thus, the expected cost rate with a profit term is given

$$C_3(T_p) = \frac{(c_1 - c_2) \int_0^\infty \overline{G}(t; T_p) dF(t) + c_2 - c \cdot SD(T_a)}{\int_0^\infty \overline{G}(t; T_p) F(t) dt}. \tag{4}$$

It is difficult to compare the three expected cost rates analytically and we investigate the three expected cost rates and the optimal PR times numerically in case that the system failure distribution is a Weibull distribution with scale parameter, $\lambda = 1$ and $c_2 = 1$.

1. *Exponential distribution case*

If the actual preventive replacement time follows an Exponential distribution of which the mean is the planned PR time, T_p, the expected cost rate with the profit term is

$$C_3(T_p) = \frac{(c_1 - c_2) \int_0^\infty \alpha t^{\alpha-1} e^{-t/T_p - t^\alpha} \, dt + c_2 - cT_p}{\int_0^\infty e^{-t/T_p - t^\alpha} \, dt}. \tag{5}$$

The optimal PR time T_p minimizing the expected cost rate is obtained numerically. For numerical examples, $c_1 = 4$ and $c_2 = 1$. Firstly, we obtain the expected cost rate without the profit rate term and from Fig. 1, we know the optimal PR interval decreases as the shape parameter of Weibull distribution increases.

Secondly, we consider the expected cost rate with the profit term and Table 2 gives the optimal T_p^* and expected cost rate $C(T_p^*)$ for different shape parameter α and cost ratio $c_2/(c_1 - c_2)$. It indicates that T_p^* increases with ratio $c_2/(c_1 - c_2)$, $C(T_p^*)$ decreases with α and ratio $c_2/(c_1 - c_2)$. For large $c_2/(c_1 - c_2)$, we do not need preventive replacement.

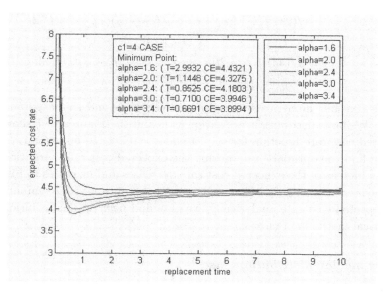

Fig. 1 Optimal T_p and expected cost rates without profit term (Exponential, $c_1 = 4$)

Table 2   Optimal time T_p^* and its cost rate $C(T_p^*)$ with $c = 1$ (random age replacement policy with Exponential distribution case)

$\dfrac{c_2}{c_1 - c_2}$	$\alpha = 2.0$		$\alpha = 2.5$		$\alpha = 3.0$		$\alpha = 5.0$	
	T_p^*	$C(T_p^*)$	T_p^*	$C(T_p^*)$	T_p^*	$C(T_p^*)$	T_p^*	$C(T_p^*)$
0.01	0.1497	27.5860	0.0856	19.8430	0.9718	15.7650	0.1250	10.4000
0.02	0.1634	20.5130	0.1160	14.9300	0.1261	12.3870	0.1510	8.8303
0.05	0.1910	12.3580	0.1803	10.1430	0.1856	8.8855	0.2031	6.9569
0.10	0.2889	8.5556	0.2686	7.4631	0.2647	6.7902	0.2707	5.6657
0.20	0.5411	5.8293	0.4540	5.3781	0.4237	5.0648	0.4000	4.4740
0.50	4.9525	3.3761	1.9169	3.3239	1.4262	3.2564	1.0420	3.0701
1.00	∞	2.2568	∞	2.2540	∞	2.2396	∞	2.1782
2.00	∞	1.6926	∞	1.6905	∞	1.6797	∞	1.6336
5.00	∞	1.3541	∞	1.3524	∞	1.3438	∞	1.3069

2. Uniform distribution case

Now we consider a different distribution case in which the actual PR time follows a uniform distribution with mean T_p and an allowed limit δ. Then, the density of T_a is given by

$$g(t; T_p) = \frac{1}{2\delta}, \quad T_p - \delta \le t \le T_p + \delta.$$

Thus, the expected cost rate is

$$C_U(T) = \frac{\begin{array}{c} 1/(2\delta) \int_{T_p-\delta}^{T_p+\delta} [(c_1 + c_2)\delta + (c_1 - c_2)(T_p - t)] f(t) dt \\ + c_1 F(T_p - \delta) + c_2 \overline{F}(T_p + \delta) - c\delta/\sqrt{3} \end{array}}{\begin{array}{c} 1/(4\delta) \int_{T_p-\delta}^{T_p+\delta} [2t(T_p + \delta) - t^2 - (T_p - \delta)^2] f(t) dt \\ + \int_0^{T_p-\delta} t f(t) dt + T_p \overline{F}(T_p + \delta) \end{array}}. \tag{6}$$

We consider numerical examples and the scale and shape parameters of the Weibull distribution are given as 1 and 2. Figures 2–4 show the expected cost rate function for $c_1 = 4$ and $c = 0.5, 1$. In numerical results, we can find the optimal δ minimizing the expected cost rate. Additionally, when δ increases, the expected cost rate decreases and increases in Figs. 2–4. Thus, even though it is not easy to find the optimal (δ, T) minimizing the expected cost rate analytically, we may find near optimal solutions by numerical methods in most cases.

3. Triangular distribution case

In this subsection, we assume that the actual preventive replacement time follows a triangular distribution with mean T_p and an allowed limit δ. Thus,

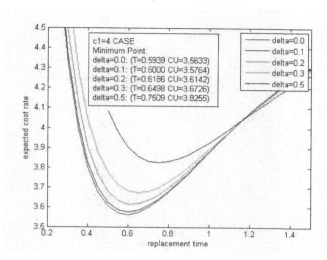

Fig. 2   Expected cost rates without profit term ($c_1 = 4$) : uniform distribution case

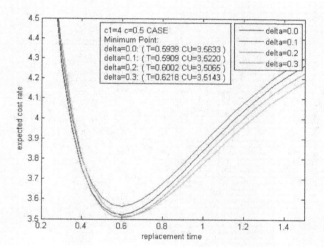

Fig. 3   Expected cost rates with different allowed limits ($c_1 = 4, c = 0.5$): uniform distribution case

the density of T_a is given by

$$g(t; T_p) = \begin{cases} \frac{x - T_p + \delta}{\delta^2}, & T_p - \delta \leq t \leq T_p \\ \frac{T_p - x + \delta}{\delta^2}, & T_p \leq t \leq T_p + \delta \end{cases} \tag{7}$$

Thus, the expected cost rate is obtained from Equation (4) as in the uniform case. From the numerical examples, we get the optimal preventive

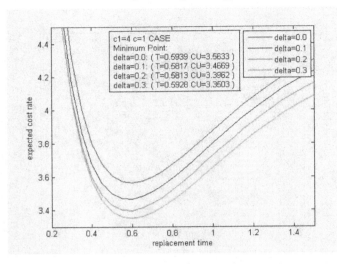

Fig. 4   Expected cost rates with different allowed limits ($c_1 = 4, c = 1$): uniform distribution case

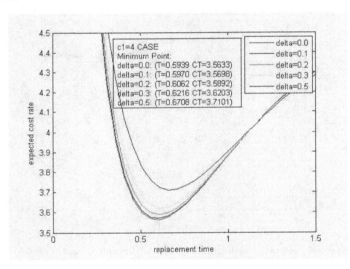

Fig. 5   Expected cost rates without profit ($c_1 = 4$): triangular distribution case

replacement time T_p^* which minimizes the expected cost rate when the scale and shape parameters of Weibull distribution are given by 1 and 2 respectively and PR cost is 1. Figures 5–7 show that when δ increases, the expected cost rate decreases.

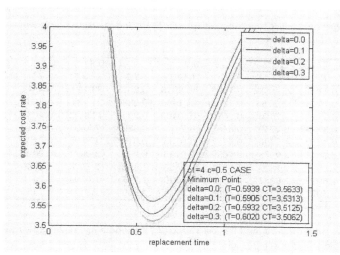

Fig. 6   Expected cost rates with different allowed limits ($c_1 = 4, c = 0.5$): triangular distribution case

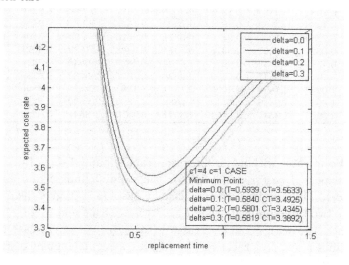

Fig. 7   Expected cost rates with different allowed limits ($c_1 = 4, c = 1$): triangular distribution case

2.3   *Numerical Examples*

In this section, we study more numerical examples, T_p^* and assume that the system failure follows a Weibull distribution and the scale and shape parameters are given by 1 and 2 respectively. Firstly, we consider the

Table 3   Optimal time T_p^* and its cost rate $C(T_p^*)$ (random age replacement policy with uniform distribution case)

$\dfrac{c_2}{c_1 - c_2}$	$\delta = 0.1$		$\delta = 0.2$		$\delta = 0.3$		$\delta = 0.5$	
	T_p^*	$C(T_p^*)$	T_p^*	$C(T_p^*)$	T_p^*	$C(T_p^*)$	T_p^*	$C(T_p^*)$
0.01	0.1161	23.0725	0.1556	30.3015	0.2078	39.1303	0.3374	57.0392
0.02	0.1538	15.2763	0.1863	18.1425	0.2330	21.9377	0.3565	30.1348
0.05	0.2331	9.2626	0.2578	10.0400	0.2964	11.1622	0.4087	13.8245
0.10	0.3259	6.4740	0.3465	6.7477	0.3802	7.1593	0.4842	8.1951
0.20	0.4610	4.5796	0.4797	4.6703	0.5107	4.8098	0.6104	5.1729
0.50	0.7444	2.9578	0.7639	2.9755	0.7967	3.0027	0.9027	3.0733
1.00	1.0987	2.1829	1.1225	2.1865	1.1621	2.1920	1.2860	2.2054
2.00	1.6999	1.6887	1.7333	1.6890	1.7871	1.6894	1.9419	1.6902
5.00	3.4071	1.3541	3.4668	1.3541	3.5504	1.3541	3.7434	1.3541

Table 4   Optimal time T_p^* and its cost rate $C(T_p^*)$ (random age replacement policy with triangular distribution case)

$\dfrac{c_2}{c_1 - c_2}$	$\delta = 0.1$		$\delta = 0.2$		$\delta = 0.3$		$\delta = 0.5$	
	T_p^*	$C(T_p^*)$	T_p^*	$C(T_p^*)$	T_p^*	$C(T_p^*)$	T_p^*	$C(T_p^*)$
0.01	0.1084	21.6024	0.1304	25.6137	0.1612	30.7542	0.2380	41.9786
0.02	0.1478	14.7329	0.1651	16.2810	0.1911	18.3915	0.2610	23.2546
0.05	0.2289	9.1236	0.2415	9.5311	0.2616	10.1468	0.3208	11.6240
0.10	0.3224	6.4262	0.3328	6.5674	0.3498	6.7899	0.4021	7.3748
0.20	0.4579	4.5639	0.4672	4.6104	0.4826	4.6846	0.5314	4.8977
0.50	0.7411	2.9547	0.7509	2.9638	0.7671	2.9784	0.8189	3.0205
1.00	1.0948	2.1822	1.1066	2.1841	1.1264	2.1872	1.1896	2.1957
2.00	1.6943	1.6886	1.7112	1.6888	1.7393	1.6890	1.8270	1.6897
5.00	3.3963	1.3541	3.4291	1.3541	3.4810	1.3541	3.6279	1.3541

expected cost rates without profit term. Tables 3 and 4 show the optimum T_p^* and expected cost rate $C(T_p^*)$ for the allowed limit δ and cost ratio $c_2/(c_1 - c_2)$. The standard age replacement cases ($\delta = 0$) give the minimum cost. The two tables indicate that T_p^* increases with δ and $c_2/(c_1 - c_2)$, and $C(T_p^*)$ increases with δ but decreases with $c_2/(c_1 - c_2)$. In case that the allowed limit δ and cost ratio $c_2/(c_1 - c_2)$ are same, triangular distribution cases give less costs than uniform cases.

Secondly, we consider the expected cost rate with profit term. Tables 5 and 6 present the optimal δ^* and T_p^* and minimal expected cost rate C^* for different values of $c_2/(c_1 - c_2)$ in uniform and triangular distribution cases, respectively ($c = 1, 2$). It indicates that T_p^* and C^* decrease with $c_2/(c_1 - c_2)$. From the tables, we can find that the random age replacement policy with profit term obtained by flexible PR times can be better than the deterministic age replacement policy (standard age replacement policy).

Table 5 Optimal δ^* and T_p^* with $c_2 = 1$ and $c = 1$ (random age replacement policy with profit term)

c_2	Uniform distribution case			Triangular distribution case		
$\overline{c_1 - c_2}$	δ^*	T_p^*	$C(T_p^*, \delta^*)$	δ^*	T_p^*	$C(T_p^*, \delta^*)$
0.01	0.01	0.1000	19.9919	0.01	0.1000	19.9923
0.02	0.02	0.1413	14.1301	0.03	0.1414	14.1309
0.05	0.05	0.2234	8.9225	0.06	0.2233	8.9223
0.10	0.10	0.3165	6.2875	0.14	0.3164	6.2873
0.20	0.22	0.4545	4.4003	0.31	0.4542	4.3998
0.50	1.30	1.3019	2.2594	0.79	0.7904	2.6008
1.00	1.55	1.5514	1.1301	1.21	1.2104	1.6087
2.00	1.64	1.6487	0.5697	1.73	1.7369	0.8752
5.00	1.69	1.7175	0.2327	2.16	2.1616	0.3538

Table 6 Optimal δ^* and T_p^* with $c_2 = 1$ and $c = 2$ (random age replacement policy with profit term)

c_2	Uniform distribution case			Triangular distribution case		
$\overline{c_1 - c_2}$	δ^*	T_p^*	$C(T_p^*, \delta^*)$	δ^*	T_p^*	$C(T_p^*, \delta^*)$
0.01	0.02	0.0996	19.9173	0.02	0.0996	19.9189
0.02	0.04	0.1404	14.0225	0.05	0.1403	14.0200
0.05	0.09	0.2197	8.7405	0.13	0.2197	8.7402
0.10	0.19	0.3077	6.0081	0.28	0.3082	6.0068
0.20	0.44	0.4456	3.9152	0.42	0.4233	3.9944
0.50	0.62	0.6247	1.9397	0.62	0.6239	2.2053
1.00	0.73	0.7358	1.0433	0.80	0.8032	1.3154
2.00	0.79	0.8139	0.5484	0.96	0.9722	0.7369
5.00	0.83	0.8905	0.2325	1.10	1.1346	0.3202

3 Preventive Replacement Policies with Random Opportunity Times

In this section, we assume that preventive replacements are allowed at opportunity times with PR cost, c_2 and the opportunity times occur according to a renewal process with a distribution $H(t)$ in which the renewal function is $M(t)$. The renewal process of opportunities is assumed to be reset after corrective and preventive replacement.

3.1 *Opportunity Age Replacement Model*

We consider an age replacement policy in which the system is replaced preventively at the first opportunity time after T_p. The age replacement policy is studied by [Dekker and Dijkstra (1992), Chen *et al.* (2010) and Zhao *et al.* (2012)].

Firstly, we derive the residual life in the renewal process of opportunities. Let the residual life be $\gamma(T_p)$ and $D(t; T_p)$ be the distribution of $\gamma(T_p)$. Then the distribution of the residual life is given [Nakagawa (2011)]

$$P(\gamma(T_p) \leq t) = D(t; T_p) = H(T_p + t) - \int_0^{T_p} \overline{H}(T_p + t - u) dM(u). \quad (8)$$

Then the actual PR time is a random variable and $T_a = T_p + \gamma(T_p)$. The expected cost of one cycle is

$$c_1 P(X \leq T_a) + c_2 P(X > T_a) = c_1 - (c_1 - c_2) P(X > T_a)$$

$$= \int_0^\infty [-(c_1 - c_2) P(X > T_p + t) + c_1] dD(t)$$

$$= c_1 - (c_1 - c_2) \int_0^\infty \overline{F}(T_p + t) dD(t; T_p).$$

The expected duration of one cycle is

$$E[\min(X, T_p)] = \int_0^\infty E[\min(X, T_p + t)] dD(t; T_p)$$

$$= \int_0^\infty \int_0^{t+T_p} \overline{F}(x) dx dD(t; T_p)$$

$$= \int_0^{T_p} \overline{F}(x) dx + \int_0^\infty \overline{F}(x + T_p) \overline{D}(x) dx.$$

The expected cost rate is given as

$$C_4(T_p) = \frac{c_1 - (c_1 - c_2) \int_0^\infty \overline{F}(T_p + t) dD(t; T_p)}{\int_0^{T_p} \overline{F}(x) dx + \int_0^\infty \overline{F}(x + T_p) \overline{D}(t; T_p) dt}, \quad (9)$$

where $D(\cdot)$ is the distribution function of the residual life time. This model is similar with the age replacement with overtime policy introduced by [Zhao et al. (2012)].

3.2   Number-based Replacement Policy

In this section, we consider a preventive replacement policy based on the number of opportunities. The system is replaced preventively before failure at the Nth opportunity time. The actual PR time is a random variable and the Nth opportunity time. The distribution function of Nth opportunity time is the j−fold Stieltjes convolution, $H^{(N)}(t)$ of $H(t)$ with itself.

$$c_1 P(X \le T_a) + c_2 P(X > T_a) = c_1 - (c_1 - c_2) P(X > T_a)$$
$$= \int_0^\infty [-(c_1 - c_2) P(X \le t) + c_1] dH^{(N)}(t)$$
$$= c_1 - (c_1 - c_2) \int_0^\infty \overline{F}(t) dH^{(N)}(t).$$

The expected duration of a cycle is

$$E[\min(X, T_a)] = \int_0^\infty E[\min(X, t)] dH^{(N)}(t) = \int_0^\infty \int_0^t \overline{F}(x) dx dH^{(N)}(t)$$
$$= \int_0^\infty \overline{F}(x)[1 - H^{(N)}(t)] dt.$$

Then the expected cost rate (Chen, 2013) is

$$C_5(N) = \frac{c_1 - (c_1 - c_2) \int_0^\infty \overline{F}(t) dH^{(N)}(t)}{\int_0^\infty \overline{F}(x)[1 - H^{(N)}(t)] dt}. \tag{10}$$

3.3 *Numerical Examples*

We consider numerical examples in which the system failure distribution is a Weibull distribution in which the scale and shape parameters are given by 1 and 2. A Poisson process is assumed and inter-arrival times of opportunities follows an Exponential distribution with a finite mean $1/\theta$.

Then, $D(t; T_p) = 1 - e^{-\theta t}$ and $H^{(N)}(t) = 1 - \sum_{j=0}^{N-1} (\theta t)^j / j! e^{-\theta t}$, the expected cost rate in (9) is

$$C_D(T_p) = \frac{c_1 - (c_1 - c_2) \int_0^\infty \theta e^{-\theta t - (t + T_p)^2} dt}{\int_0^{T_p} e^{-t^2} dt + \int_0^\infty e^{-\theta t - (t + T_p)^2} dt}, \tag{11}$$

and the expected cost rate in (10) is

$$C_O(N) = \frac{c_1 - (c_1 - c_2) \int_0^\infty e^{-t^2} d[1 - \sum_{j=0}^{N-1} (\theta t)^j / j! e^{-\theta t}]}{\int_0^\infty \sum_{j=0}^{N-1} (\theta t)^j / j! e^{-\theta t - t^2} dt}. \tag{12}$$

Table 7 presents the optimal T_p^* and expected cost rate $C(T_p^*)$ for different $c_2/(c_1 - c_2)$ and $1/\theta$. It indicates that T_p^* increases with $c_2/(c_1 - c_2)$ and θ, and $C(T_p^*)$ decreases with $c_2/(c_1 - c_2)$ and θ. Table 8 presents the optimal N^* and expected cost rate $C(N^*)$ for different $c_2/(c_1 - c_2)$ and $1/\theta$. It indicates that N^* increases with $c_2/(c_1 - c_2)$ and θ, and $C(T^*)$ decreases with $c_2/(c_1 - c_2)$ and θ. Additionally, age-based replacement policy

Table 7   Optimal time T_p^* and its cost rate $C(T_p^*)$ (random age replacement policy with exponentially distributed opportunity time)

$\dfrac{c_2}{c_1 - c_2}$	$1/\theta = 0.1$		$1/\theta = 0.2$		$1/\theta = 0.3$		$1/\theta = 0.5$	
	T_p^*	$C(T_p^*)$	T_p^*	$C(T_p^*)$	T_p^*	$C(T_p^*)$	T_p^*	$C(T_p^*)$
0.01	0.0428	27.6715	0.0260	40.1860	0.0191	54.2915	0.0153	66.4960
0.02	0.0755	17.0471	0.0543	22.3580	0.0372	28.7645	0.0301	34.5100
0.05	0.1493	9.7223	0.1166	11.2430	0.0869	13.2859	0.0730	15.2420
0.10	0.2392	6.6295	0.1932	7.1619	0.1586	7.9378	0.1368	8.7189
0.20	0.3727	4.6285	0.3143	4.7996	0.2785	5.0580	0.2489	5.3306
0.50	0.6558	2.9664	0.6302	2.9894	0.5554	3.0420	0.5214	3.0909
1.00	1.0112	2.1845	1.0062	2.1393	0.9178	2.1978	0.8827	2.2061
2.00	1.6148	1.6888	1.5766	1.6892	1.5389	1.6896	1.5086	1.6901
5.00	3.3259	1.3541	3.3109	1.3541	3.1966	1.3541	3.1460	1.3541

Table 8   Optimal time N^* and its cost rate $C(N^*)$ (Number-based replacement policy with exponentially distributed opportunity time)

$\dfrac{c_2}{c_1 - c_2}$	$1/\theta = 0.1$		$1/\theta = 0.2$		$1/\theta = 0.3$		$1/\theta = 0.5$	
	N^*	$C(N^*)$	N^*	$C(N^*)$	N^*	$C(N^*)$	N^*	$C(N^*)$
0.01	1	29.4627	1	40.6208	1	54.3889	1	66.5357
0.02	2	19.3927	1	22.9867	1	28.9488	1	34.5873
0.05	2	10.8717	1	12.4063	1	13.6848	1	15.4183
0.10	3	7.2568	2	7.8622	1	8.5968	1	9.0286
0.20	5	4.9208	3	5.2009	2	5.4474	2	5.7622
0.50	8	3.0618	5	3.1288	3	3.1923	3	3.2364
1.00	13	2.2112	8	2.2262	6	2.2373	5	2.2445
2.00	22	1.6914	14	1.6921	10	1.6924	9	1.6925
5.00	32	1.3541	19	1.3541	14	1.3541	11	1.3541

outperforms number-based policy but the gap between two policies in the expected cost rate is not big.

4   Extended Models

In this section, we consider two extended models:

Case 1: In this case we consider some extended models. In Section 3, we assume a renewal process for opportunity times. First, we assume that the opportunity times occur according to a non-homogeneous Poisson process (NHPP) with intensity function $\lambda(t)$ and let $\Lambda(t) = \int_0^t \lambda(x)\mathrm{d}x$.

Then, the residual life after T_p has the following distribution function

$$P(\gamma(T_p) \le t) = D(T_p, t)$$
$$= 1 - P\{N(T_p + t) - N(T_p) = 0\} = 1 - \mathrm{e}^{-(\Lambda(T_p+t)-\Lambda(T_p))}. \tag{13}$$

Thus, the expected cost rate under random age replacement policy can be obtained from Equation (9). For NHPP case,

$$H^{(N)}(t) = P(S_n \leq t)1 - \sum_{k=0}^{n-1} \frac{[\Lambda(t)]^k e^{-\Lambda(t)}}{k!}.$$

Thus, the expected cost rate of number-based replacement policy can be obtained easily from Equation (10).

Case 2: we have two types of preventive replacement (random and deterministic) with different PR costs. Then we consider a new age replacement policy with two parameters in which the failed system is replaced correctively before PR times. If the random PR time T_r is less than T_2 and greater than T_1, $T_1 < T_r < T_2$, then we replace the system at T_r preventively. Otherwise, the PR time is T_2. Then the expected cost rate is

$$C_R(T_1, T_2) = \frac{\begin{array}{c}[F(T_2)c_1 + \overline{F}(T_2)c_2][G(T_1) + \overline{G}(T_2)] \\ + \int_0^{T_1}[F(t)c_1 + \overline{F}(t)c_3]\mathrm{d}G(t)\end{array}}{[G(T_1) + \overline{G}(T_2)]\int_0^{T_2}\overline{F}(t)\mathrm{d}t + \int_{T_1}^{T_2}[\int_0^t \overline{F}(x)\mathrm{d}x]\mathrm{d}G(t)}. \quad (14)$$

For further studies, we can assume that the system is maintained at every opportunity and the state of the system is improved by regular maintenance. The effect of regular maintenance can be modeled by imperfect repair concept. Then we can apply age-based and number-based replacement (perfect repair) policies to overhaul the system.

5   Conclusion

In this paper, we consider random replacement models in which the actual preventive maintenance times are random variables. The actual PR time has the mean value equal to the planned PR time in the first case and preventive replacement can be done only on opportunity time points that occur according to stochastic processes in the second case. We consider age-based and number-based replacement policies in two actual PR time models. The standard age replacement and random replacement models are summarized and random age replacement models with profit terms are studied. Corrective and preventive replacement costs, and benefit for allowing random replacement are included. In particular, the benefit is dependent on the standard deviation of the random replacement. We derived the expected cost rate and optimal polices were obtained numerically. For some special cases, we found that random age replacement policy outperforms the standard age replacement policy. Time-based and number-based replacement models under random opportunistic times are reviewed and some extended models are discussed. In extended models, we derived the expected cost rates but more analysis is required for further studies.

Acknowledgement

This research was supported by Basic Science Research Program through the National Research Foundation of Korea (NRF) funded by the Ministry of Education, Science and Technology (No.2010-0025084). The first author is very grateful to Professor Toshio Nakagawa who stimulated his research interest in maintenance models and policies through many papers and four books in reliability and maintenance.

References

Barlow, R. E. and Proschan, F. (1965). *Mathematical Theory of Reliability* (Wiley, New York).

Chen, M., Mizutani, S. and Nakagawa, T. (2010). Random and age replacement policies, *International Journal of Reliability, Quality and Safety Engineering* **17**, pp. 27–39.

Chen, M., Nakamura, S. and Nakagawa, T. (2010). Replacement and preventive maintenance models with random working times, *IEICE Trans. Fundamentals* **E93-A**, pp. 500–507.

Chen, M. (2013). Optimal random replacement models with continuously processing jobs, *Applied Stochastic Models in Business and Industry* **29**, pp. 118–126.

Dekker, R. and Dijkstra, M. C. (1992) Opportunity-based age replacement: Exponentially distribution times between opportunities, *Naval Research Logistics*, **39**, pp. 175–190.

Nakagawa, T. (2005). *Maintenance Theory of Reliability* (Springer, London).

Nakagawa, T. (2011). *Stochastic Process with Applications to Reliability Theory* (Springer, London).

Nakagawa, T., Zhao, X. and Yun, W. Y. (2011). Optimal age replacement and inspection policies with random failure and replacement times, *International Journal of Reliability, Quality and Safety Engineering*, **18**, pp.405–416.

Zhao, X. and Nakagawa, T. (2012). Optimization problems of replacement first or last in reliability theory, *European Journal of Operational Research* **223**, pp. 141–149.

Zhao, X., Qian, C. and Nakagawa, T. (2012). Age replacement with overtime policy, in *The 5th Asia-Pacific International Symposium on Advanced Reliability and Maintenance Modeling V* (Nanjing, China), pp. 661–668.

Chapter 4

Optimal Replacement Interval of a Dual System

Satoshi Mizutani

*Department of Media Informations, Aichi University of Technology,
50-2 Manori, Nishihazama-cho, Gamagori 443-0047, Japan*

1 Introduction

In this chapter, we consider optimal replacement policies in which a system operates as a dual system when it is replaced. The system usually works by a aging unit system which is in random failure period or wearout failure period. When it is replaced to a fresh unit which is in initial failure period, the system works as a dual system for a while. That is, in order to avoid loss cost of the initial and wearout failure, the system works operates as a dual system from introducing the fresh unit to stopping the aging unit.

In recent years, information system with network has been greatly developed and become widely used, and hence failures of the system cause changing of the design and a social confusion. For example, when a system newly begins to operate, some troubles such as software failure or confusion of using the new control interface might be occurs. On the other hand, we delay a time of replacement, then risk of occurrence of wearout failure become large. We should devise a countermeasure, and propose that the system operates as a dual system for a while when the system is replaced. However, the cost of operating as a dual system will incur more than the one of operating as a single system. Accordingly, it is greatly important and indispensably necessary to plan suitably maintenance from the viewpoints of reliability and economics.

There have been many studies of maintenance and replacement policies using reliability theory [Barlow (1965); Osaki (1992); Nakagawa

(2011)]. Periodic replacement with minimal repair was considered in [Barlow (1965)]. Periodic replacement with minimal repair is summarized in [Nakagawa (1981)]. The periodic replacement model with two unit is considered in [Nakagawa (1987)]. In the model, when unit-1 fails, it undergoes minimal repair, and when unit-2 fails, the system is replaced without repairing. Minimal repair model in which the failure occurs at a non-homogeneous Poisson process are described in [Murthy (1991)]. There have been also many studies of maintenance policies for multi-unit systems. Nakagawa considered analytically optimal policy to decide the number of units [Nakagawa (1984)]. In the models, the system fails when all units have failed. Each unit fails from shocks, independently of the other units. Murthy and Nguyen proposed the model that the units fail with interaction [Murthy, Nguyen (1985a,b)]. Yasui and Nakagawa summarized optimum policies for a parallel system [Yasui, Nakagawa, Osaki (1988)]. Mizutani, Koike and Nakagawa considered the replacement policy that the system operates as a dual system from the beginning of fresh unit to the stopping of aging unit. In the model, the aging unit is in wearout failure period [Mizutani (2010)].

We consider a system which consists of one unit, and the system have been operated already before time 0, and we call the unit *aging unit*. In this paper, we extend the models [Mizutani (2010)] by introducing two cases that the aging unit is in random failure period or wearout failure period. We replace the aging unit with new one which is called *fresh unit*. Then, we propose that the system operates as a dual system from the beginning of fresh unit to the stopping of aging unit. Especially, we treat two cases: (i) aging unit is in random failure period, and (ii) in wearout failure period. We derive analytically optimal times of the stopping of aging unit, which minimize the expected cost. Further, for the model that the aging unit is in wearout failure period, we derive optimal time of the introducing of the fresh unit. Further, we consider a problem to obtain optimal time set which consists of a time of stopping aging unit and one of introducng fresh unit. Numerical examples are given when the failure distributions of the fresh unit is a Weibull distribution, and the aging unit is an exponential distribution or Weibull distribution.

2 Modeling

We make two models: (1) the system is replaced when the aging unit operate in random failure period, (2) the system is replaced when aging unit operate in wearout failure period. Firstly, we show common assumption for two models:

(1) Fresh unit has a failure distribution

$$F_f(t) = 1 - e^{-H_f(t)},$$

and the failure rate $h_f(t) \equiv H_f'(t)$ is a monotonic decreasing function, where $H_f'(t)$ is a differential function of $H_f(t)$, i.e., $H_f(t) = \int_0^t h_f(u)du$.

(2) When the system fails, a minimal repair is done: That is, the failure rate remains undisturbed after the repair, i.e., the system after the repair has the same failure rate as before the failure [Nakagawa (2005)]. It is assumed that the repair times are negligible.

(3) Cost c_a is a minimal repair cost when the aging unit operates as a dual system, and α_a is an additional repair cost when it operates as a single system. Cost c_f ($c_f \geq c_a > 0$) is a minimal repair cost when the fresh unit operates as a dual system, and α_f is an additional repair cost when the fresh unit operates as a single system. Furthermore, cost c_d is the operating cost per unit of time, and c_r is the replacement cost.

(1) Replacement in Random Failure Period

We consider an aging unit which is replaced at time t_a (> 0) when it is in a random failure period and has an exponential distribution $(1 - e^{-\lambda t})$. In this case, this model is a case that we can predict the time at which the state of the aging unit change to a wearout failure period, and can replace it before a wearout failure period. We can adopt this model to another case that the system is replaced for a reason beside the system wearout. For examples, we can consider the reasons such as a technological innovation.

We assume the following assumptions for this model:

(1) The fresh unit is an introduced at time 0, and is in initial failure period from time 0 to T ($T > 0$).
(2) The stopping time of an aging unit is t_a ($0 \leq t_a \leq T$).
(3) The aging unit has an exponential distribution

$$F_a(t) = 1 - e^{-\lambda t}$$

with finite mean $1/\lambda$.

The expected number of failure of the fresh unit from 0 to time T is $H_f(T)$. Especially, when the fresh unit operates as a single unit from time t_a

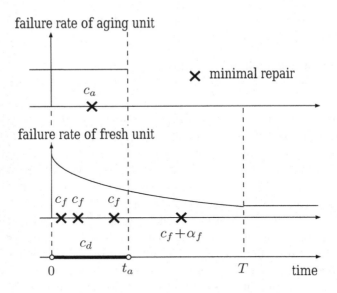

Fig. 1   Replacement model in random failure period

to time T, the expected number of failure of the fresh unit is $H_f(T) - H_f(t_a)$. The expected number of failure of aging unit from 0 to time t_a is λt_a.

Thus, the expected cost $C_1(t_a)$ from time 0 to time T is given by

$$C_1(t_a) = c_f H_f(T) + \alpha_f [H_f(T) - H_f(t_a)] + (c_a \lambda + c_d)t_a + c_r. \qquad (1)$$

(2) Replacement in Wearout Failure Period

We consider an aging unit which is replaced in a wearout failure period (Fig. 2). We make following assumptions:

(1) The aging unit is in a wearout failure period at time 0, and the fresh unit is introduced at time t_f. The fresh unit is in an initial failure period from t_f to $t_f + T$, and its state changes to random failure period. The aging unit is stopped at time t_a $(t_f \le t_a \le t_f + T)$.
(2) The aging unit has a failure distribution $F_a(t) = 1 - e^{-H_a(t)}$, and the failure rate $h_a(t) \equiv H_a'(t)$ is a monotonic increasing function.

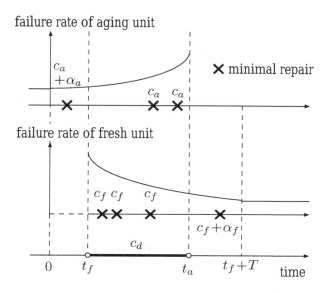

Fig. 2   Replacement in wearout failure period

The failure distribution $F(t)$ of a duplex system at time t $(0 \leq t \leq T)$ is

$$F(t) \equiv \begin{cases} 1 - e^{-H_f(t)} & t < t_f, \\ 1 - e^{-H_f(t) - H_a(t-t_f)}, & t_f \leq t \leq t_a, \\ 1 - e^{-H_a(t-t_f)}. & t_a < t < T. \end{cases}$$

The failure rate is

$$r(t) \equiv \begin{cases} h_f(t) & t < t_f, \\ h_f(t) + h_a(t - t_f), & t_f \leq t \leq t_a, \\ h_a(t - t_f). & t_a < t < T. \end{cases}$$

The density function $f(t)$ of the failure time distribution is

$$f(t) = [h_f(t) + h_a(t - t_f)]e^{-H_f(t) - H_a(t-t_f)}.$$

Therefore, the failure rate at time t $(t_f \leq t \leq t_a)$ $r(t_f, t_a)$ is

$$r(t_f, t_a) = \frac{f(t)}{1 - F(t)} = h_f(t) + h_a(t - t_f). \tag{2}$$

The expected number of the failure of the fresh unit from t_a to time $t_f + T$ is $H_f(T) - H_f(t_a)$. Thus, the expected cost rate $C_2(t_a, t_f)$ from 0 to $t_f + T$ is given by

$$C_2(t_a, t_f) = \frac{\begin{array}{c} c_a H_a(t_a) + \alpha_a H_a(t_f) + c_f H_f(T) \\ + \alpha_f [H_f(T) - H_f(t_a - t_f)] + c_d(t_a - t_f) + c_r \end{array}}{t_f + T}. \tag{3}$$

The equation (1) is the expected cost from time 0 to time T, and is not the expected cost rate, because the interval time T is given as a constant value. However, the equation (3) is the expected cos rate, because the interval time $t_f + T$ is not given as a constant value.

3   Optimal Policies

3.1   *Replacement in Random Failure Period*

We seek an optimal t_a^* $(0 \leq t_a^* \leq T)$ which minimizes the expected cost rate $C_1(t_a)$. Differentiating $C_1(t_a)$ with respect to t_a and setting it equal to 0,

$$h_f(t_a) = \frac{\lambda c_a + c_d}{\alpha_f}. \tag{4}$$

Because $h_f(t_a)$ is a monotonic decreasing function, we have the following optimal policy:

(i) If $(\lambda c_a + c_d)/\alpha_f \geq h_f(0)$ then $t_a^* = 0$
(ii) If $h_f(0) > (\lambda c_a + c_d)/\alpha_f > h_f(T)$ then there exists a finite and unique t_a^* $(0 < t_a^* < T)$ which satisfies (4).
(iii) If $h_f(T) \geq (\lambda c_a + c_d)/\alpha_f$ then $t_a^* = T$.

3.2   *Replacement in Wearout Failure Period*

3.2.1   *Analysis of Optimal t_a*

We seek optimal t_a^* that minimize the expected cost rate $C_2(t_a, t_f)$. Differentiating $C_2(t_a, t_f)$ with respect to t_a^* and setting it equal to 0,

$$\alpha_f h_f(t_a - t_f) - c_a h_a(t_a) = c_d. \tag{5}$$

Letting $L_{2a}(t_a)$ be the left-hand of above equation. Because $h_f(t)$ $(t > 0)$ is a monotonic increasing function and $h_a(t)$ $(t > 0)$ is a monotonic decreasing function, $L_{2a}(t)$ $(t > 0)$ is a monotonic increasing function. Therefore, we have following optimal policy:

(i) If $c_d \geq L_{2a}(t_f)$ then $t_a^* = 0$
(ii) If $L_{2a}(t_f) > c_d > L_{2a}(t_f + T)$ then there exits a finite and unique t_a^* which satisfies (5).
(iii) If $L_{2a}(t_f + T) \geq c_d$ then $t_a^* = t_f + T$

3.2.2 *Analysis of Optimal* t_f

Next, we seek optimal t_f^* that minimize the expected cost rate $C_2(t_a, t_f)$. Differentiating $C_2(t_a, t_f)$ with t_f and setting it equal to 0,

$$\alpha_a[h_a(t_f)(t_f + T) - H_a(t_f)] + \alpha_f[h_f(t_a - t_f)(t_f + T) + H_f(t_a - t_f)$$
$$= c_a H_a(t_a) + (c_f + \alpha_f)H_f(T) + c_d(t_a + T) + c_r. \tag{6}$$

Letting $L_{2f}(t_f)$ be the left-hand of above equation. Because

$$L'_{2f}(t_f) = \alpha_a h'_a(t_f)(t_f + T) - \alpha_f h'_f(t_a - t_f) > 0,$$

$L_{2f}(t_f)$ is a monotonic increasing function. Therefore, we have following optimal policy:

(i) If $L_{2f}(0) \geq c_a H_a(t_a) + (c_f + \alpha_f)H_f(T) + c_d(t_a + T) + c_r$ then $t_f^* = 0$
(ii) If $L_{2f}(t_a) > c_a H_a(t_a) + (c_f + \alpha_f)H_f(T) + c_d(t_a + T) + c_r > L_{2f}(0)$ then there exists a finite and unique t_f^* which satisfies (6).
(iii) If $c_a H_a(t_a) + (c_f + \alpha_f)H_f(T) + c_d(t_a + T) + c_r \geq L_{2f}(t_a)$ then $t_f^* = t_a$

3.2.3 *Analysis of Optimal* t_a *and* t_f

We seek optimal set (t_a^*, t_f^*) that minimize the expected cost rate $C_2(t_a, t_f)$. For the purpose, we can use following well-known theorem as "Second Derivative Test": Suppose that $z = f(x, y)$ has continuous first and second partial derivatives, and $\partial z/\partial x = 0$ and $\partial z/\partial y = 0$. Let $\Delta \equiv \partial^2 z/\partial x \partial y - \partial^2 z/\partial x \partial x \cdot \partial^2 z/\partial y \partial y < 0$. If $\Delta > 0$ and $\partial^2 z/\partial x \partial x < 0$, then (x, y) is a pont that z is a local maximum. If $\Delta > 0$ and $\partial^2 z/\partial x \partial x > 0$, then (x, y) is a point that z is a local minimum. If $\Delta < 0$, then (x, y) is a saddle point. To obtain optimal t_a^* and t_f^*, we have

$$\frac{\partial C_2(t_a, t_f)}{\partial t_a} = \frac{c_a h_a(t_a) - \alpha_f h_f(t_a - t_f) + c_d}{t_f + T},$$

$$\frac{\partial C_2(t_a, t_f)}{\partial t_f} = \frac{\alpha_a h_a(t_f) + \alpha_f h_f(t_a - t_f) - c_d}{t_f + T} - \frac{C_2(t_a, t_f)}{t_f + T},$$

$$\frac{\partial^2 C_2(t_a, t_f)}{\partial t_a \partial t_f} = \frac{\alpha_f [h_f'(t_a - t_f)(t_f + T) + h_f(t_a - t_f)] - c_a h_a(t_a) - c_d}{(t_f + T)^2},$$

$$\frac{\partial^2 C_2(t_a, t_f)}{\partial t_a^2} = \frac{c_a h_a'(t_f) - \alpha_f h_f'(t_a - t_f)}{t_f + T},$$

$$\frac{\partial^2 C_2(t_a, t_f)}{\partial t_f^2} = \frac{\alpha_a h_a'(t_a) - \alpha_f h_f'(t_a - t_f) - 2\partial C_2(t_a, t_f)}{t_f + T}.$$

When we set $\partial C_2(t_a, t_f)/\partial t_a = 0$ and $\partial C_2(t_a, t_f)/\partial t_f = 0$,

$$\frac{\partial^2 C_2(t_a, t_f)}{\partial t_a \partial t_f} = \frac{\alpha_f h_f'(t_a - t_f)}{(t_f + T)^2},$$

$$\frac{\partial^2 C_2(t_a, t_f)}{\partial t_a^2} = \frac{\partial^2 C_2(t_a, t_f)}{\partial t_f^2} = \frac{\alpha_a h_a'(t_a) - \alpha_f h_f'(t_a - t_f)}{t_f + T} > 0,$$

$$\frac{\partial^2 C_2(t_a, t_f)}{\partial t_a \partial t_f} - \frac{\partial^2 C_2(t_a, t_f)}{\partial t_a^2} \cdot \frac{\partial^2 C_2}{\partial t_f^2}$$
$$= \frac{2\alpha_a \alpha_f h_a'(t_a) h_f'(t_a - t_f) - [\alpha_a h_a'(t_a)]^2}{(t_f + T)^2} < 0.$$

Therefore, when $t_a > t_f$, $C_2(t_a, t_f)$ has a local minimal point.

Thus, we can say that if there exists (t_a, t_f) $(t_a > t_f)$ satisfies $\partial C_2(t_a, t_f)/\partial t_a = 0$ and $\partial C_2(t_a, t_f)/\partial t_f = 0$, if condition $t_f + T > t_a > t_f > 0$ is satisfied, then (t_a, t_f) is optimal (t_a^*, t_f^*) which minimizes the expected cost $C_2(t_a^*, t_f^*)$. Further, if (t_a, t_f) is not satisfied the condition, optimal (t_a^*, t_f^*) exists on additonal condition $t_f + T = t_a$ or $t_a = 0$ or $t_f = 0$.

4   Numerical Examples

We compute numerically optimal replacement times. Especially, we suppose that the fresh unit has a Weibull distribution with cumulative hazard function $H_f(t) = \mu t^s$ $(0 < s < 1)$, and $1/\mu = 10$, $1/\lambda = 100$, $c_a = c_f = 1$, $c_d = 2$, $T = 100$.

Table 1   Optimal t_a^* in random failure period model

α_f	s		
	0.7	0.8	0.9
10	0.0297	0.0100	0.0003
15	0.1148	0.0758	0.0187
20	0.2995	0.3196	0.3317
25	0.6302	0.9754	3.0893
30	1.1572	2.4270	19.1283
35	1.9345	5.2458	89.3604
40	3.0191	10.2275	100.0000
45	4.4708	18.4303	100.0000
50	6.3519	31.2119	100.0000

4.1   *Replacement in Random Failure Period*

We compute numerically optimal t_a^* which satisfies the equation (4), that is, t_a^* minimizes the expected cost $C_1(t_a)$ in (1). Table 1 gives optimal t_a^* for $\alpha_f = 10, 15, 20, 25, 30, 35, 40, 45, 50$ and $s = 0.7, 0.8, 0.9$. We can see that we should take a large interval of dual system, when additional cost α_f in single system is large. When α_f is small, optimal t_a^* decrease when shape parameter s increase. On the other hand, when α_f is large, optimal t_a^* increase when shape parameter s increase. This reason is considered as follows. When α_f is small t_a^* is small as mentioned above. If t_a^* is larger than 1, $h_f(t_a^*) = \mu s t_a^{*s-1}$ is large when s is large. On the other hand, if t_a^* is smaller than 1, $h_f(t_a^*)$ is smaller when s is large. We can consider this tendency give the numerical results.

4.2   *Replacement in Wearout Failure Period*

We compute numerically optimal t_a^* and t_f^* which minimize the expected cost rate $C_2(t_a, t_s)$. Especially, we suppose that the aging unit has a Weibull distribution with cumulative hazard function $H_a(t) = \lambda t^m$ $(m > 1)$.

Table 2 gives optimal t_a^* for the same parameter of Table 1 when $t_f = 10$. Because $t_f = 10$, t_a^* is larger than 10, optimal t_a^* is larger than one in Table 1. We can see $t_a^* - t_f$ in Table 2 is similar with and larger than one in Table 1.

Table 3 give optimal t_f^* for $\alpha_a = 10, 15, 20, 25, 30, 35, 40, 40, 50$ and $m = 1.1, 1.2, 1.3, 1.4, 1.5$, when $\alpha_f = 50$, $s = 0.7$, $t_a = 50$. We can see that t_a^* when α_a is large, t_f^* should be small, i.e., the interval of single unit should be small.

Table 2   Optimal t_a^* in wearout failure period when $t_f = 10$

α_f	s		
	0.7	0.8	0.9
10	10.0302	10.0102	10.0003
15	10.1167	10.0778	10.0196
20	10.3046	10.3277	10.3487
25	10.6408	11.0000	13.2473
30	11.1766	12.4883	30.1066
35	11.9669	15.3782	103.9303
40	13.0697	20.4858	110.0000
45	14.5457	28.8957	110.0000
50	16.4584	42.0000	110.0000

Table 3   Optimal t_f^* in wearout failure period when $\alpha_f = 50$, $s = 0.7$, $t_a = 50$

α_a	m				
	1.1	1.2	1.3	1.4	1.5
10	40.22793	38.82731	36.21407	31.37484	24.01530
15	39.06968	36.54794	31.62958	23.49385	15.30535
20	37.72920	33.73476	26.08212	16.47542	10.02834
25	36.17296	30.32267	20.24973	11.40427	6.92544
30	34.36208	26.32469	15.01229	8.03251	5.02004
35	32.25319	21.90113	10.87539	5.81527	3.78689
45	26.96803	13.21125	5.72344	3.31093	2.35916
50	23.73486	9.70127	4.23442	2.58987	1.92752

5   Conclusions

We have considered optimal replacement policies with an interval of dual system for two cases: (1) Aging unit is in random failure period, and (2) an aging unit is in wearout failure period. We have obtained the expected cost for a interval that the fresh unit is in initial failure period. Further, we have derived analytically optimal policies of stopping aging unit which minimizes the expected cost. Numerical examples have been given when the times to failure of an aging unit are exponential or Weibull distributions. These formulations and results would be applied to other real systems such as digital circuits or server systems by suitable modifications.

References

Barlow, R. E. and Proschan, F. (1965). *Mathematical Theory of Reliability* (John Wiley & Sons, New York).

Lotfi Tadj, Mohamed-Salah Ouali, Soumaya Yacout and Daoud Ait-Kadi (2011). *Replacement Models With Minimal Repair* (Springer Verlag, London).

Mizutani, S. and Nakagawa, T. (2010). "Optimal Maintenance Policy with an Interval of Duplex System", in: S. Chukova, J. Haywood and T. Dohi (eds), *Advanced Reliability Modeling IV* (McGraw-Hill, Taiwan), pp. 496–503.

Murthy, D. N. P. (1991). "A note on minimal repair", *IEEE Transactions Reliability*, **40**, pp. 245–246.

Murthy, D. N. P. and Nguyen, D. G. (1985a). "Study of two-component system with failure interaction", *Naval Research Logistics* **32**, pp. 239–248.

Murthy, D. N. P. and Nguyen, D. G. (1985b). "Study of a multi-component system with failure interaction", *European Journal of Operations Research* **21**, pp. 330–338.

Nakagawa, T. (1981). "A Summary of periodic replacement with minimal repair at failure", *Journal Operations Research Society Japan* **24**, pp. 213–227.

Nakagawa, T. (1984). "Optimal number of units for a parallel system", *Journal Applied Probability* **21**, pp. 431–436.

Nakagawa, T. (1987). "Optimum replacement policies for systems with two types of units", in: S. Osaki and J. H. Cao (eds), *Reliability Theory and Applications Proceedings of the China-Japan Reliability Symposium* (Shanghai, China).

Nakagawa, T. (2005). *Maintenance Theory of Reliability* (Springer Verlag, London).

Nakagawa, T. (2011). *Stochastic Processes with Applications to Reliability Theory* (Springer Verlag, London).

Osaki, S. (1992). *Applied Stochastic System Modeling* (Springer Verlag, Berlin).

Pham, H., Suprasad, A. and Misra, B. (1996). "Reliability and MTTF prediction of k-out-of-n complex systems with components subjected to multiple stages of degradation", *International Journal System Science* **27**, pp. 995–1000.

Yasui, K., Nakagawa, T. and Osaki, S. (1988). "A summary of optimum replacement policies for a parallel redundant system", *Microelectron Reliability* **28**, pp. 635–641.

Chapter 5

Cumulative Damage Models with Random Working Times

Xufeng Zhao[1,2], Cunhua Qian[1] and Shey-Huei Sheu[3]

[1]*School of Economics and Management,*
Nanjing University of Technology,
30 Puzhu Road, Nanjing 211816, China
[2]*Graduate School of Management and Information Sciences,*
Aichi Institute of Technology,
1247 Yachigusa, Yakusa-cho, Toyota 470-0392, Japan
[3]*Department of Statistics and Informatics Science,*
Providence University,
200 Chung Chi Rd., Shalu, Taichung 43301, Taiwan

1 Introduction

Most operating units are repaired or replaced when they have failed. However, it may require much time and suffer higher cost to repair a failed unit or replace it with a new one, so that it is necessary to replace it to prevent failures. Replacement after failure and before failure are called corrective replacement (CR) and preventive replacement (PR), respectively. The recent published books [Osaki (2002); Pham (2003); Nakagawa (2005); Wang and Pham (2007); Kobbacy and Murthy (2008); Manzini, Regattieri, Pham and Ferrari (2010); Nakagawa (2011)] collected many PR models or other maintenance models in theory and their applications in industrial systems. Especially, Nakagawa (2007) summarized sufficiently maintenance policies and their optimization problems for shock and damage models which are called cumulative damage models. These models, which play an important role in reliability theory, are considered as a sequence of shocks that occur randomly in time and give some amount of damage to a unit. The damage is accumulated to the current damage level, weakens the unit gradually, and makes it failure when the total damage exceeds a failure level.

Some units in offices and industries execute jobs or computer procedures with random working times successively. For such units, it would be impossible or impractical to maintain or replace them in a strict regular fashion, e.g., a planned time T, even though the maintenance or replacement time comes, because sudden suspension of the job may suffer losses of production in different degrees if there is no sufficient preparation in advance [Barlow and Proschan (1965), p. 72; Nakagawa (2005), p. 245]. On the other hand, we sometimes are interested in certain quantities of units in which to analyze their reliabilities and maintenances that are usually observed by a unique time scale such as age or operating time, but they are often measured by some combined scales in reliability applications. Alternative time scales were investigated and how to select effective ones to analyze mean time to failure (MTTF) was discussed [Duchesne and Lawless (2000)]. Replacement policies with two time scales, such as age and number of uses, were proposed [Nakagawa (2005), p. 83]. Taking parts of an aircraft as an example, appropriate maintenance and replacement policies are usually scheduled at a total hours of operations or at a specified number of flights since the last major overhaul [Nakagawa (2008), p. 149].

By considering the factors of working times in operations, the reliability quantities of the random age replacement policy were obtained [Barlow and Proschan (1965), p. 72]. Several schedules of jobs that have random processing times were summarized [Pinedo (2002)]. When a job has a variable working cycle or processing time, it would be better to do maintenance or replacement after the job is just completed even though the maintenance time has arrived [Sugiura, Mizutani and Nakagawa (2004); Nakagawa (2005), p. 245]. From such a viewpoint, age and periodic replacement policies which are done at the first completion of some working time over a planned time T was proposed [Chen, Mizutani and Nakagawa (2010); Chen, Nakamura and Nakagawa (2010)], and optimal age replacement with overtime policy was derived in [Zhao, Qian and Nakagawa (2012)].

Furthermore, by combining a planned replacement with working times, age and periodic replacement policies, where the unit is replaced at a planned time T and at the Nth random working time, were discussed [Chen, Mizutani and Nakagawa (2010); Chen, Nakamura and Nakagawa (2010)]. By considering the cases when replacement costs suffered for failures would be estimated to be not so high and the factor of working times, age and periodic replacement policies, where the unit is replaced at a planned operating time T or at a random working cycle Y, whichever occurs last, were proposed [Zhao and Nakagawa (2012)]. This also indicated that policies proposed in [Chen, Mizutani and Nakagawa (2010); Chen, Nakamura and Nakagawa (2010)] would cause frequent and unnecessary replacement which may incur production losses under the assumption of "whichever occurs first". The notion "whichever occurs last" was applied in a cumulative damage model, where the unit is replaced before failure at a planned time T, at a shock number N, or at a damage level Z, was discussed [Zhao, Nakayama and Nakamura (2011); Zhao, Qian and Nakagawa (2013)].

From above considerations, this chapter considers an operating unit which works at successive random times for jobs and its replacement policies, using the cumulative damage models [Nakagawa (2007)] by replacing *shock* with *work*: Each work, which has a variable working time, causes some damage to the unit and these damage are additive, and the system fails when the total damage has exceeded a failure level K and corrective replacement is made, or fails with probability $p(x)$ when the total damage is x and minimal repair is made. As the preventive replacement policy, (1) the unit is replaced before failure at Nth working time for the first model; (2) the unit is replaced at the first completion of working time after T has arrived; (3) the unit is replaced at the first completion of working time when the total damage has exceeded a threshold level Z; (4) the unit is replaced before failure at time T of operations or at number N of working times, whichever occurs first and last. Two cases where failure level K and failure probability $p(x)$ have been considered in (2) and (3). Expected cost rates of each model are obtained and optimal replacement policies are discussed analytically and computed numerically.

2 Modeling and Optimization

It is assumed that X_j $(j = 1, 2, \cdots)$ are successive working times of the unit and are independent and have an identical distribution $F(t) \equiv \Pr\{X_j \leq t\}$ with a finite mean $1/\lambda$. That is, the system works at a renewal process with its distribution $F(t)$. Suppose that each work of the job incurs some damage to the operating unit and the total damage is additive, which is called a *cumulative damage model* [Nakagawa (2007), p.16]. That is, the jth work causes some damage to the unit in the amount Y_j $(j = 1, 2, \cdots)$ according to an identical distribution $G(x) \equiv \Pr\{Y_j \leq x\}$ with a finite mean $1/\mu$.

Then, the probability that j times of works are completed in $(0, t]$ is, from [Nakagawa (2007), p. 17],

$$\Pr\{N(t) = j\} = F^{(j)}(t) - F^{(j+1)}(t) \quad (j = 0, 1, 2, \cdots),$$

and the distribution of the total damage $Z(t)$ at time t is

$$\Pr\{Z(t) \leq x\} = \sum_{j=0}^{\infty} G^{(j)}(x)[F^{(j)}(t) - F^{(j+1)}(t)],$$

where $\Phi^{(j)}(t)$ denotes the j-fold Stieltjes convolution of $\Phi(t)$ with itself and $\Phi^{(0)}(t) \equiv 1$ for $t > 0$ for any function $\Phi(t)$, and $M(x) \equiv \sum_{j=1}^{\infty} G^{(j)}(x)$ which is the renewal function of $G(x)$.

2.1 *Working Number N*

It is assumed that the operating unit fails when the total damage has exceeded a failure level K, and its failure is detected and replacement is made at the completion of working time. As the preventive replacement policy, the unit is replaced before failure at Nth ($N = 1, 2, \cdots$) working time. Then, the expected cost rate is [Nakagawa (2007), p. 44]

$$\frac{C(N)}{\lambda} = \frac{c_K - (c_K - c_N)G^{(N)}(K)}{\sum_{j=0}^{N-1} G^{(j)}(K)} \quad (N = 1, 2, \cdots), \tag{1}$$

where c_K and c_N are the respective replacement costs at failure and the Nth working time with $c_K > c_N$.

From the inequality $C(N+1) - C(N) \geq 0$, an optimal number N^* which minimizes the expected cost rate $C(N)$ in (1) satisfies

$$L(N) \sum_{j=0}^{N-1} G^{(j)}(K) - [1 - G^{(N)}(K)] \geq \frac{c_N}{c_K - c_N} \quad (N = 1, 2, \cdots), \tag{2}$$

where

$$L(N) \equiv \frac{G^{(N)}(K) - G^{(N+1)}(K)}{G^{(N)}(K)}.$$

If $L(N)$ increases strictly with N, then the left-hand side of (2) also increases strictly with N. Therefore, if $L(\infty)[1 + M(K)] > c_K/(c_K - c_N)$, then there exists a unique and minimum N^* ($1 \leq N^* < \infty$) which satisfies (2).

In particular, when $G(x) = 1 - e^{-\mu x}$, i.e., $G^{(j)}(x) = \sum_{i=j}^{\infty}[(\mu x)^i/i!]e^{-\mu x}$, from Example 2.2 [Nakagawa (2007), p. 24], $L(N) = [(\mu K)^N/N!]/\sum_{i=N}^{\infty}[(\mu K)^i/i!]$ increases from $e^{-\mu K}$ to 1. Then, the resulting cost rate is

$$(c_K - c_N)L(N^*) < \frac{C(N^*)}{\lambda} \leq (c_K - c_N)L(N^* + 1). \tag{3}$$

Table 1 presents optimal N^* which satisfy (2) and $C(N^*)/\lambda$ for different μK and $c_N/(c_K - c_N)$. Clearly, N^* increase with both μK and $c_N/(c_K - c_N)$. That is, to control effectively a high cost suffered for failure, we must make the replacement time earlier as a failure level K is lower or a failure replacement cost c_K is higher.

Table 1   Optimal N^* and $C(N^*)/\lambda$ for μK and $c_N/(c_K - c_N)$

$\dfrac{c_N}{c_K - c_N}$	$\mu K = 15$		$\mu K = 20$		$\mu K = 30$	
	N^*	$C(N^*)/\lambda$	N^*	$C(N^*)/\lambda$	N^*	$C(N^*)/\lambda$
0.01	6	0.0021	9	0.0013	15	0.0007
0.03	7	0.0054	10	0.0035	17	0.0020
0.05	7	0.0082	10	0.0055	17	0.0032
0.07	8	0.0110	11	0.0074	18	0.0043
0.10	8	0.0148	11	0.0101	19	0.0059
0.30	10	0.0372	13	0.0262	21	0.0160
0.50	11	0.0569	14	0.0407	22	0.0253
0.70	12	0.0753	15	0.0542	23	0.0341
1.00	12	0.1009	16	0.0734	24	0.0469

2.2   Replacement Over Time T or Level Z

2.2.1   Over time T

1. Policy I

A unit is replaced before a planned time T $(0 \leq T \leq \infty)$ when the total damage has exceeded a failure level K, and after T, it is replaced at the first completion of working time. Then, the expected cost rate is [Nakagawa (2007), p. 54]

$$\frac{C_1(T)}{\lambda} = \frac{c_K - (c_K - c_T)\sum_{j=0}^{\infty}[F^{(j)}(T) - F^{(j+1)}(T)]G^{(j+1)}(K)}{\sum_{j=0}^{\infty} F^{(j)}(T)G^{(j)}(K)}, \quad (4)$$

where c_T is replacement cost at the completion of working time after T.

We find an optimal time T_1^* which minimizes the expected cost rate $C_1(T)$ in (4) when $F(t) = 1 - e^{-\lambda t}$ and $G(x) = 1 - e^{-\mu x}$, i.e., $F^{(j)}(t) = \sum_{i=j}^{\infty}[(\lambda t)^i/i!]e^{-\lambda t}$ and $G^{(j)}(x) = \sum_{i=j}^{\infty}[(\mu x)^i/i!]e^{-\mu x}$. Differentiating $C_1(T)$ with respect to T and setting it equal to zero,

$$Q(T) \sum_{j=0}^{\infty} F^{(j)}(T)G^{(j)}(K) - \sum_{j=0}^{\infty} \frac{(\lambda T)^j}{j!}e^{-\lambda T}[1 - G^{(j+1)}(K)] = \frac{c_T}{c_K - c_T},$$

$$(5)$$

where

$$Q(T) \equiv \frac{\sum_{j=0}^{\infty}[(\lambda T)^j/j!][(\mu K)^{j+1}/(j+1)!]e^{-\mu K}}{\sum_{j=0}^{\infty}[(\lambda T)^j/j!]G^{(j+1)}(K)}.$$

Because $Q(T)$ increases strictly with T from $\mu K/(e^{\mu K} - 1)$ to 1, the left-hand side of (5) also increases strictly from

$$Q \equiv \frac{\mu K - 1 + e^{-\mu K}}{e^{\mu K} - 1} \leq \frac{\mu K}{2}$$

to μK. Therefore, we have the following optimal policies:

1. If $Q \geq c_T/(c_K - c_T)$, then $T_1^* = 0$, i.e., the unit is replaced at the first completion of working time, and the expected cost rate is

$$\frac{C_1(0)}{\lambda} = c_K - (c_K - c_T)G(K). \tag{6}$$

2. If $Q < c_T/(c_K - c_T) < \mu K$, then there exists a finite and unique T_1^* ($0 < T_1^* < \infty$) which satisfies (5), and the resulting cost rate is

$$\frac{C_1(T_1^*)}{\lambda} = (c_K - c_T)Q(T_1^*). \tag{7}$$

3. If $\mu K < c_T/(c_K - c_T)$, then $T_1^* = \infty$, i.e., the unit is replaced only at failure, and the resulting cost rate is

$$\frac{C_1(\infty)}{\lambda} = \frac{c_K}{1 + M(K)}. \tag{8}$$

Table 2 presents optimal λT_1^* which satisfy (5) and $C_1(T_1^*)/\lambda$ for different μK and $c_T/(c_K - c_T)$. It is shown that T_1^* have the same tendencies with N^* for the same parameters and $\lambda T_1^* \approx N^*$ in Table 1 for large $c_T/(c_K - c_T)$. From the economical point, the policy made at the Nth working time is better than that over time T. However, from the convenient point, the policy over time T would be easier than that at working number N, because it is not necessary to count the number of working times.

2. Policy II

A unit fails with probability $p(x)$ when the total damage becomes x at the end of each work and undergoes minimal repair at failures, i.e., its damage remains undisturbed by minimal repairs. It is assumed that the unit is not replaced at a planned time T, and after T, it is replaced at the first

Table 2   Optimal λT_1^* and $C_1(T_1^*)/\lambda$ for μK and $c_T/(c_K - c_T)$

$\dfrac{c_T}{c_K - c_T}$	$\mu K = 15$		$\mu K = 20$		$\mu K = 30$	
	λT_1^*	$C_1(T_1^*)/\lambda$	λT_1^*	$C_1(T_1^*)/\lambda$	λT_1^*	$C_1(T_1^*)/\lambda$
0.01	4.04	0.0032	6.37	0.0019	11.74	0.0010
0.03	5.16	0.0075	7.75	0.0047	13.57	0.0025
0.05	5.79	0.0111	8.52	0.0072	14.56	0.0040
0.07	6.26	0.0144	9.08	0.0094	15.27	0.0053
0.10	6.81	0.0190	9.73	0.0126	16.09	0.0072
0.30	8.96	0.0440	12.22	0.0306	19.15	0.0185
0.50	10.29	0.0647	13.74	0.0460	20.96	0.0284
0.70	11.35	0.0833	14.92	0.0601	22.35	0.0377
1.00	12.67	0.1087	16.37	0.0795	24.03	0.0508

completion of working time. Then, the expected cost rate is [Nakagawa (2007), p. 98]

$$C_2(T) = \frac{c_M \sum_{j=1}^{\infty} F^{(j)}(T)\{1 - G^*(\theta)^j\} + c_T}{T + \int_T^{\infty} \overline{F}(t)\mathrm{d}t + \sum_{j=1}^{\infty} \int_0^T [\int_{T-t}^{\infty} \overline{F}(u)\mathrm{d}u]\mathrm{d}F^{(j)}(t)}, \qquad (9)$$

where c_M is cost suffered for minimal repair at each failure, and

$$\int_0^{\infty} p(x)\mathrm{d}G^{(j)}(x) = \int_0^{\infty} (1 - \mathrm{e}^{-\theta x})\mathrm{d}G^{(j)}(x) = 1 - [G^*(\theta)]^j,$$

which means the probability that the unit fails at the end of jth work. Here, $G^*(\theta)$ denotes the Laplace-Stietjes transform of $G(x)$, i.e., $G^*(\theta) \equiv \int_0^{\infty} \mathrm{e}^{-\theta x}\mathrm{d}G(x) < 1$ for $\theta > 0$.

Differentiating $C_2(T)$ with respect to T and setting it equal to zero,

$$1 - (1 + \lambda T)G^*(\theta)\mathrm{e}^{-\lambda T[1 - G^*(\theta)]} + \frac{G^*(\theta)}{1 - G^*(\theta)}\{1 - \mathrm{e}^{-\lambda T[1 - G^*(\theta)]}\} = \frac{c_T}{c_M}. \tag{10}$$

The left-hand side of (10) increases strictly with T from $1 - G^*(\theta)$ to $1/[1 - G^*(\theta)]$. Therefore, we have the following optimal policies:

1. If $1 - G^*(\theta) \geq c_T/c_M$, then $T_2^* = 0$.
2. If $1 - G^*(\theta) < c_T/c_M < 1/[1 - G^*(\theta)]$, then there exists a finite and unique T_2^* $(0 < T_2^* < \infty)$ which satisfies (10), and the resulting cost rate is

$$C_2(T_2^*) = \lambda c_M\{1 - G^*(\theta)\mathrm{e}^{-\lambda T[1 - G^*(\theta)]}\}. \tag{11}$$

3. If $1/[1 - G^*(\theta)] \leq c_T/c_M$, then $T_2^* = \infty$.

Suppose that $G(x) = 1 - e^{-\mu x}$, Table 3 presents optimal λT_2^* which satisfies (10) and $C_2(T_2^*)$ for different θ and c_T/c_M when $\mu = 1.0$. It is shown that λT_2^* decreases with θ because the operating unit has a higher failure probability when θ becomes larger, and λT_2^* increases with c_T/c_M because minimal repair cost suffers less than before so that the unit could be operating for a longer time.

Table 3   Optimal λT_2^* and $C_2(T_2^*)$ for θ and c_T/c_M when $\mu = 1.0$.

c_T/c_M	$\theta = 0.01$		$\theta = 0.03$		$\theta = 0.05$	
	λT_2^*	$C_2(T_2^*)$	λT_2^*	$C_2(T_2^*)$	λT_2^*	$C_2(T_2^*)$
0.1	3.46	0.0430	1.48	0.0699	0.84	0.0846
0.3	6.95	0.0752	3.63	0.1263	2.60	0.1585
0.5	9.38	0.0970	5.13	0.1635	3.83	0.2060
0.7	11.38	0.1146	6.37	0.1932	4.85	0.2435
1.0	13.94	0.1366	7.98	0.2299	6.17	0.2896
3.0	25.94	0.2326	15.82	0.3868	12.86	0.4831
5.0	34.79	0.2965	22.01	0.4877	18.46	0.6041
7.0	42.39	0.3471	27.66	0.5653	23.93	0.6947
10.0	52.54	0.4091	35.85	0.6573	32.64	0.7982

2.2.2   *Over level Z*

1. Policy I

A unit is replaced when the total damage has exceeded a failure level K, and is also replaced preventively at the next working time when the damage is between Z and K. Then, the expected cost rate is [Nakagawa (2007), p. 56]

$$\frac{C_1(Z)}{\lambda} = \frac{c_K - (c_K - c_Z)\{\int_Z^K G(K-x)\mathrm{d}G(x) + \int_0^Z [\int_{Z-x}^{K-x} G(K-x-y)\mathrm{d}G(y)]\mathrm{d}M(x)\}}{1 + G(K) + \int_0^Z G(K-x)\mathrm{d}M(x)}, \qquad (12)$$

where c_Z is replacement cost at the completion of working time when the damage is between Z and K.

When $G(x) = 1 - e^{-\mu x}$, differentiating $C_1(Z)$ with respect to Z and setting it equal to zero,

$$e^{-\mu(K-Z)}\left[\frac{\mu(1+\mu Z)(K-Z)}{1 - e^{-\mu(K-Z)}} - 1\right] = \frac{c_Z}{c_K - c_Z}. \qquad (13)$$

The left-hand side of (13) increases strictly with Z from Q to μK. Therefore, we have the following optimal policies:

1. If $Q \geq c_Z/(c_K - c_Z)$, then $Z_1^* = 0$, and the expected cost rate is

$$\frac{C_1(0)}{\lambda} = \frac{c_K - (c_K - c_Z)[1 - (1 + \mu K)e^{-\mu K}]}{2 - e^{-\mu K}}. \tag{14}$$

2. If $Q < c_Z/(c_K - c_Z) < \mu K$, then there exists a finite and unique Z_1^* ($0 < Z_1^* < K$) which satisfies (13), and the resulting cost rate is

$$\frac{C_1(Z_1^*)}{\lambda} = \frac{(c_K - c_Z)\mu(K - Z_1^*)e^{-\mu(K - Z_1^*)}}{1 - e^{-\mu(K - Z_1^*)}}. \tag{15}$$

3. If $\mu K \leq c_Z/(c_K - c_Z)$, then $Z_1^* = K$, and the expected cost rate is given in (8).

Table 4 presents optimal Z_1^* which satisfies (13) and $C_1(Z_1^*)/\lambda$ for different K and $c_Z/(c_K - c_Z)$. It is shown that Z_1^* have the same tendencies with T_1^* for the same parameters and $C_1(Z_1^*) < C_1(T_1^*)$ in Table 2, which means that, from the economical point, the policy made over level Z is better than that over time T because such a over level Z policy could monitor the damage level more accurately.

Table 4   Optimal Z_1^* and $C_1(Z_1^*)/\lambda$ for K and $c_Z/(c_K - c_Z)$ when $\mu = 1.0$

$\dfrac{c_Z}{c_K - c_Z}$	$K = 15.0$		$K = 20.0$		$K = 30.0$	
	Z_1^*	$C_1(Z_1^*)/\lambda$	Z_1^*	$C_1(Z_1^*)/\lambda$	Z_1^*	$C_1(Z_1^*)/\lambda$
0.01	6.35	0.0014	10.71	0.0009	15.34	0.0006
0.03	7.44	0.0036	11.85	0.0024	16.50	0.0017
0.05	7.96	0.0057	12.39	0.0038	17.04	0.0028
0.07	8.31	0.0076	12.75	0.0051	17.41	0.0038
0.10	8.68	0.0105	13.13	0.0071	17.80	0.0054
0.30	9.87	0.0281	14.34	0.0198	19.02	0.0151
0.50	10.44	0.0445	14.92	0.0318	19.60	0.0245
0.70	10.83	0.0604	15.31	0.0435	19.99	0.0337
1.00	11.25	0.0834	15.73	0.0606	20.41	0.0472

2. Policy II

It is assumed that a unit fails with probability $p(x)$ when the total damage becomes x at the end of each work and undergoes minimal repair at failures and the unit is replaced at the first completion of working time when the total damage has exceeded a threshold level Z. Then, the expected cost rate is [Nakagawa (2007), p. 99]

$$\frac{C_2(Z)}{\lambda} = \frac{c_M\{\int_0^\infty p(x)dG(x) + \int_0^Z [\int_0^\infty p(x + y)dG(y)]dM(x)\} + c_Z}{2 + M(Z)}. \tag{16}$$

Differentiating $C_2(Z)$ with respect to Z and setting it equal to zero,

$$[2 + M(Z)] \int_0^\infty p(Z + x)\mathrm{d}G(x) - \int_0^Z \left[\int_0^\infty p(x + y)\mathrm{d}G(y) \right] \mathrm{d}M(x)$$

$$- \int_0^\infty p(x)\mathrm{d}G(x) = \frac{c_Z}{c_M}. \tag{17}$$

Denote $D(Z)$ be the left-hand side of (17), then $D(Z)$ increases strictly with Z from $\int_0^\infty p(x)\mathrm{d}G(x)$ to $D(\infty)$. Therefore, we have the following optimal policies:

1. If $\int_0^\infty p(x)\mathrm{d}G(x) \geq c_Z/c_M$, then $Z_2^* = 0$, and

$$\frac{C_2(0)}{\lambda} = \frac{c_M \int_0^\infty p(x)\mathrm{d}G(x) + c_Z}{2}. \tag{18}$$

2. If $\int_0^\infty p(x)\mathrm{d}G(x) < c_Z/c_M < D(\infty)$, then there exists a finite and unique Z_2^* $(0 < Z_2^* < \infty)$ which satisfies (17), and the resulting cost rate is

$$\frac{C_2(Z_2^*)}{\lambda} = c_M \int_0^\infty p(Z_2^* + x)\mathrm{d}G(x). \tag{19}$$

3. If $D(\infty) \leq c_Z/c_M$, then $Z_2^* = \infty$.

Suppose that $G(x) = 1 - \mathrm{e}^{-\mu x}$, Table 5 presents optimal Z_2^* which satisfies (17) and $C_2(Z_2^*)/\lambda$ for different θ and c_Z/c_M when $\mu = 1.0$. It is shown that Z_2^* have the same tendencies with Z_1^* for the same parameters and $C_2(Z_2^*) < C_2(T_2^*)$ in Table 3, the reason could be given as that for comparison between Z_1^* and T_1^*.

Table 5   Optimal Z_2^* and $C_2(Z_2^*)/\lambda$ for θ and c_Z/c_M when $\mu = 1.0$

c_T/c_M	$\theta = 0.01$		$\theta = 0.03$		$\theta = 0.05$	
	Z_2^*	$C_2(Z_2^*)/\lambda$	Z_2^*	$C_2(Z_2^*)/\lambda$	Z_2^*	$C_2(Z_2^*)/\lambda$
0.1	2.70	0.0367	0.98	0.0573	0.49	0.0704
0.3	5.98	0.0683	2.81	0.1088	1.86	0.1332
0.5	8.29	0.0902	4.17	0.1452	2.92	0.1791
0.7	10.22	0.1080	5.32	0.1748	3.83	0.2162
1.0	12.70	0.1303	6.82	0.2120	5.02	0.2630
3.0	24.36	0.2283	14.24	0.3735	11.19	0.4640
5.0	32.97	0.2936	20.12	0.4782	16.39	0.5914
7.0	40.35	0.3454	25.51	0.5591	21.47	0.6877
10.0	50.24	0.4090	33.27	0.6550	29.50	0.7979

2.3   *Replacement First and Last*

2.3.1   *Replacement first*

It is assumed that an operating unit fails when the total damage exceeds a threshold level K $(0 < K < \infty)$, its failure is immediately detected at the end of working time, and then corrective replacement is done. As preventive replacement policies, the unit is replaced before failure at time T $(0 < T \leq \infty)$ of operations or at number N $(N = 1, 2, \cdots)$ of working times, whichever occurs first. Then, the expected cost rate is [Nakagawa (2007), p. 42]

$$\frac{C_F(T, N)}{\lambda} = \frac{c_P + (c_K - c_P) \sum_{j=0}^{N-1} F^{(j+1)}(T)[G^{(j)}(K) - G^{(j+1)}(K)]}{\sum_{j=0}^{N-1} F^{(j+1)}(T)G^{(j)}(K)},$$

(20)

where c_P is the preventive replacement cost at T or N with $c_K > c_P$.

Differentiating $C_F(T, N)$ with respect to T for N and setting it equal to zero,

$$R_F(T, N) \sum_{j=0}^{N-1} F^{(j+1)}(T)G^{(j)}(K) - \sum_{j=0}^{N-1} F^{(j+1)}(T)[G^{(j)}(K) - G^{(j+1)}(K)]$$
$$= \frac{c_P}{c_K - c_P},$$

(21)

where

$$R_F(t, N) \equiv \frac{\sum_{j=0}^{N-1} f^{(j+1)}(t)[G^{(j)}(K) - G^{(j+1)}(K)]}{\sum_{j=0}^{N-1} f^{(j+1)}(t)G^{(j)}(K)}.$$

From [Zhao, Qian and Nakagawa (2013)], when $G^{(j)}(K) = \sum_{i=j}^{\infty}[(\mu K)^i / i!]e^{-\mu K}$, it is approved that $R_F(t, N) = 1 - G(K)$ for $N = 1$ and increases strictly with t for $2 \leq N < \infty$ from $1 - G(K)$ to $1 - G^{(N+1)}(K)/G^{(N)}(K)$. Denoting the left-hand side of (21) by $V_F(T, N)$ and $r_F(t, N) \equiv dR_F(t, N)/dt$,

$$\frac{dV_F(T, N)}{dT} = r_F(T, N) \sum_{j=0}^{N-1} F^{(j+1)}(T)G^{(j)}(K) \geq 0,$$

(22)

which follows that $V_F(T, N)$ increases with T from 0 to

$$V_F(\infty, N) = \frac{G^{(N)}(K) - G^{(N+1)}(K)}{G^{(N)}(K)} \sum_{j=0}^{N-1} G^{(j)}(K) - \overline{G}^{(N)}(K).$$

Therefore, if $V_F(\infty, N) > c_P/(c_K - c_P)$, then there exists a finite and unique T_F^* $(0 < T_F^* < \infty)$ that satisfies (21), and the resulting cost rate is

$$\frac{C_F(T_F^*, N)}{\lambda} = (c_F - c_P)R_F(T_F^*, N). \qquad (23)$$

If $V_F(\infty, N) \leq c_P/(c_K - c_P)$, then $T_F^* = \infty$, and the resulting cost rate is given by (1).

Next, we discuss an optimal N_F^* for T to minimize $C_F(T, N)$ which is given by (20). Forming the inequality $C_F(T, N + 1) - C_F(T, N) \geq 0$,

$$\sum_{j=0}^{N-1} F^{(j+1)}(T)G^{(j)}(K)[L(N) - L(j)] \geq \frac{c_P}{c_K - c_P}, \qquad (24)$$

where $L(j)$ $(j = 0, 1, 2, \cdots)$ is given in (2).

Denoting the left-hand side of (24) by $V_F(N, T)$, then $V_F(N + 1, T) - V_F(N, T)$ is

$$[L(N + 1) - L(N)] \sum_{j=0}^{N-1} F^{(j+1)}(T)G^{(j)}(K) > 0, \qquad (25)$$

which follows that $V_F(N, T)$ increases strictly with N to

$$N(K) = \sum_{j=1}^{\infty} F^{(j)}(T)G^{(j)}(K).$$

Therefore, if $N(K) > c_P/(c_K - c_P)$, then there exists a unique and minimum N_F^* $(1 \leq N_F^* < \infty)$ which satisfies (24), and the resulting cost rate is

$$(c_K - c_P)L(N_F^* - 1) < \frac{C_F(T, N_F^*)}{\lambda} \leq (c_K - c_P)L(N_F^*). \qquad (26)$$

If $N(K) \leq c_P/(c_K - c_P)$, then $N_F^* = \infty$ and the resulting cost rate is given by

$$\frac{C_F(T, \infty)}{\lambda} = \frac{c_K - (c_K - c_P)\sum_{j=0}^{\infty}[F^{(j)}(T) - F^{(j+1)}(T)]G^{(j)}(K)}{\sum_{j=0}^{\infty} F^{(j+1)}(T)G^{(j)}(K)}. \qquad (27)$$

2.3.2 *Replacement last*

Suppose that the unit is replaced before failure at time T $(0 \leq T \leq \infty)$ of operations or at number N $(N = 0, 1, 2, \cdots)$ of working times, whichever occurs last. Then, the expected cost rate is [Zhao, Qian and Nakagawa (2013)]

$$\frac{C_L(T, N)}{\lambda} = \frac{c_P + (c_K - c_P)\{1 - \sum_{j=N}^{\infty} \overline{F}^{(j+1)}(T)[G^{(j)}(K) - G^{(j+1)}(K)]\}}{\sum_{j=0}^{N-1} \overline{F}^{(j+1)}(T)G^{(j)}(K) + \sum_{j=0}^{\infty} F^{(j+1)}(T)G^{(j)}(K)}.$$

(28)

Differentiating $C_L(T, N)$ with respect to T for N and setting it equal to zero,

$$R_L(T, N)\left[\sum_{j=0}^{N-1} \overline{F}^{(j+1)}(T)G^{(j)}(K) + \sum_{j=0}^{\infty} F^{(j+1)}(T)G^{(j)}(K)\right]$$

$$+ \sum_{j=N}^{\infty} \overline{F}^{(j+1)}(T)[G^{(j)}(K) - G^{(j+1)}(K)] - 1 = \frac{c_P}{c_K - c_P},$$

(29)

where

$$R_L(t, N) \equiv \frac{\sum_{j=N}^{\infty} f^{(j+1)}(t)[G^{(j)}(K) - G^{(j+1)}(K)]}{\sum_{j=N}^{\infty} f^{(j+1)}(t)G^{(j)}(K)},$$

and $f^{(j+1)}(t) \equiv [\lambda(\lambda t)^j/j!]e^{-\lambda t}$. When $G^{(j)}(K) = \sum_{i=j}^{\infty}[(\mu K)^i/i!]e^{-\mu K}$, it is proved from [Zhao, Qian and Nakagawa (2013)], that $R_L(t, N)$ increases strictly with t from $1 - G^{(N+1)}(K)/G^{(N)}(K)$ to 1. Denoting the left-hand side of (29) by $V_L(T, N)$ and $r_L(t, N) \equiv dR_L(t, N)/dt$, then $dV_L(T, N)/dT$ is

$$r_L(T, N)\left[\sum_{j=0}^{N-1} \overline{F}^{(j+1)}(T)G^{(j)}(K) + \sum_{j=0}^{\infty} F^{(j+1)}(T)G^{(j)}(K)\right] > 0,$$

(30)

which follows that $V_L(T, N)$ increases strictly with T from

$$V_L(0, N) = L(N)\sum_{j=0}^{N-1} G^{(j)}(K) - \overline{G}^{(N)}(K)$$

to $V_L(\infty, N) = M(K)$.

Therefore, if $V_L(0, N) < c_P/(c_K - c_P) < M(K)$, then there exists a finite and unique T_L^* $(0 < T_L^* < \infty)$ that satisfies (29), and the resulting cost rate is

$$\frac{C_L(T_L^*, N)}{\lambda} = (c_K - c_P)R_L(T_L^*, N).$$

(31)

If $M(K) \leq c_P/(c_K - c_P)$, then $T_L^* = \infty$, and the resulting cost rate is given by (8). If $V_L(0, N) \geq c_P/(c_K - c_P)$, then $T_L^* = 0$, and the resulting cost rate is given by (1).

Forming the inequality $C_L(N+1) - C_L(N) \geq 0$,

$$\sum_{j=N}^{\infty} \overline{F}^{(j+1)}(T)G^{(j)}(K)[L(j) - L(N)] + L(N)\sum_{j=0}^{\infty} G^{(j)}(K) - 1 \geq \frac{c_P}{c_K - c_P},$$

(32)

where $L(j)$ $(j = 0, 1, 2, \cdots)$ is given in (2).

From [Nakagawa (2007), p. 24], $G^{(j+1)}(K)/G^{(j)}(K)$ decreases strictly with j when $G^{(j)}(K) = \sum_{i=j}^{\infty}[(\mu K)^i/i!]e^{-\mu K}$, and $L(j)$ increases strictly with j from $e^{-\mu K}$ to 1. Denoting the left-hand side of (32) by $V_L(N, T)$, $V_L(N+1, T) - V_L(N, T)$ is

$$[R(N+1) - R(N)]\left[\sum_{j=N}^{\infty} F^{(j+1)}(T)G^{(j)}(K) + \sum_{j=0}^{N-1} G^{(j)}(K)\right] > 0, \quad (33)$$

which follows that $V_L(N, T)$ increases strictly with N from

$$\sum_{j=0}^{\infty} F^{(j+1)}(T) \int_0^K [\overline{G}(K) - \overline{G}(K - x)]\mathrm{d}G^{(j)}(x) < 0$$

to $M(K)$.

Therefore, if $M(K) > c_P/(c_K - c_P)$, then there exists a unique and minimum N_L^* $(1 \leq N_L^* < \infty)$ which satisfies (32), and the resulting cost rate is

$$(c_K - c_P)L(N_L^* - 1) < \frac{C_L(T, N_L^*)}{\lambda} \leq (c_K - c_P)L(N_L^*). \quad (34)$$

If $M(K) \leq c_P/(c_K - c_P)$, then $N_L^* = \infty$ and the resulting cost rate is given by (8).

2.3.3 *Comparisons*

First, we compare T_F^* in (23) with T_L^* in (31) to decide which policy is better for a given N. Denoting the left-hand side of (2) by $V(N)$, it is easily shown that $V(N) = V_L(0, N) = V_F(\infty, N)$. Then, the following comparison results between replacement first and last can be given:

1: If a predetermined $N < N^*$, then $T_F^* = \infty$, and $0 < T_L^* < \infty$ when $M(K) > c_P/(c_K - c_P)$, and $T_L^* = \infty$ when $M(K) \leq c_P/(c_F - c_P)$, i.e., replacement last should be adopted when $N < N^*$.

2: If a predetermined $N \geq N^*$, then $T_L^* = 0$, and $0 < T_F^* < \infty$ when $V(N) > c_P/(c_K - c_P)$ and $T_F^* = \infty$ when $V(N) \leq c_P/(c_K - c_P)$, i.e., replacement first should be adopted when $N \geq N^*$ and $V(N) > c_P/(c_K - c_P)$, and the policy in Section Working Number N should be adopted when $N \geq N^*$ and $V(N) \leq c_P/(c_K - c_P)$.

Table 6 Optimal T_L^*, T_F^*, and their cost rates when $\lambda = 1$, $\mu = 1$ and $K = 10$

c_P	$N = 5$				$N = 8$				
$c_K - c_P$	T_L^*	$C_L(T_L^*, N)$	T_F^*	$C_F(T_F^*, N)$	T_L^*	$C_L(T_L^*, N)$	T_F^*	$C_F(T_F^*, N)$	N^*
0.1	0.00	0.026	∞	0.026	0.00	0.041	4.75	0.033	5
0.2	1.28	0.046	∞	0.046	0.00	0.054	6.84	0.052	6
0.3	3.56	0.065	∞	0.066	0.00	0.067	9.49	0.066	6
0.4	5.01	0.084	∞	0.086	0.00	0.080	14.09	0.079	7
0.5	6.09	0.100	∞	0.106	0.00	0.093	28.35	0.092	7
0.6	6.97	0.115	∞	0.126	0.00	0.106	∞	0.106	8
0.7	7.75	0.129	∞	0.146	0.00	0.119	∞	0.119	8
0.8	8.44	0.142	∞	0.166	0.00	0.131	∞	0.131	8
0.9	9.07	0.155	∞	0.186	0.00	0.144	∞	0.144	8
1.0	9.68	0.167	∞	0.206	2.62	0.157	∞	0.157	9

Table 6 indicates:

- There exist three cases between T_L^* and T_F^* according to N^*: $0 < T_L^* < \infty$ and $T_F^* = \infty$ for $N < N^*$, $0 < T_F^* < \infty$ and $T_L^* = 0$ for $N > N^*$, and $T_L^* = 0$ and $T_F^* = \infty$ for $N = N^*$. That is, replacement last should be adopted for $N < N^*$, e.g., when $N = 5$ and $c_P/(c_K - c_P) = 0.5$, $C_L(T_L^*, N) = 0.100 < C_F(T_F^*, N) = 0.106$; replacement first should be adopted for $N > N^*$, e.g., when $N = 8$ and $c_P/(c_K - c_P) = 0.1$, $C_F(T_F^*, N) = 0.033 < C_L(T_L^*, N) = 0.041$; the policy in (1) should be adopted when $N = N^*$, e.g., when $N = 5$ and $c_P/(c_K - c_P) = 0.1$, $C_F(T_F^*, N) = C_L(T_L^*, N) = 0.026$.

- When $0 < T_L^* < \infty$ or $0 < T_F^* < \infty$, both T_L^* and T_F^* increase with the replacement cost ratio $c_P/(c_K - c_P)$, i.e., decrease with c_K/c_P. When c_K/c_P increases, preventive replacement should be advanced to prevent a higher corrective replacement cost. In other words, the unit can be operating for a longer time as c_K/c_P becomes smaller, e.g., when $N = 5$ and $c_P/(c_K - c_P) = 0.2, 0.5$, $T_L^* = 1.28, 6.09$; when $N = 8$ and $c_P/(c_K - c_P) = 0.1, 0.5$, $T_F^* = 4.75, 28.35$.

- When a predetermined N becomes smaller, replacement last shows more superior cases than replacement first, not only because of comparison results between optimal cost rates in Table 6, but also because replacement last can let the unit work as longer as possible. For example, $\mu K = 10$ means that the unit will be suffered failure around 10 shocks, when $N = 5$, preventive replacement is done at 5 working times for replacement first, no matter whether c_K/c_P is large or small; however, preventive replacement would be done around from $\lambda T_L^* \approx 0$ to $\lambda T_L^* \approx 10$ working times according to c_K/c_P. It seem more reasonable when replacement last is adopted for such cases to avoid unnecessary replacements.

Next, we compare N_F^* in (26) with N_L^* in (34) to decide which policy is better for a given T. When $N(K) > c_P/(c_K - c_P)$, compare the left-hand side of (24) and (32) by denoting

$$A(N) \equiv V_L(N, T) - V_F(N, T).$$

It can be easily shown that $A(N + 1) - A(N)$ is

$$[L(N + 1) - L(N)] \left[\sum_{j=0}^{N-1} \overline{F}^{(j+1)}(T)G^{(j)}(K) + \sum_{j=N}^{\infty} F^{(j+1)}(T)G^{(j)}(K) \right] > 0,$$

which follows that $A(N)$ increases with N strictly to

$$A(\infty) = \sum_{j=0}^{\infty} \overline{F}^{(j)}(T)G^{(j)}(K) > 0.$$

Thus, there exists a unique and minimum N_A^* ($1 \leq N_A^* < \infty$) which satisfies $A(N) \geq 0$.

From (24), denoting that

$$H(N_A^*) \equiv \sum_{j=0}^{N_A^*-1} F^{(j+1)}(T)G^{(j)}(K)[L(N_A^*) - L(j)]. \tag{35}$$

Then, the following comparison results can be given:

1: If $H(N_A^*) < c_P/(c_K - c_P)$, then $N_L^* \leq N_F^*$, and $C_L(T, N_L^*) \leq C_F(T, N_F^*)$, i.e., replacement last should be adopted.
2: If $H(N_A^* - 1) > c_P/(c_K - c_P)$, then $N_F^* \leq N_L^*$, i.e., replacement first should be adopted.
3: If $H(N_A^* - 1) \leq c_P/(c_K - c_P) \leq H(N_A^*)$, then either replacement last or first may be better than the other, or the same with each other.

Table 7 indicates:

- There exist three cases between N_L^* and N_F^* according to $H(N_A^*-1)$ and $H(N_A^*)$: When $H(N_A^*) < c_P/(c_K-c_P)$, then $N_L^* \leq N_F^*$, i.e., replacement

Table 7   Optimal N_L^*, N_F^*, and their cost rates when $\lambda = 1$, $\mu = 1$ and $K = 10$

$\dfrac{c_P}{c_K - c_P}$	$T = 5$				$T = 8$				N^*
	N_L^*	$C_L(T, N_L^*)$	N_F^*	$C_F(T, N_F^*)$	N_L^*	$C_L(T, N_L^*)$	N_F^*	$C_F(T, N_F^*)$	
0.1	5	0.032	5	0.029	6	0.049	5	0.026	5
0.2	6	0.048	6	0.052	6	0.063	6	0.049	6
0.3	6	0.064	7	0.074	7	0.075	6	0.064	6
0.4	7	0.078	7	0.095	7	0.088	7	0.080	7
0.5	7	0.093	8	0.116	7	0.100	7	0.096	7
0.6	8	0.106	8	0.137	8	0.112	8	0.111	8
0.7	8	0.119	9	0.157	8	0.124	8	0.126	8
0.8	8	0.132	9	0.178	8	0.136	8	0.141	8
0.9	9	0.145	10	0.198	9	0.148	9	0.156	8
1.0	9	0.156	10	0.219	9	0.159	9	0.169	9
N_A^*	6				8				
$H(N_A^*)$	0.270				0.821				
$H(N_A^* - 1)$	0.142				0.547				

last should be adopted, e.g., when $T = 5$, $H(N_A^*) = 0.270$ which is less than $c_P/(c_K - c_P) = 0.3$, then $C_L(T, N_L^*) < C_F(T, N_F^*)$; when $H(N_A^* - 1) > c_P/(c_K - c_P)$, then $N_F^* \leq N_L^*$, i.e., replacement first should be adopted, e.g., when $T = 5$, $H(N_A^* - 1) = 0.142 > c_P/(c_K - c_P) = 0.1$, then $C_F(T, N_F^*) = 0.029 < C_L(T, N_L^*) = 0.032$; when $H(N_A^* - 1) \leq c_P/(c_K - c_P) \leq H(N_A^*)$, then $N_L^* = N_F^*$, and either replacement first or last may be better than the other, e.g., when $T = 5$ and $c_P/(c_K - c_P) = 0.2$, then $N_L^* = N_F^* = 6$ but $C_L(T, N_L^*) = 0.048 < C_F(T, N_F^*) = 0.052$, and when $T = 8$ and $c_P/(c_K - c_P) = 0.6$, then $N_L^* = N_F^* = 8$ but $C_F(T, N_F^*) = 0.111 < C_L(T, N_L^*) = 0.112$, it is interesting that such a case number will increase when T becomes larger.

• N_L^* and N_F^* and their cost rates $C_L(T, N_L^*)$ and $C_F(T, N_F^*)$ show some different properties from those in 6. For example, N_L^* and N_F^* increase with the replacement cost ratio $c_P/(c_K - c_P)$, N_L^* increases with T from N^* and N_F^* decreases with T to N^*. So that analyses could be obtained in a similar way, e.g., $\mu K = 10$, when $T = 5$, preventive replacement is done around $\lambda T = 5$ working times for replacement first and at from $N_L^* = 5$ to $N_L^* = 9$ working times for replacement last.

• Compare the cost rates in Tables 6 and 7 for the same $\lambda T = N$ and $c_P/(c_K - c_P)$ when $\mu = 1$ and $K = 10$, replacement last done at N_L^* is better than that done at T_L^* in most cases, when a predetermined T or N is in a moderate size, e.g., 5; however, when T or N is large enough, e.g., 8, preventive replacement done at T_L^* is better than that done at N_L^*.

3   Conclusions

We have discussed four replacement policies for an operating unit which works at successive random times for jobs, where the unit fails due to damage that can be additive caused by jobs. Using the technique of cumulative damage models, expected cost rates have been obtained, and the optimal replacement policies have been discussed analytically. Numerical examples have been computed for all models and some useful explanations have been given.

The policies in Sections 2.1 and 2.2 are made only at the end of some work completion, although the performances for these three policies are different. It seems more practical to perform replacement when the unit has completed the job without stops. In Section 2.1, we could consider the replacement at working number N as a standard policy, which is a single policy that has been only considered the factor of job. When we need to take the factors, e.g., total operation time and damage level, into consideration, the models proposed in Sections 2.2.1 and 2.2.2 are more practical to perform and two cases where failure level K and failure probability $p(x)$ have been considered. From numerical comparisons in these two sections, the over level Z policies are better than those in over time T from the economical point, because over level Z policies could monitor the damage level more accurately that would cause the failure of the unit.

The policies in Section 2.3 are combined a planned replacement with working times, i.e., time T of operations and number N of working times. Two performance mechanisms "whichever occurs first" and "whichever occurs last" are modeled, we called them replacement first and last, respectively. As a newly proposed notion "whichever occurs last", the most interesting point is that we have already found the cases when replacement last should be adopted or not.

Finally, we give a potential application of such models to maintain a database in computer science: (i) Normally, the database is maintained at periodic times such as day, week, month, etc. However, it is necessary to guarantee ACID (atomicity, consistency, isolation, durability) properties of database transactions [Haerder and Reuter (1983); Gray and Reuter (1992); Lewis, Bernstein and Kifer (2002)], so that it is not advisable to suspend any transaction when it is under busy state. (ii) Cumulative damage models have been successfully formulated the incremental processes of updated data in a database, [Qian, Pan and Nakagawa (2002)]. In other words, we can monitor the cumulative updated data at any time.

References

Barlow, R. E. and Proschan, F. (1965). *Mathematical Theory of Reliability* (Wiley, New York).

Chen, M., Mizutani, S. and Nakagawa, T. (2010). Random and age replacement policies, *International Journal of Reliability, Quality and Safety Engineering* **17**, pp. 27–39.

Chen, M., Nakamura, S. and Nakagawa, T. (2010). Replacement and preventive maintenance models with random working times, *IEICE Trans. Fundamentals* **E93-A**, pp. 500–507.

Duchesne, T. and Lawless, J. (2000). Alternative time scales and failure time models, *Lifetime Data Analysis* **6**, pp. 157–179.

Gray, J. and Gray, A. (1992). *Transaction Processing: Concepts and Techniques* (Morgan Kaufmann, USA)

Haerder, T. and Reuter, A. (1983). Principles of transaction-oriented database recovery, *ACM Computing Surveys* **15**, pp. 287–317.

Kobbacy, K. A. H. and Murthy, D. N. P. (2008). *Complex System Maintenance Handbook* (Springer, London).

Lewis, M. L., Bernstein, B. and Kifer, M. (2002). *Databases and Transaction Processing: An Application-Oriented Approach* (Addison Wesley, USA).

Manzini, R., Regattieri, A., Pham, H. and Ferrari, E. (2010). *Maintenance for Industrial Systems* (Springer, London).

Nakagawa, T. (2005). *Maintenance Theory of Reliability* (Springer, London).

Nakagawa, T. (2007). *Shock and Damage Models in Reliability Theory* (Springer, London).

Nakagawa, T. (2008). *Advanced Reliability Models and Maintenance Policies* (Springer, London).

Nakagawa, T. (2011). *Stochastic Process with Applications to Reliability Theory* (Springer, London).

Osaki, S. (2002). *Stochastic Models in Reliability and Maintenance* (Springer, Berlin).

Pinedo, M. (2002). *Scheduling Theory, Algorithms and Systems* (Prentice Hall, Englewood Cliffs, NJ).

Pham, H. (2003). *Handbook of Reliability Engineering* (Springer, London).

Qian, C., Pan, Y. and Nakagawa, T. (2002). Optimal policies for a database system with two backup schemes. *RAIRO-Operations Research* **36**, pp. 227–235.

Sugiura, T., Mizutani, S. and Nakagawa, T. (2004). Optimal random replacement policies, in *Proceedings of the Tenth ISSAT International Conference on Reliability and Quality in Design* (Las Vegas, Nevada), pp. 99–103.

Wang, H. and Pham, H. (2007). *Reliability and Optimal Maintenance* (Springer, London).

Zhao, X., Nakayama, K. and Nakamura, S. (2011). Cumulative damage models with replacement last, in *International Conferences ASEA/DRBC/EL,*

Communications in Computer and Information Science 257 (Jeju Island, Korea), pp. 338–345.

Zhao, X., Qian, C. and Nakagawa, T. (2012). Age replacement with overtime policy, in *The 5th Asia-Pacific International Symposium on Advanced Reliability and Maintenance Modeling V* (Nanjing, China), pp. 661–668.

Zhao, X. and Nakagawa, T. (2012). Optimization problems of replacement first or last in reliability theory, *European Journal of Operational Research* **223**, pp. 141–149.

Zhao, X., Qian C. and Nakagawa, T. (2013). Optimal policies for cumulative damage models with maintenance last and first, *Reliability Engineering & System Safety* **110**, pp. 50–59.

PART 2
Reliability Analysis

Chapter 6

Modules of Multi-State Coherent Systems — Order Theoretical Relations

Fumio Ohi[1]

Omohi College, Nagoya Institute of Technology, Gokiso-cho, Showa-ku, Nagoya, 466-8555, Japan
email: ohi.fumio@nitech.ac.jp

1 Introduction

A basic problem of reliability theory is to explain relationships among the reliability performances of systems and the components. Using Boolean functions, [Mine (1959)] introduced the concept of monotone systems, where all the state spaces were assumed to be $\{0, 1\}$, so called binary state systems, where 0 and 1 denote the failure and the functioning states, respectively. Mathematical aspects of binary state systems have been fully examined by [Birnbaum and Esary (1965)], [Birnbaum *et al.* (1961)] and [Esary and Proschan (1963)]. [Barlow and Proschan (1975)] have summarized the reliability studies of the binary state monotone systems.

In many practical situations, however, systems and their components could take many other performance levels from the perfectly functioning state to the complete failure state and sometimes for some two states we cannot say which state is better/worse than another state, therefore we need multi-state reliability models with partially ordered state spaces to understand and solve practical problems as reliability estimation and risk analysis.

---

[1]Corresponding Author, Tel.: +81-52-735-5393

Multi-state systems were introduced in the context of cannibalization by [Hirsch, Meisner and Boll (1968)] and [Hochberg (1973)]. After this work, mathematical studies of multi-state systems were carried out by many authors; first making use of the minimal path and cut sets of binary state systems, [Barlow and Alexander (1978)] defined multi-state coherent systems. [El-Neweihi, Proschan and Sethuraman (1978)] defined the multi-state systems assuming all the state spaces to be commonly $\{0, 1, \cdots, M\}$. [Huang, Zuo and Fang (2003)] introduced the multi-state consecutive k-out-of-n systems and provided algorithms to evaluate the performance probabilities of the systems, assuming the state spaces to be the same finite totally ordered sets as [El-Neweihi, Proschan and Sethuraman (1978)].

[Ohi and Nishida (1983)], [Ohi (2010)] have defined multi-state systems, of which state spaces are finite totally ordered sets having not necessarily the same cardinal number, where they have fully examined order and stochastic properties of the systems. [Yu, Koren and Guo (1994)] presented a model of multi-state systems under an assumption for the state spaces to be partially ordered sets.

A mathematical generalization to partially ordered set cases, aiming at giving a theoretical framework of multi-state systems is given by [Ohi (2011)] showing an existence theorem of series and parallel systems and a decomposition of systems by series and parallel systems, which is called in the binary case as "max − min" formulae with minimal path or cut sets, see [Barlow and Proschan (1975)].

A concept of modules or sub-systems plays a crucial role for evaluating reliability performances of systems, since practical systems are usually composed of modules and the components are arranged to form these modules. A concept of module of multi-state systems for the case of partially ordered sets has been given by [Ohi (2012)], a continuation of our recent work of [Ohi (2011)], where some basic order theoretical properties are presented without proofs. In this article, we show perfect proofs of the propositions of [Ohi (2012)].

In this paper, we present a definition of modules of multi-state systems for the case of partially ordered state spaces and show some order theoretical properties of them, which are thought to be basic for further examinations of modules as giving stochastic bounds for the systems by modules. By these properties, it is suggested that normal property of systems play an important role for the definition of modules. Since if the normal property is not assumed in the definition of systems, any subset of the components could be a module in the context of partially ordered sets and no brake is applied to the defining modules, in other words, the definition of modules without normal property is meaningless.

It is well known in the case of binary state series and parallel systems that any subset of components consisting the system can be a module. We show the similar result in the case of totally ordered state spaces, following the definition of series and parallel systems of [Ohi (2010)]. We also present only definitions of a k-out-of-n and a consecutive k-out-of-n multi-state systems, of which precise examinations are remained for future work.

Notations

In this paper we use the following notations. A finite set $C = \{1, 2, \cdots, n\}$, Ω_i ($i \in C$) and S are interpreted respectively as the set of the components, the state spaces of the ith component and the system. The definition of a system is presented in Definition 6.1.

1) For $A \subset C$, $\Omega_A = \prod_{i \in A} \Omega_i$ is the product set of Ω_i ($i \in A$).
2) An element of Ω_A ($A \subset C$) is denoted by $\boldsymbol{x}^A$ and also simply $\boldsymbol{x}^A = \boldsymbol{x}$ if there is no confusion. When $A = C$, $\boldsymbol{x} \in \Omega_C$ is precisely written as $\boldsymbol{x} = (x_1, \cdots, x_n)$, $x_i \in \Omega_i$ ($i = 1, \cdots n$).
3) Let $\{\, B_j \mid 1 \leq j \leq m \,\}$ be a partition of $A \subset C$. Then for $\boldsymbol{x}_j \in \prod_{i \in B_j} \Omega_i$ ($1 \leq j \leq m$), $\boldsymbol{x} = (\boldsymbol{x}_1, \cdots, \boldsymbol{x}_m)$ is an element of Ω_A such that $P_{\Omega_{B_j}} \boldsymbol{x} = \boldsymbol{x}_j$. Then for every $\boldsymbol{x} \in \Omega_A$ ($A \subset C$), $\boldsymbol{x} = (\boldsymbol{x}^{B_1}, \cdots, \boldsymbol{x}^{B_m})$, where $\boldsymbol{x}^{B_i} = P_{B_i}(\boldsymbol{x})$ ($i = 1, \cdots, m$). P_{B_i} is the projection from Ω_A to Ω_{B_i}.
4) $(k_i, \boldsymbol{x}) \in \Omega_A$ ($A \subset C$, $i \in A$) is an element of Ω_A such that $k \in \Omega_i$ and $\boldsymbol{x} \in \prod_{j \in A \setminus \{i\}} \Omega_j$.
5) For $\boldsymbol{x} \in \Omega_C$, we sometimes use

$$(k_i, \boldsymbol{x}) = (x_1, \cdots, x_{i-1}, k, x_{i+1}, \cdots, x_n)$$

to stress that the state of the ith element of $\boldsymbol{x}$ is k.

2 Multi-state Coherent Systems

Definition 6.1. (Definition of a Multi-state System) A system composed of n components is a triplet (Ω_C, S, φ) satisfying the following conditions.

(1) $C = \{1, \cdots, n\}$ is the set of integers from 1 to n, where each number is the index of each unit.

(2) Ω_i ($i \in C$) is a finite lattice set having the least and the greatest elements denoted by m_i and M_i, respectively.

(3) $\Omega_C = \prod_{i=1}^n \Omega_i$ is the product lattice set of Ω_i $(i \in C)$. An element $\boldsymbol{x} = (x_1, \cdots, x_n) \in \Omega_C$, called a state vector, means a combination of states of the components as $x_i \in \Omega_i$ is the state of the ith component.

(4) S is a finite lattice set having the least and the greatest elements denoted by m and M, respectively.

(5) φ is a surjection from Ω_C to S, which is also called a structure function. For a state vector $\boldsymbol{x} \in \Omega_C$, $\varphi(\boldsymbol{x})$ is the state of the system.

The least and the greatest elements mean the full failure and the perfect functioning states, respectively.

The orders on Ω_i $(i \in C), S$ are denoted by the common symbol $\leqq$, and $a < b$ is used for a and b of Ω_i $(i \in C)$ or S to mean $a \leqq b$ and $a \neq b$. For $\boldsymbol{x} = (x_1, \cdots, x_n)$ and $\boldsymbol{y} = (y_1, \cdots, y_n)$ of Ω_C, we use the symbols $\boldsymbol{x} \leqq \boldsymbol{y}$ for $x_i \leqq y_i$ $(\forall i \in C)$, $\boldsymbol{x} < \boldsymbol{y}$ for $\boldsymbol{x} \leqq \boldsymbol{y}$ and $\boldsymbol{x} \neq \boldsymbol{y}$, $\boldsymbol{x} << \boldsymbol{y}$ for $x_i < y_i$ $(\forall i \in C)$.

A system (Ω_C, S, φ) is simply called a system φ when there is no confusion. For a system φ and an element $s \in S$, we use the following notations;

$$\varphi^{-1}(s \leqq) \overset{def}{=} \{\boldsymbol{x} \in \Omega_C \mid \varphi(\boldsymbol{x}) \geqq s\},$$

$$\varphi^{-1}(\leqq s) \overset{def}{=} \{\boldsymbol{x} \in \Omega_C \mid \varphi(\boldsymbol{x}) \leqq s\},$$

$$\varphi^{-1}(s) \overset{def}{=} \{\boldsymbol{x} \in \Omega_C \mid \varphi(\boldsymbol{x}) = s\}.$$

Generally $MI(A)$ and $MA(A)$ are the sets of all the minimal and maximal elements of a finite ordered set A, respectively. For s and t of A such that $s < t$, a path of length k from s to t is a series $(s_0, s_1, \cdots, s_k)$ satisfying that $s_0 = s < s_1 < \cdots < s_k = t$ and there is no element between s_i and s_{i+1} $(i = 0, \cdots, k-1)$.

Definition 6.2. (Increasing Property) A system φ is called increasing, when $\varphi(\boldsymbol{x}) \leqq \varphi(\boldsymbol{y})$ holds for every $\boldsymbol{x}$ and $\boldsymbol{y} \in \Omega_C$ such that $\boldsymbol{x} \leqq \boldsymbol{y}$.

Definition 6.3. (Normal Property) A system φ is called normal when for every $s \in S$,

$$\forall \boldsymbol{x} \in MI\left(\varphi^{-1}(s \leqq)\right), \; \varphi(\boldsymbol{x}) = s, \tag{1}$$

$$\forall \boldsymbol{x} \in MA\left(\varphi^{-1}(\leqq s)\right), \; \varphi(\boldsymbol{x}) = s, \tag{2}$$

in other words, (1) is equivalent to $MI\left(\varphi^{-1}(s \leqq)\right) = MI\left(\varphi^{-1}(s)\right)$ and (2) is to $MA\left(\varphi(\leqq s)\right) = MA\left(\varphi^{-1}(s)\right)$.

Definition 6.4. (Strong Relevant Property) (1) The component $i \in C$ is said to be strong relevant when the following is satisfied.

$$\forall r, \; \forall s \in S \; (r \neq s), \; \exists (k_i, \boldsymbol{x}), \; \exists (l_i, \boldsymbol{x}), \; \varphi(k_i, \boldsymbol{x}) = r, \; \varphi(l_i, \boldsymbol{x}) = s.$$

(2) A system φ is called strong relevant when every component is relevant to the system.

The strong relevant property of the component i means that for arbitrary assigned two different states r and s of the system, the component i can shift the state from r to s only by changing the states of the component, remaining the other components' states unchanged. The strong relevant, which is called relevant in [Ohi (2011)], is a condition apparently stronger than that of Definition 6.6.

Definition 6.5. (Strong Coherent System) A system φ is called strong coherent, when φ is increasing, normal and strong relevant.

Definition 6.6. (Relevant Property) (1) The component $i \in C$ is said to be relevant when the following is satisfied.

$$\forall k, \ \forall l \in \Omega_i \ (k \neq l), \ \exists(k_i, \boldsymbol{x}), \ \exists(l_i, \boldsymbol{x}), \ \varphi(k_i, \boldsymbol{x}) \neq \varphi(l_i, \boldsymbol{x}).$$

(2) A system φ is called relevant when every component is relevant.

The condition (1) of Definition 6.6 has essentially no practical restriction on a system, since if the condition does not hold for a component $i \in C$, we have

$$\exists k, \ \exists l \in \Omega_i \ (k \neq l), \ \forall(k_i, \boldsymbol{x}), \ \forall(l_i, \boldsymbol{x}), \ \varphi(k_i, \boldsymbol{x}) = \varphi(l_i, \boldsymbol{x}),$$

which means that the states k and l equivalently contribute to the system states and therefore are not necessarily defined to be different states, in other words, we may combine the two states k and l into one state.

Definition 6.7. (Coherent System) A system φ is called coherent, when φ is increasing, normal and relevant.

Series and parallel systems play important roles in the theory and practical situation as series and parallel decomposition of systems, deriving stochastic bounds for reliability, and so on. See [Ohi (2010)], [Ohi (2011)].

Definition 6.8. (Series and Parallel Systems) Let φ be increasing.
   (1) φ is called a series system when for every $s \in S$, $\varphi^{-1}(s \leqq)$ has the least element which is simply written as $\boldsymbol{m}_s$.
   (2) φ is called a parallel system when for every $s \in S$, $\varphi^{-1}(\leqq s)$ has the greatest element which is simply written as $\boldsymbol{M}_s$.

Proposition 6.1. (1) Let (Ω_C, S, φ) be a strong coherent series system and S is a totally ordered set. Then we have

$$\forall s, \ \forall t \in S \ (s < t), \ \boldsymbol{m}_s << \boldsymbol{m}_t.$$

(2) Let (Ω_C, S, φ) be a strong coherent parallel system and S is a totally ordered set. Then we have

$$\forall s, \ \forall t \in S \ (s < t), \ \boldsymbol{M}_s << \boldsymbol{M}_t.$$

Proof. The proof of (2) is similar to (1), then we only prove (1). Let $s, t \in S$ and $s < t$. Since φ is normal, we have $\boldsymbol{m}_s < \boldsymbol{m}_t$. We suppose that for $i \in C$, $(\boldsymbol{m}_s)_i = (\boldsymbol{m}_t)_i$. The system φ is strong coherent, then

$$\exists (k_i, \boldsymbol{x}), \ \exists (l_i, \boldsymbol{x}), \ \ \varphi(k_i, \boldsymbol{x}) = s, \ \varphi(l_i, \boldsymbol{x}) = t.$$

Since $\boldsymbol{m}_s$ and $\boldsymbol{m}_t$ are the least elements of $\varphi^{-1}(s)$ and $\varphi^{-1}(t)$, respectively, we have two inequalities $\boldsymbol{m}_s \leqq (k_i, \boldsymbol{x})$ and $\boldsymbol{m}_t \leqq (l_i, \boldsymbol{x})$. From the assumption $(\boldsymbol{m}_s)_i = (\boldsymbol{m}_t)_i$, we have $\boldsymbol{m}_t \leqq (k_i, \boldsymbol{x})$ and then $\varphi(k_i, \boldsymbol{x}) = t$, contradicting to $\varphi(k_i, \boldsymbol{x}) = t$. ∎

From Proposition 6.1, we have the next corollary asserting the necessary and sufficient condition for the existence of series and parallel strong coherent systems.

Corollary 6.1. (Existence of Strong Coherent Series and Parallel Systems) For a system (Ω_C, S, φ) to be a strong coherent series (parallel) system, it is necessary and sufficient that for every $i \in C$, there is a length $|S|$ path from the least element to the greatest element of Ω_i.

Proof. (only if part) Let (Ω_C, S, φ) be a strong coherent series system. From Proposition 6.1, $\boldsymbol{m}_0 << \boldsymbol{m}_1 << \cdots << \boldsymbol{m}_M$, where the state space S of the system is assumed to be $\{0, 1, \cdots, M\}$ without generality, since S is a totally ordered set. Then we have

$$(\boldsymbol{m}_0)_i < (\boldsymbol{m}_1)_i < \cdots < (\boldsymbol{m}_M)_i$$

for each component $i \in C$, meaning that there should be a length $|S|$ path from the least element to the greatest element of Ω_i.

(if part) Following [Ohi (2010)], we may construct the least element $\boldsymbol{m}_s$ for each $s \in S$, and then a strong coherent series system is given. ∎

Using the definition of series systems we give a definition of a k-out-of-n system, which is a straight generalization of that of [Ohi (2010)].

Definition 6.9. (A definition of a k-out-of-n system) A system φ is called a k-out-of-n system, when for each subset A of C such that $|A| = k$, there exists a series system (Ω_A, φ_A, S) satisfying

$$\forall \boldsymbol{x} \in \Omega_C, \ \varphi(\boldsymbol{x}) = \inf_{A \subset C, \ |A| = k} \varphi_A(\boldsymbol{x}^A).$$

Definition 6.10. (A definition of a consecutive k-out-of-n system) A system φ is called a consecutive k-out-of-n system, when for each subset A of C such that $A = \{i, i+1, \cdots, i+k-1\}$, which is called a k-consecutive subset of Ω, there exists a series system (Ω_A, φ_A, S) satisfying

$$\forall \boldsymbol{x} \in \Omega_C, \ \varphi(\boldsymbol{x}) = \inf_{A: \ \text{a } k\text{-consecutive subset of } C} \varphi_A(\boldsymbol{x}^A),$$

where if $j \in A$ is greater than n, j is considered to be $j - n$, then this definition is a generalization of so-called circular consecutive k-out-of-n system.

When the state spaces are binary sets, the structure functions of these definitions are easily shown to be usual binary state k-out-of-n and consecutive k-out-of-n systems. In this paper we present only definitions of these systems of which precise examinations as order theoretic and probabilistic examinations are remained for future work.

3   Modules

Following [Ohi (2010)], we examine the concept of modules by using the equivalent relation $\overset{\varphi}{=}$.

Letting (Ω_C, S, φ) be a system, we define a pseudo-order $\overset{\varphi}{\leq}$ on a product set Ω_A of Ω_i ($i \in A$), where A is a non-empty subset of C. For $\boldsymbol{x}, \boldsymbol{y} \in \Omega_A$

$$\boldsymbol{x} \overset{\varphi}{\leq} \boldsymbol{y} \overset{def}{\iff} \forall \boldsymbol{z} \in \Omega_{A'}, \ \varphi(\boldsymbol{x}, \boldsymbol{z}) \leq \varphi(\boldsymbol{y}, \boldsymbol{z}),$$

where $A' = C \backslash A$. Furthermore an equivalent relation $\overset{\varphi}{=}$ is defined on Ω_A as for $\boldsymbol{x}, \boldsymbol{y} \in \Omega_A$,

$$\boldsymbol{x} \overset{\varphi}{=} \boldsymbol{y} \overset{def}{\iff} \forall \boldsymbol{z} \in \Omega_{A'}, \ \varphi(\boldsymbol{x}, \boldsymbol{z}) = \varphi(\boldsymbol{y}, \boldsymbol{z}) \iff \boldsymbol{x} \overset{\varphi}{\leq} \boldsymbol{y} \ \text{and} \ \boldsymbol{y} \overset{\varphi}{\leq} \boldsymbol{x}.$$

$\Omega_A |\overset{\varphi}{=}$ is the quotient space of Ω_A with respect to the equivalent relation $\overset{\varphi}{=}$. Each element of $\Omega_A |\overset{\varphi}{=}$ is called an equivalent class. Defining an order $\overset{\varphi}{\leq}$ on $\Omega_A |\overset{\varphi}{=}$ as

$$D, E \in \Omega_A |\overset{\varphi}{=}, \ D \overset{\varphi}{\leq} E \overset{def}{\iff} \boldsymbol{x} \overset{\varphi}{\leq} \boldsymbol{y} \ (\boldsymbol{x} \in D, \boldsymbol{y} \in E),$$

we have an ordered set $(\Omega_A| \overset{\varphi}{=}, \overset{\varphi}{\leqq})$. This definition is not depend on the choice of $\boldsymbol{x}$ and $\boldsymbol{y}$, since D and E are equivalent classes.

For $\boldsymbol{x} \in \Omega_A$, $[\boldsymbol{x}]$ is the equivalent class to which $\boldsymbol{x}$ belongs. We define a mapping $\varphi_A : (\Omega_A| \overset{\varphi}{=}, \overset{\varphi}{\leqq}) \times \Omega_{A'} \to S$ as

$$([\boldsymbol{x}], \boldsymbol{y}) \in \Omega_A| \overset{\varphi}{=} \times \Omega_{A'}, \quad \varphi_A([\boldsymbol{x}], \boldsymbol{y}) \overset{def}{=} \varphi(\boldsymbol{x}, \boldsymbol{y}),$$

and $\chi_A : \Omega_A \to (\Omega_A| \overset{\varphi}{=}, \overset{\varphi}{\leqq})$ as

$$\boldsymbol{x} \in \Omega_A, \quad \chi_A(\boldsymbol{x}) = [\boldsymbol{x}].$$

These mappings are related with each other as

$$\boldsymbol{x} \in \Omega_C, \quad \varphi(\boldsymbol{x}) = \varphi_A\left(\chi_A(\boldsymbol{x}^A), \boldsymbol{x}^{A'}\right). \tag{3}$$

Proposition 6.2. φ : increasing $\iff \chi_A, \varphi_A$: increasing.

Proof. (only if part) Let $\varphi : \Omega_C \to S$ be increasing.

For $\boldsymbol{x}, \boldsymbol{y} \in \Omega_A$, $\boldsymbol{x} \leqq \boldsymbol{y}$, we have

$$\forall \boldsymbol{z} \in \Omega_{A'}, \ (\boldsymbol{x}, \boldsymbol{z}) \leqq (\boldsymbol{y}, \boldsymbol{z}) \text{ then } \varphi(\boldsymbol{x}, \boldsymbol{z}) \leqq \varphi(\boldsymbol{y}, \boldsymbol{z}),$$

which means

$$[\boldsymbol{x}] \overset{\varphi}{\leqq} [\boldsymbol{y}] \ \text{ i.e. } \ \chi_A(\boldsymbol{x}) \leqq \chi_A(\boldsymbol{y}).$$

For $\boldsymbol{x}, \boldsymbol{y} \in \Omega_A$ such that $[\boldsymbol{x}] \overset{\varphi}{\leqq} [\boldsymbol{y}]$ and $\boldsymbol{x}', \boldsymbol{y}' \in \Omega_{A'}$ such that $\boldsymbol{x}' \leqq \boldsymbol{y}'$ we have the following chain of inequalities.

$$\varphi_A([\boldsymbol{x}], \boldsymbol{x}') = \varphi(\boldsymbol{x}, \boldsymbol{x}') \leqq \varphi(\boldsymbol{y}, \boldsymbol{x}') \leqq \varphi(\boldsymbol{y}, \boldsymbol{y}') = \varphi_A([\boldsymbol{y}], \boldsymbol{y}'),$$

where the first inequality is from $\boldsymbol{x} \overset{\varphi}{\leqq} \boldsymbol{y}$, and the second is from the increasing property of φ.

(if part) For $\boldsymbol{x}, \boldsymbol{y} \in \Omega_A, \boldsymbol{x}', \boldsymbol{y}' \in \Omega_{A'}$ such that $(\boldsymbol{x}, \boldsymbol{x}') \leqq (\boldsymbol{y}, \boldsymbol{y}')$,

$$\varphi(\boldsymbol{x}, \boldsymbol{x}') = \varphi_A(\chi_A(\boldsymbol{x}), \boldsymbol{x}') \leqq \varphi_A(\chi_A(\boldsymbol{y}), \boldsymbol{y}') = \varphi(\boldsymbol{y}, \boldsymbol{y}'),$$

since $\boldsymbol{x} \leqq \boldsymbol{y}$ and the increasing property of χ_A imply $\chi_A(\boldsymbol{x}) \leqq \chi_A(\boldsymbol{y})$, and furthermore $\boldsymbol{x}' \leqq \boldsymbol{y}'$ and the increasing property of φ_A imply the above inequality. ∎

Proposition 6.3. φ : relevant $\iff \chi_A, \varphi_A$: relevant.

Proof. (only if part) Suppose $i \in A$. i is relevant to φ, then

$$\forall k, \forall l \in \Omega_i \ (k \neq l), \ \exists (\cdot_i, \boldsymbol{x}), \ \varphi(k_i, \boldsymbol{x}) \neq \varphi(l_i, \boldsymbol{x}).$$

Therefore $(k_i, \boldsymbol{x})^A \overset{\varphi}{\neq} (l_i, \boldsymbol{x})^A$, i.e., $\left[(k_i, \boldsymbol{x})^A\right] \neq \left[(l_i, \boldsymbol{x})^A\right]$ and then

$$\chi_A \left(k_i, \boldsymbol{x}^{A \setminus \{i\}} \right) \neq \chi_A \left(l_i, \boldsymbol{x}^{A \setminus \{i\}} \right),$$

meaning the relevant property of χ_A.

The relevant property of φ_A is easily proved. Suppose $i \in A'$.

$$\forall k, \forall l \in \Omega_i \ (k \neq l), \ \exists (\cdot_i, \boldsymbol{x}), \ \varphi(k_i, \boldsymbol{x}) \neq \varphi(l_i, \boldsymbol{x}).$$

Then we have

$$\varphi_A \left(\left[\boldsymbol{x}^A \right], k_i, \boldsymbol{x}^{A' \setminus \{i\}} \right) \neq \varphi_A \left(\left[\boldsymbol{x}^A \right], l_i, \boldsymbol{x}^{A' \setminus \{i\}} \right),$$

meaning that the component $i \in A'$ is relevant to φ_A.

Next we suppose that for $\boldsymbol{x}, \boldsymbol{y} \in \Omega_A$, $[\boldsymbol{x}] \overset{\varphi}{\neq} [\boldsymbol{y}]$ holds. From the definition of the equivalent relation $\overset{\varphi}{=}$, $\varphi(\boldsymbol{x}, \boldsymbol{z}) \neq \varphi(\boldsymbol{y}, \boldsymbol{z})$ for some $\boldsymbol{z} \in \Omega_{A'}$. Noticing the relation (3), we have $\varphi_A([\boldsymbol{x}], \boldsymbol{z}) \neq \varphi_A([\boldsymbol{y}], \boldsymbol{z})$. Then the relevant property of φ_A is proved.

The if part is apparent, and then the proof is omitted. ∎

Proposition 6.4. When φ is increasing, we have the following relation about the minimal elements.

$$\forall s \in S, \ \boldsymbol{x} \in MI\left(\varphi^{-1}(s)\right) \implies \boldsymbol{x}^A \in MI\left([\boldsymbol{x}^A]\right).$$

For maximal elements, we have a similar relation.

Proof. If $\boldsymbol{z} \leqq \boldsymbol{x}^A$ and $\boldsymbol{z} \in [\boldsymbol{x}^A]$, then

$$\varphi\left(\boldsymbol{z}, \boldsymbol{x}^{A'}\right) = \varphi\left(\boldsymbol{x}^A, \boldsymbol{x}^{A'}\right) = s, \quad \left(\boldsymbol{z}, \boldsymbol{x}^{A'}\right) \leq \left(\boldsymbol{x}^A, \boldsymbol{x}^{A'}\right).$$

Since $\boldsymbol{x}$ is a minimal element of $\varphi^{-1}(s)$, $\left(\boldsymbol{z}, \boldsymbol{x}^{A'}\right) = \left(\boldsymbol{x}^A, \boldsymbol{x}^{A'}\right)$ and thus $\boldsymbol{z} = \boldsymbol{x}^A$ follows. ∎

Proposition 6.5. When φ is increasing, for every $s \in S$ and $\boldsymbol{x} \in \Omega_A$ we have the following relation .

$$([\boldsymbol{x}], \boldsymbol{y}) \in MI\left(\varphi_A^{-1}(s)\right) \text{ and } \boldsymbol{x}_m \in MI\left([\boldsymbol{x}]\right) \implies (\boldsymbol{x}_m, \boldsymbol{y}) \in MI\left(\varphi^{-1}(s)\right).$$

For maximal elements, we have a similar relation.

Proof. We suppose

$$\boldsymbol{u} \in \Omega_A, \ \boldsymbol{v} \in \Omega_{A'}, \ (\boldsymbol{u}, \boldsymbol{v}) \leqq (\boldsymbol{x}_m, \boldsymbol{y}), \ \varphi(\boldsymbol{u}, \boldsymbol{v}) = s,$$

and notice $[\boldsymbol{x}_m] = [\boldsymbol{x}]$.

Since $\boldsymbol{u} \leq \boldsymbol{x}_m$, $[\boldsymbol{u}] \overset{\varphi}{\leqq} [\boldsymbol{x}_m]$ and then $([\boldsymbol{u}], \boldsymbol{v}) \leqq ([\boldsymbol{x}_m], \boldsymbol{y})$. On the other hand $\varphi(\boldsymbol{u}, \boldsymbol{v}) = \varphi_A([\boldsymbol{u}], \boldsymbol{v}) = s$, then $([\boldsymbol{u}], \boldsymbol{v}) = ([\boldsymbol{x}_m], \boldsymbol{y})$, because $([\boldsymbol{x}], \boldsymbol{y}) = ([\boldsymbol{x}_m], \boldsymbol{y})$ is a minimal element of $\varphi_A^{-1}(s)$. Furthermore $[\boldsymbol{u}] = [\boldsymbol{x}_m]$ which means $\boldsymbol{u} \in [\boldsymbol{x}_m]$, $\boldsymbol{x}_m$ is a minimal element of $[\boldsymbol{x}]$, and $\boldsymbol{u} \leqq \boldsymbol{x}_m$, then we have $\boldsymbol{u} = \boldsymbol{x}_m$. ∎

Proposition 6.6. φ : increasing, normal $\implies \varphi_A$: normal.

Proof. We suppose

$$s \in S, \ t \in S, \ s < t, \ ([\boldsymbol{x}], \boldsymbol{y}) \in MI\left(\varphi_A^{-1}(t)\right).$$

Letting $\boldsymbol{x}_m \in MI([\boldsymbol{x}])$, we have $(\boldsymbol{x}_m, \boldsymbol{y}) \in MI\left(\varphi^{-1}(t)\right)$ by Proposition 6.5. Since φ is normal, then

$$\exists \boldsymbol{z} \in MI\left(\varphi^{-1}(s)\right), \ \boldsymbol{z} \leqq (\boldsymbol{x}_m, \boldsymbol{y}),$$

from which $\left[\boldsymbol{z}^A\right] \overset{\varphi}{\leqq} [\boldsymbol{x}_m]$ follows because $\boldsymbol{z}^A \leqq \boldsymbol{x}_m$ and φ is increasing. Noticing $\boldsymbol{z}^{A'} \leq \boldsymbol{y}$, we have

$$\left([\boldsymbol{z}^A], \boldsymbol{z}^{A'}\right) \leqq ([\boldsymbol{x}_m], \boldsymbol{y}), \ \text{ and } \varphi_A\left([\boldsymbol{z}^A], \boldsymbol{z}^{A'}\right) = \varphi(\boldsymbol{z}) = s,$$

which means that φ_A is normal. ∎

Proposition 6.7. When φ is increasing, we have

$$\chi_A, \ \varphi_A : \text{normal} \implies \varphi : \text{normal}.$$

Proof. Supposing

$$s \in S, \ t \in S, \ s < t, \ \boldsymbol{x} \in MI\left(\varphi^{-1}(t)\right),$$

we have

$$\varphi_A\left([\boldsymbol{x}^A], \boldsymbol{x}^{A'}\right) = \varphi\left(\boldsymbol{x}^A, \boldsymbol{x}^{A'}\right) = t, \ \boldsymbol{x}^A \in MI\left([\boldsymbol{x}^A]\right),$$

where the above inclusion relation is by Proposition 6.4. The normal property of φ_A implies

$$\exists ([\boldsymbol{y}], \boldsymbol{z}) \leqq \left([\boldsymbol{x}^A], \boldsymbol{x}^{A'}\right), \ \varphi_A([\boldsymbol{y}], \boldsymbol{z}) = s.$$

Since $[\boldsymbol{y}] \overset{\varphi}{\lesssim} [\boldsymbol{x}^A]$ and χ_A is normal, we have

$$\exists \boldsymbol{y}_m \in MI\left([\boldsymbol{y}]\right), \ \boldsymbol{y}_m \leq \boldsymbol{x}^A,$$

and then finally

$$(\boldsymbol{y}_m, \boldsymbol{z}) \leq \left(\boldsymbol{x}^A, \boldsymbol{x}^{A'}\right), \ \varphi(\boldsymbol{y}_m, \boldsymbol{z}) = \varphi_A\left([\boldsymbol{y}_m], \boldsymbol{z}\right) = \varphi_A([\boldsymbol{y}], \boldsymbol{z}) = s,$$

which means φ to be normal. ∎

Proposition 6.8. When φ is increasing and χ_A is normal, then we have the following relation about minimal elements.

$$\boldsymbol{x} \in MI\left(\varphi^{-1}(s)\right) \implies \left([\boldsymbol{x}^A], \boldsymbol{x}^{A'}\right) \in MI\left(\varphi_A^{-1}(s)\right)$$

The similar relation holds for maximal elements.

Proof. Let $\boldsymbol{x} \in MI\left(\varphi^{-1}(s)\right)$. An inclusion relation $\left([\boldsymbol{x}^A], \boldsymbol{x}^{A'}\right) \in \varphi_A^{-1}(s)$ is clear from $\varphi_A\left([\boldsymbol{x}^A], \boldsymbol{x}^{A'}\right) = \varphi\left(\boldsymbol{x}^A, \boldsymbol{x}^{A'}\right) = s$.

We suppose $([\boldsymbol{y}], \boldsymbol{z})$ to be

$$\varphi_A([\boldsymbol{y}], \boldsymbol{z}) = s, \ ([\boldsymbol{y}], \boldsymbol{z}) \leq \left([\boldsymbol{x}^A], \boldsymbol{x}^{A'}\right), \ ([\boldsymbol{y}], \boldsymbol{z}) \neq \left([\boldsymbol{x}^A], \boldsymbol{x}^{A'}\right),$$

noticing that $[\boldsymbol{y}] \overset{\varphi}{\lesssim} [\boldsymbol{x}^A]$ holds.

If $[\boldsymbol{y}] = [\boldsymbol{x}^A]$, then $\boldsymbol{z} \leq \boldsymbol{x}^{A'}$ and $\boldsymbol{z} \neq \boldsymbol{x}^{A'}$, therefore

$$(\boldsymbol{x}^A, \boldsymbol{z}) \leq \left(\boldsymbol{x}^A, \boldsymbol{x}^{A'}\right), \ (\boldsymbol{x}^A, \boldsymbol{z}) \neq \left(\boldsymbol{x}^A, \boldsymbol{x}^{A'}\right), \ \varphi\left(\boldsymbol{x}^A, \boldsymbol{z}\right) = s,$$

contradicting to $\boldsymbol{x} \in MI\left(\varphi^{-1}(s)\right)$.

Hence we have $[\boldsymbol{y}] \leq [\boldsymbol{x}^A]$ and $[\boldsymbol{y}] \neq [\boldsymbol{x}^A]$. Since $\boldsymbol{x}^A \in MI\left([\boldsymbol{x}^A]\right)$ by Proposition 6.4, there exists a minimal element $\boldsymbol{y}_m$ of $[\boldsymbol{y}]$ from the normal property of χ_A and thus we have

$$\varphi(\boldsymbol{y}_m, \boldsymbol{z}) = \varphi(\boldsymbol{y}, \boldsymbol{z}) = \varphi_A([\boldsymbol{y}], \boldsymbol{z}), \ (\boldsymbol{y}_m, \boldsymbol{z}) \leq \left(\boldsymbol{x}^A, \boldsymbol{x}^{A'}\right), \ (\boldsymbol{y}_m, \boldsymbol{z}) \neq \left(\boldsymbol{x}^A, \boldsymbol{x}^{A'}\right),$$

which contradicts to $\boldsymbol{x} = \left(\boldsymbol{x}^A, \boldsymbol{x}^{A'}\right) \in MI\left(\varphi^{-1}(s)\right)$. ∎

From the above propositions, increasing, normal and relevant properties of φ determine the increasing and relevant properties of φ_A and χ_A, and furthermore the normal property of φ_A, but not the normal property of χ_A. Then $A \subset C$ is a module or not, in other words, $\left(\Omega_A, \Omega_A| \overset{\varphi}{=}, \chi_A\right)$ is a coherent system or not, according to χ_A is normal or not, respectively.

Summarized relations are shown in the following.

$$\varphi : \text{increasing} \iff \varphi_A, \chi_A : \text{increasing},$$
$$\varphi : \text{relevant} \iff \varphi_A, \chi_A : \text{relevant},$$
$$\varphi : \text{increasing and normal} \implies \varphi_A : \text{normal},$$
$$\varphi : \text{normal} \impliedby \varphi_A, \chi_A : \text{normal}.$$

When φ is increasing,

$$x \in MI(\varphi^{-1}(s)) \implies x^a \in MI([x^a]),$$
$$([x], y) \in MI(\varphi_A^{-1}(s)) \text{ and } x_m \in MI([x]) \implies (x_m, y) \in MI(\varphi^{-1}(s)).$$

When φ is increasing and χ_A is normal,

$$x \in MI(\varphi^{-1}(s)) \implies \left([x^a], x^{A'}\right) \in MI(\varphi_A^{-1}(s)).$$

Here, reminding that a system is called coherent in this paper when the system is increasing, relevant and normal from Definition 6.7, we have the following theorem.

Theorem 6.1. When (Ω_C, S, φ) is a coherent system, $A \subset C$ is a module if and only if χ_A is normal.

From Propositions 6.4, 6.5 and 6.8, we have the next theorem about maximal and minimal elements.

Theorem 6.2. Let φ be a coherent system, and $A \subset C$ is assumed to be a module. For $x \in \Omega_A, y \in \Omega_{A'}$, we have the following equivalent relation;

$$(x, y) \in MI\left(\varphi^{-1}(s)\right) \iff ([x], y) \in MI\left(\varphi_A^{-1}(s)\right) \text{ and } x \in MI([x]).$$

Similar equivalent relation holds for maximal elements.

Remark 6.1. Normal property seems to play an important role for defining modules, since from the above propositions, when we do not assume the normal property on the definition of modules, every subset A of C may be a module, which means no brakes to defining the modules.

It is well known that every subset of components is a module in the case of binary state series and parallel systems. We have a similar conclusion for the case of multi-state systems. Here we prove the proposition for strong coherent systems when the state spaces are totally ordered sets.

Proposition 6.9. Suppose a system (Ω_C, S, φ) is a strong series (parallel) system and the state spaces are totally ordered sets. Then every subset A of components of the system is a module.

Proof. Without loss of generality, we may assume $S = \{0, 1, \cdots, M\}$, since S is assumed to be a finite totally ordered set. From Proposition 6.1 (1),

$$m_0 << m_1 << \cdots << m_M. \tag{4}$$

We define a system (Ω_A, S, χ_A) as

$$x \in \Omega_A, \ \chi_A(x) \overset{def}{=} \max \left\{ \ s \mid x \geq m_s^A \ \right\}.$$

By (4), we have

$$m_0^A << m_1^A << \cdots << m_M^A,$$

then the system χ_A is clearly normal and a series system.

Next we define a system $\varphi_A : S \times \Omega_{A'} \to S$ as

$$s \in S \text{ and } y \in \Omega_{A'}, \ \varphi_A(s, y) \overset{def}{=} \varphi(m_s^A, y),$$

then we have

$$x \in \Omega_A, \ y \in \Omega_{A'}, \ \varphi(x, y) = \varphi_A(\chi_A(x), y). \tag{5}$$

Since from the definition of χ_A, $\chi_A(x) = s$ is equivalent to

$$m_s^A \leq x \text{ and } m_{s+1}^A \not\leq x, \tag{6}$$

and so

$$\forall y \in \Omega_{A'}, \ \varphi(x, y) = \varphi(m_s^A, y).$$

Because, when $\varphi(x, y) = t$, $(x, y) \geq m_t$ holds, and then from (6), we have

$$m_t \leq (m_s^A, y) \leq (x, y).$$

Thus the increasing property of φ implies $\varphi(m_s^A, y) = s$.

Hence (5) holds and A is proved to be a module.

In the case of a parrallel system, the proposition is similarly proved. ∎

Remark 6.2. It is easily proved that χ_A and φ_A of the proof of Theorem 6.9 are strong coherent series systems.

4 Concluding Remarks

In this paper we presented a definition of modules and showed that the normal property plays an important role for the definition, since if we do not assume the normal property on the definition of systems, any subset of the components could be module in the context of partially ordered sets and no brake is applied to defining modules, in other words, the definition of modules without the assumption of normal property is meaningless. Furthermore we showed that any subset of components consisting a coherent series or parallel system is a module in the multi-state case of which state spaces are totally ordered sets, following the definition of series and parallel systems of [Ohi (2010)].

On the other hand, we left some problems for future works. In the case of binary state case, it is well known that any k-out-of-n system ($2 \leqq k \leqq n-1$) has no module. It is left an open problem to prove the similar assertion for the multi-state case. The second problem, which is more important, is to give stochastic bounds for systems through modules, also an open problem.

References

Barlow, R. E. and Proschan, F. (1975). *Statistical Theory of Reliability of Life Testing*, Holt, Rinehart and Winston, New York.

Barlow, R. E. and Alexander, S. Wu (1978). Coherent systems with multistate components, *Mathematics of Operations Research*, **3**, pp. 275–281.

Birnbaum, Z. W. and Esary, J. D. (1965). Modules of coherent binary systems, *SIAM J. Appl. Math.*, **13**, pp. 444–462.

Birnbaum, Z. W., Esary, J. D. and Saunder, S. C. (1961). Multi-component systems and structures and their reliability, *Technometrics*, **3**, pp. 55–77.

El-Neweihi, E., Proschan, F. and Sethuraman, J. (1978). Multistate coherent systems, *J. Appl. Probability*, **15**, pp. 675–688.

Esary, J. D. and Proschan, F. (1963). Coherent structures of non-identical components, *Technometrics*, **5**, pp. 191–209.

Esary, J. D., Marshall, A. W. and Proschan, F. (1970). Some reliability application of hazard transform, *SIAM J. Appl. Math.*, **18**, pp.331–359.

Hirsch, W. M., Meisner, M. and Boll, C. (1968). Cannibalization in multicomponent systems and the theory of reliability, *Naval Res. Logist. Quart.*, **15**, pp. 331–359.

Hochberg (1973). Generalized multistate systems under cannibalization, *Naval Res. Logistic. Quart.*, **20**, pp. 585–605.

Huang, J., Zuo, M. J. and Fang, Z. (2003). Multi-state consecutive-k-out-of-n systems, *IIE Transactions*, **35**, pp. 527–534.

Mine, H. (1959). Reliability of physical system, *IRE*, **CT-6** Special Supplement, pp. 138–151.

Ohi, F. and Nishida, T. (1983). Generalized multistate coherent systems, *J. Japan Statist. Soc.*, **13**, pp. 165–181.

Ohi, F. and Nishida, T. (1984). On Multistate Coherent Systems, *IEEE Transactions on Reliability*, **R-33**, pp. 284–288.

Ohi, F. and Nishida, T. (1984). *Multistate Systems in Reliability Theory, Stochastic Models in Reliability Theory, Lecture Notes in Economics and Mathematical Systems 235*, Springer-Verlag, pp. 12–22.

Ohi, F. (2010). Multistate Coherent Systems, in "Stochastic Reliability Modeling, Optimization and Applications", edited by S. Nakamura and T. Nakagawa, World Science, pp. 3–34.

Ohi, F. (2011). Lattice Set Theoretic Treatment of Multi-state Coherent Systems, Proceedings of The 7th International Conference on "Mathematical Method in Reliability": Theory. Methods. Applications, edited by Lirong Cui & Xian Zhao, pp. 383–389.

Ohi, F. (2012). Multi-State Coherent Systems and Modules – Basic Properties –, The 5th Asia-Pasific International Symposium on Advanced Reliability and Maintenance Modeling, in Advanced reliability and maintenance modeling v, Basis of Reliability Analysis, Nanjing, China, 1-3 November 2012/12/02, edited by Hisashi Yamamoto, Chunhua Qian, Lirong Cui and Takashi Dohi, McGrow Hill Education, pp. 374–381.

Shinmori, S., Ohi, F., Hagihara, H. and Nishida, T. (1989). Modules for Two Classes of Multi-State Systems, *The Transactions of the IEICE*, **E72**, pp. 600–608.

Yu, K., Koren, I. and Guo, Y. (1994). Generalized Multistate Monotone Coherent Systems, *IEEE Transactions on Reliability*, **43**, pp. 242–250.

Chapter 7

Calculation Algorithms for the System State Distributions of Multi-State k-out-of-n Systems

Hisashi Yamamoto[1] and Tomoaki Akiba[2]

[1]*Faculty of System Design, Tokyo Metoropolitan University,*
6-6 Asahigaoka, Hino, Tokyo 191-0065, Japan
[2]*Faculty of Social System Science, Chiba Institute of Technology,*
2-17-1, Narashino, Chiba 275-0016, Japan

1 Introduction

In traditional reliability theory, both the system and its components are considered to take only two possible states: working or failed. In the binary context, a system with n components in sequence is called the k-out-of-n:F (G) system if the system fails (works) whenever at least k components in the system work (fail). When $k = n$, k-out-of-n:F (G) system is parallel (series) system.

In many practical situations, the states of the systems and their components are considered to take more than two different levels, ranging from perfectly working to completely failed. Many researchers have extended binary system of k-out-of-n:F (G) system, for example, see, [Barlow and Prochan (1975)], [Kołowrocki (2004)], [Chang et $al.$ (2000)] and [Kuo and Zuo (2003)].

In the multi-state k-out-of-n:F (G) system, both the system and the components are allowed to be in $M + 1$ possible states, $0, 1, 2, \cdots, M$, where M is a positive integer which represents the system or component in perfectly working state, while zero represents completely failure state. The multi-state k-out-of-n system reliability model provides more flexibility for the modeling of equipment conditions. Some papers provided with the algorithms for computing system state distribution of multi-state system. Researchers have extended binary k-out-of-n system to multi-state k-out-of-n system, for example, see, [Barlow and Wu (1978)], [El-Neweihi *et al.* (1978)], [Griffith (1980)], [Kossow and Preuss (1995)], [Malinowski and Preuss (1995, 1996)], [Zuo and Liang (1994)], [Koutras (1997)], [Haim and Porat (1991)], [Huang *et al.* (2000)], [Zuo *et al.* (2003, 2006)], [Amari *et al.* (2009)] and [Tian *et al.* (2009)]. [Huang *et al.* (2000)] proposed the definition of generalized multi-state k-out-of-n:G system and an efficient algorithm for evaluating the system state distributions in each case of decreasing, increasing and constant multi-state k-out-of-n:G systems. [Yamamoto *et al.* (2006)] proposed an efficient algorithm for evaluating the system state distribution of the generalized multi-state k-out-of-n:G system in non-*i.i.d.* case. On the other hand, [Zuo *et al.* (2003)] proposed a definition of the generalized multi-state k-out-of-n:F system and an efficient algorithm for evaluating the system state distribution of a decreasing multi-state k-out-of-n:F system. [Zuo *et al.* (2006)] describes multi-state k-out-of-n:F system becomes multi-state $(n - k + 1)$-out-of-n:G system. Therefore, system state distribution of generalized multi-state k-out-of-n:F system can be evaluated by using algorithms for multi-state k-out-of-n:G system, for example, [Yamamoto *et al.* (2006)]'s algorithm. However, their algorithms are not so efficient if k_l are small ($l = 1, 2, \cdots, M$) for a multi-state multi-state k-out-of-n:F system. Accordingly, [Yamamoto *et al.* (2011)] proposed a faster algorithm using idea of virtual component states, which is efficient even in non-*i.i.d.* case, for evaluating the system state distribution of the generalized multi-state k-out-of-n:F system by extending [Yamamoto *et al.* (2009)]'s algorithm proposed for evaluating the system state distribution of a generalized multi-state k-out-of-n:F system.

In this article, we show efficient algorithms for computing system state distribution of multi-state k-out-of-n:F system and multi-state k-out-of-n:G system by using recursive algorithms and techniques of virtual component states which enable us to reduce the considered number of states. We present the order of computing time and memory capacity of proposed algorithms. We also perform numerical experiments. The results show that the proposed algorithms are more efficient than the existing algorithms for evaluating the system state distribution of multi-state k-out-of-n:F system and multi-state k-out-of-n:G system.

2 Definition for the Multi-state k-out-of-n Systems

Let $\mathbf{u}$ be vector of component states $\mathbf{u} = (u_1, u_2, \cdots, u_n)$ where u_i is state of component i, $u_i \in \{0, 1, \cdots, M\}$, for $i = 1, 2, \cdots, n$. And, $\phi(\mathbf{u})$ define system structure function representing the state of the system, $\phi(\mathbf{u}) \in \{0, 1, \cdots, M\}$. And $\mathbf{0}$ means n-dimensional zero vector.

First, we show the definition of generalized multi-state k-out-of-n:F system as follows.

Definition [Zuo *et al.* (2003)]: $\phi(\mathbf{u}) < j$ $(j = 1, 2, \cdots, M)$ if at least k_l components are in states below l for all l such that $j \leq l \leq M$. An n-component system with such a property is called a multi-state k-out-of-n:F system.

The condition in this definition can also be phrased as follows: at least k_l components are in states below j, and at least k_{j+1} components are in states below $j + 1, \cdots$, and at least k_M components are in states below M.

The multi-state k-out-of-n:F system is applicable to many practical problems. The following is an example of practical applications.

Example (A problem for functioning of a pump system of pipeline with different pump level): A pump machine for a liquid pipeline (such as an oil pipeline between cities or countries) may be labeled as being one of the following three operational classes based on the failure level of pumping attenuation A (a pump can pump toward next segment), attenuation B (a pump can pump toward two segments away), and failure (no pumping).

We consider this pump system has less than 50% capacity (state A) when at least three pumps have attenuation A. In addition, we consider the system has less than 70% capacity (state B) when at least two pumps have attenuation B.

In this case, the two-state k-out-of-n:F system when $k_A = 3$ and $k_B = 2$ applies to decision-making regarding a repair problem for this pump system.

Next, we show the definition of generalized multi-state k-out-of-n:G system as follows.

Definition [Huang *et al.* (2000)]: $\phi(\mathbf{u}) \geq j$ $(j = 1, 2, \cdots, M)$ if there exists an integer value l $(j \leq l \leq M)$ such that at least k_l components are in state l or above for all l. An n-component system with such a property is called a multi-state k-out-of-n:G system.

The condition in this definition can also be phrased as follows: at least k_j components are in states below j, or at least k_{j+1} components are in states below $j + 1$, or $\cdots$, or at least k_M components are in states below M.

Next, we assumed the following throughout this article.

Assumption:

(1) Multi-state k-out-of-n systems are multi-state monotone [Griffith (1980)]. That is, system structure function $\phi(\mathbf{u})$ satisfies,

    i. $\phi(\mathbf{u})$ is increasing function of $\mathbf{u} \geq \mathbf{0}$ and,

    ii. u_i or equivalently, $\phi(\mathbf{u}) = j$ for $j = 0, 1, \cdots, M$.

[Huang *et al.* (2000)] and [Zuo *et al.* (2003)] considered special cases of this definition of multi-state k-out-of-n:F (G) system. When $k_1 \geq k_2 \geq \cdots \geq k_M$ ($k_1 < k_2 < \cdots < k_M$) for $0 < k_l \leq n$, multi-state k-out-of-n:F (G) system was called the decreasing (increasing) multi-state k-out-of-n:F (G) system.

In addition, we note that the decreasing multi-state k-out-of-n:F (G) system includes special case when all k_l ($l = 1, 2, \cdots, M$) are constant. When k_l's are the same, i.e. $k_1 = k_2 = \cdots = k_M = k$, the structure of the system is the same for all system state levels. We called such systems constant multi-state k-out-of-n:F (G) system. The decreasing and increasing types of multi-state k-out-of-n:F (G) systems are also called monotone systems and generalized multi-state k-out-of-n:F (G) systems not including monotone systems are called non-monotone systems.

Now, we note that the following relation is evident:

$$Pr\{\phi(\mathbf{u})=j\}=\begin{cases} Pr\{\phi(\mathbf{u}) < j\} & \text{if } j=0, \\ Pr\{\phi(\mathbf{u}) < j+1\} - Pr\{\phi(\mathbf{u}) < j\} & \text{if } j=1,2,\cdots,M-1, \\ Pr\{\phi(\mathbf{u}) < M\} & \text{if } j=M. \end{cases}$$

In this article, a system state distribution means a set of probabilities of the system state j ($j = 0, 1, \cdots, M$), that is, $\{Pr\{\phi(\mathbf{u}) = j\}|j = 0, 1, \cdots, M\}$.

3 Theorem and Algorithm for the System State Distributions of the Generalized Multi-State k-out-of-n:F System

In this section, we show an algorithm for evaluating the system state distribution of the generalized multi-state k-out-of-n:F system [Yamamoto *et al.* (2011)] by unifying [Tian *et al.* (2009)]'s and [Yamamoto *et al.* (2009)]'s ideas.

Now, we define the following notations. For $i = 1, 2, \cdots, n$, and $j = 1, 2, \cdots, M$,

p_{ij} : probability that component i is in state j.

$F^{(j)}(i; \mathbf{k}, \mathbf{P})$: probability that state of the multi-state k-out-of-i:F system is below j where $\mathbf{k} = (k_1, k_2, \cdots, k_M)$ and

$$\mathbf{P} \equiv \begin{pmatrix} p_{1,0} & p_{1,1} & \cdots & p_{1,M} \\ p_{2,0} & p_{2,1} & \cdots & p_{2,M} \\ \vdots & \vdots & \ddots & \vdots \\ p_{n,0} & p_{n,1} & \cdots & p_{n,M} \end{pmatrix}.$$

And, the following are assumed throughout this article.

Assumption:

(2) State of component occurs to mutually statistically independent.

The important point of proposed algorithm is to reduce the considered number of states by using Corollary 2 which is described later. For this, we consider virtual component states, each of which consists of some actual component states. We express the virtual component states as $\bar{0}, \bar{1}, \bar{2}, \cdots, \bar{m}$ where $\bar{0} < \bar{1} < \bar{2} < \cdots < \bar{m}$ for $m = 1, 2, \cdots, M$.

Consider the case that actual component states are $0 < 1 < \cdots < j_1 < \cdots < j_2 < \cdots < j_3 < \cdots < M$ and, $j_l \in \{1, 2, \cdots, M\}$ means any actual states for $l = 1, 2, 3$, for example, virtual component state $\bar{0}$ consists of actual component states $0, 1, \cdots, j_1 - 1$, virtual component state $\bar{1}$ consists of actual component states $j_1, j_1 + 1, \cdots, j_2 - 1$, virtual component state $\bar{2}$ consists of actual component states $j_2, j_2 + 1, \cdots, j_3 - 1$, virtual component state $\bar{3}$ consists of actual component states $j_3, j_3 + 1, \cdots, M$ and then $m = 3$.

These virtual component states above are automatically decided by Corollary 7.2 as described latter.

Before stating Corollary 7.2, we introduce a Theorem (Theorem 7.1), which is the basic one in our algorithm. Before that, we define the following notations. For $i = 1, 2, \cdots, n, j = \bar{0}, \bar{1}, \cdots, \bar{m}$ and $m = 1, 2, \cdots, M$,

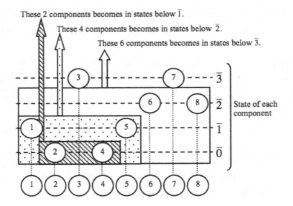

Fig. 1   An example for the relation of virtual component states with events $S(i, l; x_l)$, when $m = 3$, $(x_{\bar{1}}, x_{\bar{2}}, x_{\bar{3}}) = (2, 4, 6)$ and $i = 8$

$\rho_{i,j}$: probability that component i is in virtual states j.

$\mathbf{P}^{(\bar{m})}$: $n \times (m + 1)$ matrix of the component state probability, that is,

$$
\mathbf{P}^{(\bar{m})} \equiv
\begin{pmatrix}
\rho_{1,\bar{0}} & \rho_{1,\bar{1}} & \cdots & \rho_{1,\bar{m}} \\
\rho_{2,\bar{0}} & \rho_{2,\bar{1}} & \cdots & \rho_{2,\bar{m}} \\
\vdots & \vdots & \ddots & \vdots \\
\rho_{n,\bar{0}} & \rho_{n,\bar{1}} & \cdots & \rho_{n,\bar{m}}
\end{pmatrix}.
$$

Then $\rho_{i,j}$ can be expressed as

$$
\rho_{i,l} = \sum_{t=j_l}^{j_{l+1}-1} p_{i,t}
$$

for $l = \bar{0}, \bar{1}, \bar{2}, \cdots, \bar{m}$, where $j_{\bar{0}} \equiv 0$ and $j_{\bar{m}+1} \equiv M + 1$.

We also define notations as follows. For $i = 0, 1, \cdots, n$, $x_l = 0, 1, \cdots, k_l$, $l = \bar{1}, \bar{2}, \cdots, \bar{m}$ and $m = 1, 2, \cdots, M$,

$S(i, l; x_l)$: the event that at least x_l components in (virtual or actual) state below l occur from component 1 to i when $i = 1, 2, \cdots, n$; null event when $x_l > i$; whole event when $i = 0$.

$Q(i, (x_{\bar{1}}, x_{\bar{2}}, \cdots, x_{\bar{m}}), \mathbf{P}^{(\bar{m})})$: $Pr\{\bigcap_{l=\bar{1}}^{\bar{m}} S(i, l, x_l)\}$ when $\rho_{i,j}$ is the probability that component i is in the virtual component state j.

Figure 1 shows the relation between the virtual component states and $S(i, l; x_l)$ by using a simple example. We are now ready to present the following Theorem.

Theorem 7.1.

1) For $i = 1, 2, \cdots, n$ and $m = M - j + 1$,
$$F^{(j)}(i; \mathbf{k}, \mathbf{P}) = Q(i, (k_j, k_{j+1}, \cdots, k_M), \mathbf{P}^{(\bar{m})}) \tag{1}$$

where

$$\mathbf{P}^{(\bar{m})} \equiv \begin{pmatrix} \sum_{l=0}^{j-1} p_{1,l} & p_{1,j} & \cdots & p_{1,M} \\ \sum_{l=0}^{j-1} p_{2,l} & p_{2,j} & \cdots & p_{2,M} \\ \vdots & \vdots & \ddots & \vdots \\ \sum_{l=0}^{j-1} p_{n,l} & p_{n,j} & \cdots & p_{n,M} \end{pmatrix}.$$

2) For $i = 0, 1, \cdots, n$ and $l = \bar{1}, \bar{2}, \cdots, \bar{m}$,

$$Q(i, (x_{\bar{1}}, x_{\bar{2}}, \cdots, x_{\bar{m}}), \mathbf{P}^{(\bar{m})}) =$$
$$\begin{cases} 0 & \text{if } \exists l : x_l > i, \\ 1 & \text{if } (x_{\bar{1}}, x_{\bar{2}}, \cdots, x_{\bar{m}}) = \mathbf{0}, \\ \sum_{h=\bar{0}}^{\bar{m}} \rho_{i,h} Q(i - 1, (x_{\bar{1}}, x_{\bar{2}}, \cdots, x_h, \\ \qquad \max\{x_{h+1} - 1, 0\}, \cdots, \\ \qquad\qquad \max\{x_m - 1, 0\}), \mathbf{P}^{(\bar{m})}) \\ \qquad\qquad \text{if } x_l \leq i \text{ for } \forall l \text{ and } i > 0. \end{cases}$$

$$\tag{2}$$

Theorem 7.1 can be proven with a similar manner to Theorem 1 in [Yamamoto *et al.* (2009)]. From Theorem 7.1, we can get the following Corollary for the system state distribution of a binary k-out-of-n:F system.

Corollary 7.1. For $i = 0, 1, \cdots, n$,

$$Q(i, (x_{\bar{1}}), \mathbf{P}^{(\bar{1})}) = \begin{cases} 0 & \text{if } x_{\bar{1}} > i, \\ 1 & \text{if } x_{\bar{1}} = 0, \\ \rho_{i,\bar{0}} Q(i - 1, (\max\{x_1 - 1, 0\}), \mathbf{P}^{(\bar{1})}) & \text{if } x_{\bar{1}} \leq i \text{ and } i > 0. \end{cases} \tag{3}$$

Corollary 7.1 is the same as traditional theorem of reliability of a binary k-out-of-n:F system. (see, for example, [Rushdi (1986)] etc..)

Next, from the definition of the event $S(i, l; x_l)$, we can get the following Lemma.

Lemma 7.1. For $i = 0, 1, \cdots, n$, if $x_{\bar{t}} \geq \max\limits_{l = \bar{t}+1, \cdots, \bar{s}} \{x_l\}$ then we can get the following relation of the event $S^F(i, l; x_l)$s.

$$\bigcap_{l=\bar{t}}^{\bar{s}} S(i, l; x_l) = S(i, \bar{t}; x_{\bar{t}}). \tag{4}$$

The proof of Lemma 7.1 is provided in appendix at [Yamamoto *et al.* (2009)]. From Lemma 7.1, we can get the following Theorem.

Theorem 7.2.

We take $x_{\overline{j(l)}} \in \{x_{\bar{1}}, x_{\bar{2}}, \cdots, x_{\bar{m}}\}$ for $l = 1, 2$.

1) If $x_{\overline{j(2)}}$ exists such that $x_{\overline{j(1)}} \geq \max\limits_{l = \overline{j(1)}+1, \cdots, \overline{j(2)}-1} \{x_l\}$ and $x_{\overline{j(1)}} < x_{\overline{j(2)}}$,

$$Q(i, (x_{\bar{1}}, x_{\bar{2}}, \cdots, x_{\overline{j(1)}}, x_{\overline{j(2)}}, \cdots, x_{\bar{m}}), \mathbf{P}^{\overline{(m-(j(2)-j(1))+1)}}) =$$
$$Q(i, (x_{\bar{1}}, x_{\bar{2}}, \cdots, x_{\overline{j(1)}}, x_{\overline{j(1)+1}}, \cdots, x_{\overline{j(2)}} \cdots, x_{\bar{m}}), \mathbf{P}^{(\overline{m})}) \tag{5}$$

for $i = 0, 1, \cdots, n$, where

$$\mathbf{P}^{\overline{(m-(j(2)-j(1))+1)}} = \begin{pmatrix} \rho_{1,\bar{0}} \ \rho_{1,\bar{1}} \ \cdots \ \sum\limits_{l=\overline{j(1)}}^{\overline{j(2)}-1} \rho_{1,l} \ \cdots \ \rho_{1,\overline{j(2)}} \ \cdots \ \rho_{1,\bar{m}} \\ \rho_{2,\bar{0}} \ \rho_{2,\bar{1}} \ \cdots \ \sum\limits_{l=\overline{j(1)}}^{\overline{j(2)}-1} \rho_{2,l} \ \cdots \ \rho_{2,\overline{j(2)}} \ \cdots \ \rho_{2,\bar{m}} \\ \vdots \\ \rho_{n,\bar{0}} \ \rho_{n,\bar{1}} \ \cdots \ \sum\limits_{l=\overline{j(1)}}^{\overline{j(2)}-1} \rho_{n,l} \ \cdots \ \rho_{n,\overline{j(2)}} \ \cdots \ \rho_{n,\bar{m}} \end{pmatrix}.$$

2) If $x_{\overline{j(1)}} \geq \max\limits_{l = \overline{j(1)}+1, \cdots, \bar{m}} \{x_l\}$,

$$Q(i, (x_{\bar{1}}, x_{\bar{2}}, \cdots, x_{\overline{j(1)}}), \mathbf{P}^{(\overline{j(1)})}) =$$
$$Q(i, (x_{\bar{1}}, x_{\bar{2}}, \cdots, x_{\overline{j(1)}}, x_{\overline{j(1)+1}}, \cdots, x_{\bar{m}}), \mathbf{P}^{(\overline{m})}) \tag{6}$$

for $i = 0, 1, \cdots, n$, where

$$
\mathbf{P}^{(\overline{j(1)})} = \begin{pmatrix}
\rho_{1,\bar{0}} & \rho_{1,\bar{1}} & \cdots & \displaystyle\sum_{l=\overline{j(1)}}^{\bar{m}} \rho_{1,l} \\[2ex]
\rho_{2,\bar{0}} & \rho_{2,\bar{1}} & \cdots & \displaystyle\sum_{l=\overline{j(1)}}^{\bar{m}} \rho_{2,l} \\[2ex]
& & \vdots & \\[1ex]
\rho_{n,\bar{0}} & \rho_{n,\bar{1}} & \cdots & \displaystyle\sum_{l=\overline{j(1)}}^{\bar{m}} \rho_{n,l}
\end{pmatrix} .
$$

The proof of Theorem 7.2 is provided in appendix at [Yamamoto *et al.* (2011)]. From Theorem 7.2, we can get the following Corollary.

Corollary 7.2. We take $\{x_{\bar{1}}, x_{\overline{j(1)}}, \cdots, x_{\overline{j(r)}}\} \subseteq \{x_{\bar{1}}, x_{\bar{2}}, \cdots, x_{\bar{m}}\}$. If $x_{\overline{j(t)}} \geq \max\limits_{l=j(t)+1,\cdots,j(t+1)-1}\{x_l\}$, $x_{\overline{j(t)}} < x_{\overline{j(t+1)}}$ $(t = 0, 1, \cdots, r-1)$ and $x_{\overline{j(r)}} \geq \max\limits_{l=j(r)+1,\cdots,\bar{m}}\{x_l\}$, then, for $i = 0, 1, \cdots, n$,

$$
Q(i, (x_{\bar{1}}, x_{\overline{j(1)}}, x_{\overline{j(2)}}, \cdots, x_{\overline{j(r)}}), \mathbf{P}^{(\overline{r+1})}) = Q(i, (x_{\bar{1}}, x_{\bar{2}}, \cdots, x_{\bar{m}}), \mathbf{P}^{(\overline{m})}) \quad (7)
$$

where

$$
\mathbf{P}^{(\overline{r+1})} = \begin{pmatrix}
\rho_{1,\bar{0}} & \displaystyle\sum_{l=\bar{1}}^{\overline{j(1)}-1} \rho_{1,l} & \displaystyle\sum_{l=\overline{j(1)}}^{\overline{j(2)}-1} \rho_{1,l} & \cdots & \displaystyle\sum_{l=\overline{j(r-1)}}^{\overline{j(r)}-1} \rho_{1,l} & \displaystyle\sum_{l=\overline{j(r)}}^{\bar{m}} \rho_{1,l} \\[2ex]
\rho_{2,\bar{0}} & \displaystyle\sum_{l=\bar{1}}^{\overline{j(1)}-1} \rho_{2,l} & \displaystyle\sum_{l=\overline{j(1)}}^{\overline{j(2)}-1} \rho_{2,l} & \cdots & \displaystyle\sum_{l=\overline{j(r-1)}}^{\overline{j(r)}-1} \rho_{2,l} & \displaystyle\sum_{l=\overline{j(r)}}^{\bar{m}} \rho_{2,l} \\[1ex]
& & & \vdots & & \\[1ex]
\rho_{1,\bar{0}} & \displaystyle\sum_{l=\bar{1}}^{\overline{j(1)}-1} \rho_{n,l} & \displaystyle\sum_{l=\overline{j(1)}}^{\overline{j(2)}-1} \rho_{n,l} & \cdots & \displaystyle\sum_{l=\overline{j(r-1)}}^{\overline{j(r)}-1} \rho_{n,l} & \displaystyle\sum_{l=\overline{j(r)}}^{\bar{m}} \rho_{n,l}
\end{pmatrix} ,
$$

and $x_{\overline{j(0)}} \equiv x_{\bar{1}}$.

Corollary 7.2 can be proven directly from Theorem 7.2.

We show the following example for the sake of understanding the meanings of Corollary 7.2. When $m = 5$, and $(x_{\bar{1}}, x_{\bar{2}}, x_{\bar{3}}, x_{\bar{4}}, x_{\bar{5}}) =$

$(3, 2, 1, 5, 4)$, we can get $Q(i, (3, 5), \mathbf{P}^{(\bar{2})}) = Q(i, (3, 2, 1, 5, 4), \mathbf{P}^{(\bar{5})})$ from $x_{\bar{1}} > \max\{x_{\bar{2}}, x_{\bar{3}}\}, x_{\bar{4}} > x_{\bar{5}}$ and $x_{\bar{1}} < x_{\bar{4}}$, where

$$\mathbf{P}^{(\bar{5})} \equiv \begin{pmatrix} \rho_{1,\bar{0}} & \rho_{1,\bar{1}} & \rho_{1,\bar{2}} & \rho_{1,\bar{3}} & \rho_{1,\bar{4}} & \rho_{1,\bar{5}} \\ \rho_{2,\bar{0}} & \rho_{2,\bar{1}} & \rho_{2,\bar{2}} & \rho_{2,\bar{3}} & \rho_{2,\bar{4}} & \rho_{2,\bar{5}} \\ & & \vdots & & & \\ \rho_{n,\bar{0}} & \rho_{n,\bar{1}} & \rho_{n,\bar{2}} & \rho_{n,\bar{3}} & \rho_{n,\bar{4}} & \rho_{n,\bar{5}} \end{pmatrix},$$

and

$$\mathbf{P}^{(\bar{2})} \equiv \begin{pmatrix} \rho_{1,\bar{0}} & \rho_{1,\bar{1}} + \rho_{1,\bar{2}} + \rho_{1,\bar{3}} & \rho_{1,\bar{4}} + \rho_{1,\bar{5}} \\ \rho_{2,\bar{0}} & \rho_{2,\bar{1}} + \rho_{2,\bar{2}} + \rho_{2,\bar{3}} & \rho_{2,\bar{4}} + \rho_{2,\bar{5}} \\ & \vdots & \\ \rho_{n,\bar{0}} & \rho_{n,\bar{1}} + \rho_{n,\bar{2}} + \rho_{n,\bar{3}} & \rho_{n,\bar{4}} + \rho_{n,\bar{5}} \end{pmatrix}.$$

We demonstrate how to obtain $F^{(j)}(n; \mathbf{k}, \mathbf{P})$ $(j = 1, 2)$ in the case of $M = 2$ by using the above Theorems and Corollaries.

Example :

1) When $j = 2$, $F^{(j)}(n; \mathbf{k}, \mathbf{P})$ can be expressed from Equation 1 as follows.
   $F^{(2)}(n; (k_1, k_2), \mathbf{P}) = Q(n, (k_2), \mathbf{P}^{(\bar{1})})$
   where

$$\mathbf{P}^{(\bar{1})} = \begin{pmatrix} \rho_{1,\bar{0}} & \rho_{1,\bar{1}} \\ \rho_{2,\bar{0}} & \rho_{2,\bar{1}} \\ \vdots & \\ \rho_{n,\bar{0}} & \rho_{n,\bar{1}} \end{pmatrix} = \begin{pmatrix} p_{1,0} + p_{1,1} & p_{1,2} \\ p_{2,0} + p_{2,1} & p_{2,2} \\ \vdots & \\ p_{n,0} + p_{n,1} & p_{n,2} \end{pmatrix}.$$

And, we can get

$$Q(i, (x_{\bar{1}}), \mathbf{P}^{(\bar{1})}) = \begin{cases} 0 & \text{if } x_{\bar{1}} \neq 0, \\ 1 & \text{if } x_{\bar{1}} = 0, \end{cases}$$

for $i = 0$ and
$Q(i, (x_{\bar{1}}), \mathbf{P}^{(\bar{1})}) =$

$$\begin{cases} 0 & \text{if } x_{\bar{1}} > i > 0, \\ 1 & \text{if } x_{\bar{1}} = 0, \\ (p_{i,0} + p_{i,1})Q(i - 1, (x_{\bar{1}} - 1), \mathbf{P}^{(\bar{1})}) & \\ \qquad + p_{i,2}Q(i - 1, (x_{\bar{1}}), \mathbf{P}^{(\bar{1})}) & \text{if } i \geq x_{\bar{1}} > 0, \end{cases}$$

for $i = 1, 2, \cdots, n$, from Equation 2.

2) When $j = 1$, $F^{(j)}(n; \mathbf{k}, \mathbf{P})$ can be expressed from Equation 1 as follows.
   $F^{(1)}(n; (k_1, k_2), \mathbf{P}) = Q(n, (k_1, k_2), \mathbf{P}^{(\bar{2})})$
   where

$$\mathbf{P}^{(\bar{2})} = \begin{pmatrix} \rho_{1,\bar{0}} & \rho_{1,\bar{1}} & \rho_{1,\bar{2}} \\ \rho_{2,\bar{0}} & \rho_{2,\bar{1}} & \rho_{2,\bar{2}} \\ & \vdots & \\ \rho_{n,\bar{0}} & \rho_{n,\bar{1}} & \rho_{n,\bar{2}} \end{pmatrix} = \begin{pmatrix} p_{1,0} & p_{1,1} & p_{1,2} \\ p_{2,0} & p_{2,1} & p_{2,2} \\ & \vdots & \\ p_{n,0} & p_{n,1} & p_{n,2} \end{pmatrix}.$$

From Equation 2,

$$Q(i, (x_1, x_2), \mathbf{P}^{(\bar{2})}) = \begin{cases} 0 & \text{if } (x_1, x_2) \neq (0, 0), \\ 1 & \text{if } (x_1, x_2) = (0, 0), \end{cases}$$

for $i = 0$ and
$$Q(i, (x_1, x_2), \mathbf{P}^{(\bar{2})}) =$$

$$\begin{cases} 0 & \text{if } x_1 > i \text{ or } x_2 > i, \\ 1 & \text{if } x_1 = 0 \text{ and } x_2 = 0, \\ \begin{aligned} & p_{i,0} Q(i-1, (x_1-1, x_2-1), \mathbf{P}^{(\bar{2})}) \\ & + p_{i,1} Q(i-1, (x_1, x_2-1), \mathbf{P}^{(\bar{2})}) \\ & + p_{i,2} Q(i-1, (x_1, x_2), \mathbf{P}^{(\bar{2})}) \end{aligned} & \\ & \text{if } i \geq x_2 > x_1 > 0, \\ \begin{aligned} & p_{i,0} Q(i-1, (0, x_2-1), \mathbf{P}^{(\bar{2})}) \\ & + p_{i,1} Q(i-1, (0, x_2-1), \mathbf{P}^{(\bar{2})}) \\ & + p_{i,2} Q(i-1, (0, x_2), \mathbf{P}^{(\bar{2})}) \end{aligned} & \\ & \text{if } x_1 = 0 \text{ and } i \geq x_2 > 0, \\ \begin{aligned} & p_{i,0} Q(i-1, (x_1-1), \mathbf{P}^{(\bar{1})}) \\ & + (p_{i,1} + p_{i,2}) Q(i-1, (x_1), \mathbf{P}^{(\bar{1})}) \end{aligned} & \\ & \text{if } i \geq x_1 \geq x_2 > 0, \end{cases}$$

for $i = 1, 2, \cdots, n$ and $x_l = 0, 1, \cdots, k_l$ $(l = 1, 2)$.

Using Theorem 7.1 and Corollaries 7.1 and 7.2, we obtain the following algorithm steps for computing $F^{(j)}(n; \mathbf{k}, \mathbf{P})$ for $j + 1, 2, \cdots, M$.

STEP 0 : (Setting initial value)
Set $i = 0$, $m = M - j + 1$ and

$$Q(i, (x_{\bar{1}}, x_{\bar{2}}, \cdots, x_{\bar{m}}), \mathbf{P}^{(\bar{m})}) = \begin{cases} 0 & \text{if } (x_{\bar{1}}, x_{\bar{2}}, \cdots, x_{\bar{m}}) \neq \mathbf{0}, \\ 1 & \text{if } (x_{\bar{1}}, x_{\bar{2}}, \cdots, x_{\bar{m}}) = \mathbf{0}, \end{cases}$$

from Equation 2.

STEP 1 : Set $i = i + 1$.

STEP 2 : Enumerate all $(x_{\bar{1}}, x_{\bar{2}}, \cdots, x_{\bar{m}}) \in A$, where
$A \equiv \{(x_{\bar{1}}, x_{\bar{2}}, \cdots, x_{\bar{m}}) | x_{\bar{l}} = 0, 1, \cdots, k_{l+j-1} (l = 1, 2, \cdots, m)\}$.

STEP 3 :   Pick up $(x_{\bar{1}}, x_{\bar{2}}, \cdots, x_{\bar{m}}) \in A$. For the $(x_{\bar{1}}, x_{\bar{2}}, \cdots, x_{\bar{m}})$, suppose we obtain
$$Q(i, (x_{\bar{1}}, x_{\overline{j(1)}}, x_{\overline{j(2)}}, \cdots, x_{\overline{j(r)}}), \mathbf{P}^{(\overline{r+1})}) \;=\; Q(i, (x_{\bar{1}}, x_{\bar{2}}, \cdots, x_{\bar{m}}), \mathbf{P}^{(\bar{m})})$$
from Corollary 7.2.

STEP 4 :   Obtain $Q(i, (x_{\bar{1}}, x_{\overline{j(1)}}, x_{\overline{j(2)}}, \cdots, x_{\overline{j(r)}}), \mathbf{P}^{(\overline{r+1})})$ using Theorem 7.1 and memorize $Q(i, (x_{\bar{1}}, x_{\overline{j(1)}}, x_{\overline{j(2)}}, \cdots, x_{\overline{j(r)}}), \mathbf{P}^{(\overline{r+1})})$.
Go to **STEP 5** if all $(x_{\bar{1}}, x_{\bar{2}}, \cdots, x_{\bar{m}})$'s $\in A$ are picked up. When otherwise, go to **STEP 3**.

STEP 5 :   Go to **STEP 1** when $i < n$, and go to **STEP 6** when $i = n$.

STEP 6 :   Obtain $F^{(j)}(n; \mathbf{k}, \mathbf{P})$ for $j = 1, 2, \cdots, M$ using Equation 1.

Note that we can get $F^{(2)}(n; \mathbf{k}, \mathbf{P}), F^{(3)}(n; \mathbf{k}, \mathbf{P}), \cdots, F^{(M)}(n; \mathbf{k}, \mathbf{P})$ at the same time to obtain $F^{(1)}(n; \mathbf{k}, \mathbf{P})$ using Equation 1. The similar method is used in [Yamamoto *et al.* (2009)].

4   Theorem and Algorithm for the System State Distributions of the Generalized Multi-State k-out-of-n:G System

In this section, we propose the algorithm for evaluating the system state distribution of the generalized multi-state k-out-of-n:G system by using similar manner from the previous section.

First, we add definitions as following notations. For $i = 0, 1, \cdots, n$, $j = 1, 2, \cdots, M$, $x_l = 0, 1, \cdots, k_l$, $l = \bar{1}, \bar{2}, \cdots, \bar{m}$ and $m = 1, 2, \cdots, M$,

$R^{(j)}(i; \mathbf{k}, \mathbf{P})$:   probability that state of the multi-state k-out-of-i:G system is more than or equal to j where $\mathbf{k} = (k_1, k_2, \cdots, k_M)$ and

$$\mathbf{P} \equiv \begin{pmatrix} p_{1,0} & p_{1,1} & \cdots & p_{1,M} \\ p_{2,0} & p_{2,1} & \cdots & p_{2,M} \\ \vdots & \vdots & \ddots & \vdots \\ p_{n,0} & p_{n,1} & \cdots & p_{n,M} \end{pmatrix}.$$

$S^F(i, l; x_l)$:   the event that less than components in state more than l occur from component 1 to i when $i = 1, 2, \cdots, n$; null event when $x_l = 0$; whole event when $i = 0$ and $x_l > 0$.

$F(i, (x_{\bar{1}}, x_{\bar{2}}, \cdots, x_{\bar{m}}), \mathbf{P}^{(\bar{m})})$: $Pr\{\bigcap\limits_{l=\bar{1}}^{\bar{m}} S^F(i, l, x_l)\}$.   That is, probability when the event $S^F(i, l, x_l)$ occurs for all l such that $\bar{1} \leq l \leq \bar{m}$; $F(i, (x_{\bar{1}}, x_{\bar{2}}, \cdots, x_{\bar{m}}), \mathbf{P}^{(\bar{m})}) = 0$ when $\min\limits_{l}\{x_l\} = 0$.

First, we propose the following Theorem. Some notations are same as the previous section.

Theorem 7.3.

1) For $i = 1, 2, \cdots, n$ and $j = 1, 2, \cdots, M$,

$$R^{(j)}(i; \mathbf{k}, \mathbf{P}) = 1 - F(i, (k_j, k_{j+1}, \cdots, k_M), \mathbf{P}^{(M-j+1)}) \qquad (8)$$

where

$$\mathbf{P}^{(M-j+1)} \equiv \begin{pmatrix} \displaystyle\sum_{l=0}^{j-1} p_{1,l} \ p_{1,j} \ \cdots \ p_{1,M} \\ \displaystyle\sum_{l=0}^{j-1} p_{2,l} \ p_{2,j} \ \cdots \ p_{2,M} \\ \vdots \quad \vdots \quad \ddots \quad \vdots \\ \displaystyle\sum_{l=0}^{j-1} p_{n,l} \ p_{n,j} \ \cdots \ p_{n,M} \end{pmatrix}.$$

2) For $i = 0, 1, \cdots, n$, $l = \bar{1}, \bar{2}, \cdots, \bar{m}$ and $m = 1, 2, \cdots, M$,

$$F(i, (x_{\bar{1}}, x_{\bar{2}}, \cdots, x_{\bar{m}}), \mathbf{P}^{(\bar{m})}) =$$

$$\begin{cases} 1 & \text{if } {}^{\forall}l : x_l > i, \\ 0 & \text{if } \min_l\{x_l\} = 0, \\ \displaystyle\sum_{h=\bar{1}}^{\bar{m}} \rho_{i,h} F(i-1, (y_{\bar{1}}, \cdots, y_h, y_{h+1}, \cdots, y_{\bar{m}}), \mathbf{P}^{(\bar{m})}) & \text{otherwise}, \end{cases}$$

$$(9)$$

where

$$y_l = \begin{cases} 0 & \text{if } x_l \leq 0, \\ x_l - 1 & \text{if } x_l > 0 \text{ and } l \leq h, \\ x_l & \text{if } x_l > 0 \text{ and } l > h. \end{cases}$$

3) For $i = 0, 1, \cdots, n$ and $m = 1$,

$$F(i, (x_{\bar{1}}), \mathbf{P}^{(\bar{1})}) =$$

$$\begin{cases} 1 & \text{if } x_{\bar{1}} > i, \\ 0 & \text{if } x_{\bar{1}} = 0, \\ \rho_{i,\bar{0}} F(i-1, (\max\{x_{\bar{1}} - 1, 0\}), \mathbf{P}^{(\bar{1})}) \\ \quad + \rho_{i,\bar{1}} F(i-1, (x_{\bar{1}}), \mathbf{P}^{(\bar{1})}) & \text{if } x_{\bar{1}} \leq i, i > 0. \end{cases}$$

$$(10)$$

Theorem 7.3 can be proven with a similar manner to Theorem 7.1. Equation 10 means boundary condition when the number m of virtual states of components is 1. Then, we can get the system state distribution by using equation for the reliability of a binary k-out-of-n:G system.

In addition, from the definition of the event $S^F(i, l; x_l)$, we can get the following Lemma.

Lemma 7.2. For $i = 0, 1, \cdots, n$, $s = 1, 2, \cdots, m$ and $t = 1, 2, \cdots, m$, if $x_{\bar{s}} \leq \min\limits_{l=s+1,\cdots,\bar{t}} \{x_l\}$ then we can get the following relation of the event $S^F(i, l; x_l)$s.

$$\bigcap_{l=\bar{s}}^{\bar{t}} S^F(i, l, x_l) = S^F(i, \bar{s}; x_{\bar{s}}). \tag{11}$$

From Lemma 7.2, we can get the following Lemma.

Lemma 7.3. We take $x_{\overline{j(2)}} \in \{x_{\bar{1}}, \cdots, x_{\bar{m}}\}$.

1) If $x_{\overline{j(1)}}$ exists such that $x_{\overline{j(1)}} < \min\limits_{l=\overline{j(1)+1},\cdots,\overline{j(2)-1}}$ and $x_{\overline{j(1)}} > x_{\overline{j(2)}}$,

$$F(i, (x_{\bar{1}}, x_{\bar{2}}, \cdots, x_{\overline{j(1)}}, x_{\overline{j(2)}}, \cdots, x_{\bar{m}}), \mathbf{P}^{\overline{(m-(j(2)-j(1))+1)}}) =$$
$$F(i, (x_{\bar{1}}, x_{\bar{2}}, \cdots, x_{\overline{j(1)}}, x_{\overline{j(1)+1}}, \cdots, x_{\overline{j(2)}} \cdots, x_{\bar{m}}), \mathbf{P}^{(\overline{m})}) \tag{12}$$

for $i = 0, 1, \cdots, n$, where $n \times (m - (j(2) - j(1)) + 2)$ matrix $\mathbf{P}^{(m-(j(2)-j(1))+1)}$ is shown as

$$\mathbf{P}^{\overline{(m-(j(2)-j(1))+1)}} =$$

$$\begin{pmatrix} \rho_{1,\bar{0}} & \cdots & \rho_{1,\overline{j(1)}} & \sum\limits_{l=\overline{j(1)+1}}^{\overline{j(2)}} \rho_{1,l} & \rho_{1,\overline{j(2)+1}} & \cdots & \rho_{1,\bar{m}} \\ \rho_{2,\bar{0}} & \cdots & \rho_{2,\overline{j(1)}} & \sum\limits_{l=\overline{j(1)+1}}^{\overline{j(2)}} \rho_{2,l} & \rho_{2,\overline{j(2)+1}} & \cdots & \rho_{2,\bar{m}} \\ \vdots & \ddots & \vdots & \vdots & \vdots & \ddots & \vdots \\ \rho_{n,\bar{0}} & \cdots & \rho_{n,\overline{j(1)}} & \sum\limits_{l=\overline{j(1)+1}}^{\overline{j(2)}} \rho_{n,l} & \rho_{n,\overline{j(2)+1}} & \cdots & \rho_{n,\bar{m}} \end{pmatrix}.$$

2) If $x_{\overline{j(2)}} < \min\limits_{l=\overline{1},\cdots,\overline{j(2)-1}}\{x_l\}$,

$$F(i,(x_{\overline{j(2)}},\cdots,x_{\bar{m}}),\mathbf{P}^{(\overline{m-j(2)+1})}) = F(i,(x_{\bar{1}},\cdots,x_{\overline{j(2)}}\cdots,x_{\bar{m}}),\mathbf{P}^{(\overline{m})}) \tag{13}$$

for $i = 0,1,\cdots,n$, where

$$\mathbf{P}^{(\overline{j(2)})} = \begin{pmatrix} \rho_{1,\bar{0}} & \sum\limits_{l=\bar{1}}^{\overline{j(2)}}\rho_{1,l} & \rho_{1,\overline{j(2)+1}} & \cdots & \rho_{1,\bar{m}} \\ \rho_{2,\bar{0}} & \sum\limits_{l=\bar{1}}^{\overline{j(2)}}\rho_{2,l} & \rho_{2,\overline{j(2)+1}} & \cdots & \rho_{2,\bar{m}} \\ \vdots & \vdots & \vdots & \ddots & \vdots \\ \rho_{n,\bar{0}} & \sum\limits_{l=\bar{1}}^{\overline{j(2)}}\rho_{n,l} & \rho_{1,\overline{j(2)+1}} & \cdots & \rho_{n,\bar{m}} \end{pmatrix}.$$

From the Lemmas, the following Corollary 7.3 holds.

Corollary 7.3. Now, we take $\{x_{\overline{j(1)}},x_{\overline{j(2)}},\cdots,x_{\overline{j(r)}}\} \subseteq \{x_{\bar{1}},x_{\bar{2}},\cdots,x_{\bar{m}}\}$. For $t = 1,2,\cdots,r$ and $j(r) \equiv m$, if $x_{\overline{j(t)}} \leq \min\limits_{l=\overline{j(t)+1},\cdots,\overline{j(t+1)-1}}\{x_l\}$, and $x_{\overline{j(t)}} > x_{\overline{j(t+1)}}$, then,

$$F(i,(x_{\overline{j(1)}},x_{\overline{j(2)}},\cdots,x_{\overline{j(r)}}),\mathbf{P}^{(\bar{r})}) =$$

$$F(i,(x_{\bar{1}},\cdots,x_{\overline{j(1)}},\cdots,x_{\overline{j(2)}},\cdots,x_{\overline{j(r-1)}},\cdots,x_{\bar{m}}),\mathbf{P}^{(\overline{m})}) \tag{14}$$

for $i = 0,1,\cdots,n$, where

$$\mathbf{P}^{(\bar{r})} = \begin{pmatrix} \rho_{1,\bar{0}} & \sum\limits_{l=\bar{1}}^{\overline{j(1)}}\rho_{1,l} & \sum\limits_{l=\overline{j(1)+1}}^{\overline{j(2)}}\rho_{1,l} & \cdots & \sum\limits_{l=\overline{j(r-2)+1}}^{\overline{j(r-1)}}\rho_{1,l} & \sum\limits_{l=\overline{j(r-1)+1}}^{\bar{m}}\rho_{1,l} \\ \rho_{2,\bar{0}} & \sum\limits_{l=\bar{1}}^{\overline{j(1)}}\rho_{2,l} & \sum\limits_{l=\overline{j(1)+1}}^{\overline{j(2)}}\rho_{2,l} & \cdots & \sum\limits_{l=\overline{j(r-2)+1}}^{\overline{j(r-1)}}\rho_{2,l} & \sum\limits_{l=\overline{j(r-1)+1}}^{\bar{m}}\rho_{2,l} \\ \vdots & \vdots & \vdots & \ddots & \vdots & \vdots \\ \rho_{n,\bar{0}} & \sum\limits_{l=\bar{1}}^{\overline{j(1)}}\rho_{n,l} & \sum\limits_{l=\overline{j(1)+1}}^{\overline{j(2)}}\rho_{n,l} & \cdots & \sum\limits_{l=\overline{j(r-2)+1}}^{\overline{j(r-1)}}\rho_{n,l} & \sum\limits_{l=\overline{j(r-1)+1}}^{\bar{m}}\rho_{n,l} \end{pmatrix}.$$

The probability $F(i, (k_j, k_{j+1}, \cdots, k_M), \mathbf{P}^{(M-j+1)})$ can be given by the following Theorem.

Theorem 7.4. Now, we take $\{k_{\overline{j(1)}}, k_{\overline{j(2)}}, \cdots, k_{\overline{j(r)}}\} \subseteq \{k_1, k_2, \cdots, k_M\}$ where $j \leq \overline{j(1)} < \overline{j(2)} < \cdots < \overline{j(r)} = m$ for $j = 1, 2, \cdots, M$ and $m = 1, 2, \cdots, M$. For $j = 1, 2, \cdots, M$, $s = 1, 2, \cdots, r$, $t = 1, 2, \cdots, r$ and $r \equiv M$. If $k_{\bar{s}} < \min\limits_{l=\overline{s+1}, \cdots, \overline{t-1}} \{k_l\}$ and $k_{\bar{s}} > k_{\bar{t}}$, then,

$$F(i, (k_{\overline{j(1)}}, k_{\overline{j(2)}}, \cdots, k_{\overline{j(r)}}), \mathbf{P}^{(\bar{r})}) =$$
$$F(i, (k_j, \cdots, k_{\bar{1}}, \cdots, k_{\bar{2}}, \cdots, k_{\bar{r}}, \cdots, k_M), \mathbf{P}^{(M)})$$

$$(15)$$

where

$$\mathbf{P}^{(\bar{r})} =$$

$$\begin{pmatrix}
\sum\limits_{l=0}^{j-1} p_{1,l} & \sum\limits_{l=j}^{\overline{j(1)}} p_{1,l} & \sum\limits_{l=\overline{j(1)}+1}^{\overline{j(2)}} p_{1,l} & \cdots & \sum\limits_{l=\overline{j(r-2)}+1}^{\overline{j(r-1)}} p_{1,l} & \sum\limits_{l=\overline{j(r-1)}+1}^{M} p_{1,l} \\
\sum\limits_{l=0}^{j-1} p_{2,l} & \sum\limits_{l=j}^{\overline{j(1)}} p_{2,l} & \sum\limits_{l=\overline{j(1)}+1}^{\overline{j(2)}} p_{2,l} & \cdots & \sum\limits_{l=\overline{j(r-2)}+1}^{\overline{j(r-1)}} p_{2,l} & \sum\limits_{l=\overline{j(r-1)}+1}^{M} p_{2,l} \\
\vdots & \vdots & \vdots & \ddots & \vdots & \vdots \\
\sum\limits_{l=0}^{j-1} p_{n,l} & \sum\limits_{l=j}^{\overline{j(1)}} p_{n,l} & \sum\limits_{l=\overline{j(1)}+1}^{\overline{j(2)}} p_{n,l} & \cdots & \sum\limits_{l=\overline{j(r-2)}+1}^{\overline{j(r-1)}} p_{n,l} & \sum\limits_{l=\overline{j(r-1)}+1}^{M} p_{n,l}
\end{pmatrix},$$

for $i = 0, 1, \cdots, n$.

Theorem 7.4 can be proven with a similar manner to Theorem 7.2. Using the Lemmas and the Theorems, we obtain the following algorithm for computing $R^{(j)}(i; \mathbf{k}, \mathbf{P})$ for $j = 1, 2, \cdots, M$.

STEP 0 : (Setting initial value)
We take $j = 1$. First, set $i = 0$, and we calculate

$$F(i, (x_1, \cdots, x_M), \mathbf{P}^{(M)}) = \begin{cases} 0 & \text{if } \min\limits_{l} \{x_l\} = 0, \\ 1 & \text{otherwise,} \end{cases}$$

from Equation 11, and
memorize the results of $F(i, (x_1, \cdots, x_M), \mathbf{P}^{(M)})$s. Go to **STEP 1**.

STEP 1 : Set $i = i + 1$.

STEP 2 : Set $m = M$, and enumerate combinations for all element x_l $(l = \bar{1}, \bar{2}, \cdots, \bar{m}, x_l \in \{0, , \cdots, \max\{x_l\}\})$

STEP 3 : Reduced the elements for all x_ls $(l = \bar{1}, \bar{2}, \cdots, \bar{m})$ by using Lemma 7.3.

STEP 4 : For all x_ls $(l = \bar{1}, \bar{2}, \cdots, \bar{m})$ obtained in the **STEP 3**, obtain the value of y_ls $(l = \bar{1}, \bar{2}, \cdots, \bar{m})$ for all h by using the Equation 11 or 10 in Theorem 7.3. Next, we reduced the elements for all y_ls $(l = \bar{1}, \bar{2}, \cdots, \bar{m})$ by using Lemma 7.3.

STEP 5 : Obtain $F(i - 1, (y_{\bar{1}}, \cdots, y_{\bar{m}}), \mathbf{P}^{(\bar{m})})$ from the memory by using y_ls that these were calculated in **STEP 4**. Next, obtain $F(i - 1, (x_{\bar{1}}, \cdots, x_{\bar{m}}), \mathbf{P}^{(\bar{m})})$ by using Equation 11 or 10 in Theorem 7.3, and memorize the result of $F(i - 1, (x_{\bar{1}}, \cdots, x_{\bar{m}}), \mathbf{P}^{(\bar{m})})$.

STEP 6 : Go to **STEP 7** if we finished enumerating combinations for all element x_ls, and go to **STEP 2** otherwise.

STEP 7 : Go to **STEP 1** if $i < n$, and go to **STEP 8** if $i = n$.

STEP 8 : For all j $(j = 1, 2, \cdots, M)$, obtain $F(i, (k_j, k_{j+1}, \cdots, k_M), \mathbf{P}^{(M)})$ by using Equation 15 in Theorem 7.4. In this case, all $F(i, (k_{\overline{j(1)}}, k_{\overline{j(2)}}, \cdots, k_{\overline{j(r)}}), \mathbf{P}^{(\bar{r})})$s are given the values memorized in **STEP 5**.

STEP 9 : Obtain the value of $R^{(j)}(i; \mathbf{k}, \mathbf{P})$ for $j = 1, 2, \cdots, M$ by using the results of **STEP 8**.

5   Evaluations

5.1   *Orders of Computing Time and Memory Size*

First, we consider the order of computing time. For multi-state k-out-of-n:F system, the order of [Yamamoto *et al.* (2009)]'s algorithm is $O(n(2^M + \prod_{l=1}^{M} k_l))$. That is, the order of computing time is exponential of M and polynomial of n.

The order of an algorithm in section 3 is given as follows. The maximum number of $Q(i, (x_{\bar{1}}, x_{\bar{2}}, \cdots, x_{\bar{m}}), \mathbf{P}^{(\bar{m})})$'s for each i is $\prod_{l=1}^{M} k_l$. So, the order of computing time is $O(n \prod_{l=1}^{M} k_l)$. That is, the order of computing time is of exponential of M and polynomial of n, which is the same as [Yamamoto *et al.* (2009)]'s algorithm. However, as the proposed algorithm reduce the number of component states by using Corollary 7.2, the computing time seems to be less than or equal to [Yamamoto *et al.* (2009)]'s algorithm.

Next, the order of the required memory size is $O(n \prod_{l=1}^{M} k_l)$ by using [Yamamoto *et al.* (2009)]'s algorithm. That is, the order is also exponential of M, but does not depend on n. On the other hand, the algorithm for computing $Q(i, (x_{\bar{1}}, x_{\bar{2}}, \cdots, x_{\bar{m}}), \mathbf{P}^{(\bar{m})})$ needs the maximum of $2(n \prod_{l=1}^{M} k_l)^2$ entries, because we need $\prod_{l=1}^{M} k_l$ entries for even i and odd i, respectively.

Therefore, the order of the required memory size is $O((\prod_{l=1}^{M} k_l)^2)$ for the proposed algorithm.

For multi-state k-out-of-n:G system, we evaluate the orders of computing time and memory size for the proposed algorithm in section 4.

The proposed algorithm uses the similar recursive equation to [Yamamoto *et al.* (2006)]'s algorithm. In order to compute $F(i, (x_{\bar{1}}, x_{\bar{2}}, \cdots, x_{\bar{m}}), \mathbf{P}^{(\bar{m})})$'s for each i and all x_l's such that x_l takes k_l for all $l(l = \bar{1}, \bar{2}, \cdots, \bar{m})$, we must use Equation 11 a maximum of $\prod_{l=1}^{M} k_l$ times. Therefore, the order of obtaining all $F(i, (x_{\bar{1}}, x_{\bar{2}}, \cdots, x_{\bar{m}}), \mathbf{P}^{(\bar{m})})$s for $j = 1, 2, \cdots, M$ is $O(Mn \prod_{l=1}^{M} k_l)$ by Lemma 7.3. This is the same order as [Yamamoto *et al.* (2006)]'s algorithm. However, our proposed algorithm can reduce more computing time by Theorem 7.4 for redundant x_l's. Furthermore, all $F(i, (x_{\bar{1}}, x_{\bar{2}}, \cdots, x_{\bar{m}}), \mathbf{P}^{(\bar{m})})$s for $j = 2, 3, \cdots, M$ can be obtained when $j = 1$, because $F(i, (x_{\bar{1}}, x_{\bar{2}}, \cdots, x_{\bar{m}}), \mathbf{P}^{(\bar{m})})$s for $j = 2, 3, \cdots, M$ are same for each j.

Therefore, the order of obtaining $F(i, (x_{\bar{1}}, x_{\bar{2}}, \cdots, x_{\bar{m}}), \mathbf{P}^{(\bar{m})})$ for all $j = 1, 2, \cdots, M$ is $O(n \prod_{l=1}^{M} k_l)$ by proposed algorithm.

Accordingly, we estimate that the proposed algorithm can get faster or equal to computing time by using [Yamamoto *et al.* (2006)]'s algorithm. Similarly, the order of the memory required by the proposed algorithm is $O((\prod_{l=1}^{M} k_l)^2)$.

These results indicate that (1) [Yamamoto *et al.* (2009)] and [Yamamoto *et al.* (2006)]'s algorithms are advantageous over proposed algorithm with respect to memory size, (2) proposed algorithms are advantageous over [Yamamoto *et al.* (2009)] and [Yamamoto *et al.* (2006)]'s algorithms with respect to computing time.

5.2 The Result of Numerical Experiments

Though we discussed the orders of computing time and memory size, their orders are close. So, we performed a numerical experiment in order to compare proposed algorithms with [Yamamoto *et al.* (2006, 2009)]'s algorithms by actual computing times.

First, we evaluated system state distributions of increasing four-state k-out-of-n:F systems with $k_1 = 4, k_2 = 5, k_3 = 6, k_4 = 7$. The results are shown in Table 1. The proportion means the ratio of the computing time of the proposed algorithm to [Yamamoto *et al.* (2009)]'s algorithm. All the experiments were executed using a Pentium 4 (2.8GHz) computer with 1.0GBytes of RAM, MS-Windows XP, Visual C++.NET2003 and C language programming.

In Table 1, computing times are the averages of five trials for each n.

From the results of the numerical experiments, we see these two algorithms are efficient for evaluating the system state distributions of

Table 1 Computing time for the system state distribution of increasing four-state k-out-of-n:F systems (sec.)

n	Yamamoto *et al.*'s algorithm(2009)	Proposed algorithm	Proportion
10	0.109	0.015	13.8%
20	0.265	0.016	6.0%
30	0.398	0.045	11.3%
40	0.520	0.059	11.3%
50	0.718	0.079	11.0%
100	1.790	0.136	7.6%
150	2.300	0.265	11.5%
200	3.000	0.361	12.0%
300	4.980	0.631	12.7%
500	7.500	0.891	11.9%
700	13.200	1.267	9.6%
1000	17.900	1.735	9.7%

the generalized multi-state k-out-of-n:F systems, including not only the decreasing, increasing, and constant types, but also other non-monotonic types. Table 1 shows that the computing times of the proposed algorithm are less than the [Yamamoto *et al.* (2009)]'s algorithm, especially when the number of components n is large.

Similarly, we evaluated computing times for four-state k-out-of-n:G systems. All the experiments were executed using a Pentium M 1.3GHz CPU, 768MB memory, Microsoft Windows 2000, Visual C++.NET2003 and C language programming.

Though the numerical experiments, we compared computing times of the proposed algorithm with those of [Yamamoto *et al.* (2006)]'s algorithm. We calculated the system state distribution of a decreasing, increasing and generalized four-state k-out-of-n:G system as shown in Table 2, which are adverse conditions for the proposed algorithm. In Table 2, the results of computing times are the averages from five trials for each n value.

From results of the numerical experiments, these two algorithms are efficient for multi-state k-out-of-n:G system when the number of n is large. And, Table 2 shows that the proposed algorithm takes less computing time than the [Yamamoto *et al.* (2006)]'s algorithm, especially when the number of components n is large.

From both results of orders and numerical examples, we see that the proposed algorithms are efficient for computation time by reducing redundant x_l's, and these are efficient for evaluating the system state distribution for large n.

Table 2   Computing time for the system state distribution of increasing four-state k-out-of-n:G systems (sec.)

n	$k_1 = 7, k_2 = 6, k_3 = 5, k_4 = 4$		$k_1 = 4, k_2 = 5, k_3 = 6, k_4 = 7$		$k_1 = 5, k_2 = 4, k_3 = 6, k_4 = 7$	
	Proposed Algorithm	Yamamoto *et al.* (2006)	Proposed Algorithm	Yamamoto *et al.* (2006)	Proposed Algorithm	Yamamoto *et al.* (2006)
50	0.140	0.164	0.160	0.196	0.162	0.192
100	0.274	0.334	0.316	0.395	0.314	0.407
150	0.410	0.507	0.475	0.591	0.475	0.595
200	0.544	0.677	0.629	0.791	0.631	0.795
300	0.821	1.015	0.949	1.192	0.945	1.202
500	1.386	1.698	1.590	1.999	1.580	2.011
700	1.908	2.383	2.206	2.802	2.209	2.808
900	2.456	3.066	2.838	3.601	2.840	3.611
1000	2.715	3.411	3.153	4.004	3.155	4.014

However, the order of the memory shows the proposed algorithms requires too much memory capacity for large M.

6 Conclusions

In this article, we discussed efficient recursive algorithm for evaluating the system state distribution of generalized multi-state k-out-of-n:F(G) system in the non-*i.i.d.* case. We proposed algorithms are using the idea for reducing the number of the considered component states. We compared the orders of computing time and memory size requirements of proposed algorithms with other algorithms.

Though their orders are close, from the results of numerical experiments, we can conclude that the proposed algorithms can calculate the system state distributions of generalized multi-state k-out-of-n:F(G) systems faster than other algorithms in most cases, especially for large n, within the range of our experiments. Such an advantage is significant especially when n is large. However, this algorithm requires too much memory capacity to evaluate the system state distributions, when M is large. For such a case, we can recommend to use other algorithms.

Acknowledgment

This work was partially supported by Grant No.23510204, Grant-in-Aid for Scientific Research (c) from JSPS(2011-). The authors thank the JSPS for their support.

References

Amari, S. V., Zuo, M. J. and Dill, G. (2009). A Fast and Robust Reliability Evaluation Algorithm for Generalized Multi-State k-out-of-n Systems, *IEEE Transactions on Reliability*, **58**, 1, pp. 88–93.

Barlow, R. E. and Prochan, F. (1975). *Statistical Theory of Reliability and Life Testing. Probability Models* (Holt Rinehart and Winston).

Barlow, R. E. and Wu, A. S. (1978). Coherent systems with multi-state components, *Mathematics of Operations Research*, **3**, 4, pp. 275–281.

Chang, G. J., Cui, L. and Hwang, F. K. (2000). *Reliabilities of Consecutive-k Systems*, Network Theory and Applications, Volume 4 (Kluwer Academic Publishers).

El-Neweihi, E., Proschan F. and Sethuraman, J. (1978). Multi-state coherent system, *Journal of Applied Probability*, **15**, pp. 675–688.

Griffith, W. S. (1980). Multistate reliability models, *Journal of Applied Probability*, **17**, pp. 735–744.

Haim, M. and Porat, Z. (1991). Bayes reliability modeling of a multistate consecutive k-out-of-n:F system, *Proceeding Annual Reliability and Maintainability Symposium*, pp. 582–586.

Huang, J., Zuo, M. J. and Wu, Y. H. (2000). Generalized multi-state k-out-of-n:G systems, *IEEE Transactions on Reliability*, **49**, 1, pp. 105–111.

Kołowrocki, K. (2004). *Reliability of Large Systems* (Elsevier).

Kossow, A. and Preuss, W. (1995). Reliability of linear consecutively-connected systems with multistate Components, *IEEE Transactions on Reliability*, **44**, 3, pp. 518–522.

Koutras, M. V. (1997). Consecutive-k,r-out-of-n:DFM systems, *Microelectronics and Reliability*, **37**, 4, pp. 597–603.

Kuo, W. and Zuo, M. J. (2003). *Optimal Reliability Modeling*, Principles and Applications (John Wiley and Sons).

Malinowski, J. and Preuss, W. (1995). Reliability of circular consecutively-connected systems with multi-state components, *IEEE Transactions on Reliability*, **44**, 3, pp. 532–534.

Malinowski, J. and Preuss, W. (1996). Reliability of reverse-Tree-Structured systems with multi-state components, *Microelectronics and Reliability*, **36**, 1, pp. 1–7.

Rushdi, A. M. (1986). Utilization of symmetric switching functions in the computation of k-out-of-n system reliability, *Microelectronics and Reliability*, **26**, 5, pp. 973–987.

Tian, Z., Zuo, M. J. and Yam, R. CM. (2009). Multi-state k-out-of-n systems and their performance evaluation, *IIE Transactions*, **41**, pp. 32–44.

Yamamoto, H., Akiba, T. and Nagatsuka, H. (2006). Efficient methods for the system state distribution of Multi-state k-out-of-n:G Systems, *The Journal of Reliability Engineering Association of Japan*, **28**, 5, pp. 395–404. (in Japanese)

Yamamoto, H., Akiba, T. and Nagatsuka, H. (2009). Calculating method for the system state distribution of Multi-state k-out-of-n:F Systems, *IEICE Transactions on Fundamentals of Electronics, Communications and Computer Sciences*, E**92**-A, 7, pp. 1593–1599.

Yamamoto, H., Akiba, T., Yamaguchi, T. and Nagatsuka, H. (2011). An evaluating algorithm for the system state distributions of generalized multi-state k-out-of-n:F Systems, *Journal of Japan Industrial Management Association*, **61**, 6E, pp. 347–354.

Zuo, M. J. and Liang, M. (1994). Reliability of multistate consecutively-connected systems, *Reliability Engineering and System Safety*, **44**, pp. 173–176.

Zuo, M. J., Huang, J. and Kuo, W. (2003). Chapter 1: Multi-state k-out-of-n Systems, *Handbook of Reliability Engineering* (Springer), pp. 3–17.

Zuo, M. J. and Tian, Z. (2006). Performance evaluation of generalized multi-state k-out-of-n Systems, *IEEE Transactions on Reliability*, **55**, 2, pp. 319–327.

Chapter 8

Multi-state Components Assignment Problem with Optimal Network Reliability Subject to Assignment Budget

Yi-Kuei Lin and Cheng-Ta Yeh

Department of Industrial Management,
National Taiwan University of Science & Technology,
No. 43, Sec. 4, Keelung Rd., Da'an Dist., Taipei City 106, Taiwan

1 Introduction

The assignment problem (AP) is a critical issue in decision making. A typical AP is to assign a set of components to a set of locations in a system with maximal total profit or minimal total cost. For instance, a set of transmission lines, such as coaxial cables, fiber optics, etc., is ready to be assigned to the given locations of an information network. The AP has many extensions in practical applications: facility location, task assignment, personnel scheduling, and so on [Winston (1995)]. [Pentico (2007)] surveyed many typical APs and made the categorization for them, but these problems focused on maximizing the total profit or minimizing the total cost without taking the failure of the component into consideration. In practice, each component should be multistate due to failure, maintenance and partially failure. That is, each component owns several possible capacities with a probability distribution and may fail. As the set of multistate components is assigned to the arcs (i.e. locations) of a system, each arc is multistate. Such a system is therefore treated as a stochastic-flow network (SFN) with a set of arcs and nodes. Then the network reliability is defined to be the probability that the d units of homogenous commodity are transmitted successfully from the source node s to the sink node t. Many studies evaluated the network reliability in terms of minimal paths (MPs) [Lin, Jane and Yuan (1995)]. An MP is an order sequence of arcs from s to t that has no cycle.

In addition to evaluating the network reliability, several issues about the network reliability optimization have been explored recently [Hsieh and Chen (2005); Levitin and Lisnianski (2001); Lisnianski and Levitin (2003); Liu, Zhang, Ma and Zhao (2007); Painton and Campbell (1995)]. Accordingly, [Hsieh and Chen (2005)] developed an updating schema to determine the optimal multistate multi-terminal network reliability under a range of demand under the demand-dependent and demand-independent cost constraints. [Liu *et al.* (2007)] adopted genetic algorithm (GA) to evaluate the optimal multistate multi-terminal network reliability under flow assignments of multi-commodity. [Lisnianski and Levitin (2003)] classified the network reliability optimization problems into two categories: achieving the optimal network reliability under different constraints [Painton and Campbell (1995)] and minimizing the resources required subject to a network reliability level [Levitin and Lisnianski (2001)].

The preceding literatures major in the topics of commodity allocation, flow assignment, or network structure. How to assign the multistate components such that the network reliability is maximal is never discussed. Thus, this article studies the multistate components assignment problem (MCAP) with optimal network reliability subject to the assignment budget for desirable and practical requirements. In this problem, the network topology is immobile and there is a set of components able to be assigned to the network's arcs. Each component has an assignment cost, the cost of assigning/purchasing the component. Intuitively, the implicit enumeration method (IEM) may searches for the optimal components assignment with maximal network reliability. However, this method is inefficient when the network is large. GA is a powerful probabilistic search and optimization algorithm. Many researchers [Chu and Beasley (1997); Harper, De Senna, Vieira and Shahani (2005); Majumdar and Bhunia (2007); Wang (2002); Wilson (1997)] have employed GA to solve some versions of APs. Moreover, GA is applied to solve not only APs, but also the network reliability optimization problems [Levitin and Lisnianski (2001); Liu, Zhang, Ma and Zhao (2007); Painton and Campbell (1995)]. Therefore, this article refers to the standard GA from [Goldberg (1989)] and proposes a GA based algorithm to determine the optimal components assignment with maximal network reliability subject to the assignment budget. In the GA-based algorithm, a components assignment is represented as a chromosome (solution), and the fitness function is to evaluate the network reliability of a chromosome in terms of MPs and state-space decomposition. The experimental results show that the GA-based algorithm is executed in a reasonable time.

2   Assumptions

Let (N, A) be an SFN with a single source s and a single sink t, where N denotes the set of nodes and $A = \{a_i | 1 \leq i \leq n\}$ denotes the set of n directed arcs connecting nodes. Suppose there are totally m MPs, $\Lambda_1, \Lambda_2, \cdots, \Lambda_m$. Let $\Omega = \{\omega_k | 1 \leq k \leq z\}$ denote the set of components. Each component ω_k has multiple states, $1, 2, \cdots, M_k$, with corresponding capacities, $0 = h_k(1) < h_k(2) < \cdots < h_k(M_k)$ where $h_k(l)$ is the lth capacity of ω_k for $l = 1, 2, \cdots, M_k$, and $h_k(M_k)$ is the maximal capacity of ω_k Let $B = (b_1, b_2, \cdots, b_n)$ be a components assignment in which component ω_k is assigned to arc a_i if $b_i = k$. In this article, there should be several assumptions addressed as follows:

I.   No component is assigned to any node.
II. Flow in (N, A) must satisfy the flow-conservation law (Ford and Fulkerson, 1962).
III. Each component can be assigned to at most one arc and each arc must be given with exact one component.
IV. The capacities of components are statistically independent.

3   SFN under a Components Assignment

The SFN is described in terms of two vectors: state vector $X = (x_1, x_2, \cdots, x_n)$ where x_i denotes the current capacity of a_i for all i, and flow vector $F = (f_1, f_2, \cdots, f_m)$ in which f_i denotes the flow through Λ_i, $j = 1, 2, \cdots, m$. Any state vector X is said to be under the components assignment B if

$$x_i \leq h_{b_i}(M_{b_i}), \quad i = 1, 2, \cdots, n. \tag{1}$$

Constraint (1) shows the current capacity x_i can not exceed the maximal capacity of component ω_k assigned to arc a_i. Any F is said to be feasible under the components assignment B if and only if

$$f_j \leq \min_{a_i \in \Lambda_j} \{h_{b_i}(M_{b_i})\}, \quad j = 1, 2, \cdots, m, \tag{2}$$

$$\sum_{a_i \in \Lambda_j} f_j \leq \{h_{b_i}(M_{b_i})\}, \quad i = 1, 2, \cdots, n, \tag{3}$$

where $\min_{a_i \in \Lambda_j}\{h_{b_i}(M_{b_i})\}$ is the maximal capacity of Λ_j. Constraint (2) represent that the flow through Λ_j, and constraint (3) means the sum of flow through arc a_i can not exceed the maximal capacity of component ω_k assigned to arc a_i. Similarly, any F is feasible under X if and only if

$$f_j \leq \min_{a_i \in \Lambda_j} \{x_i\}, \quad j = 1, 2, \cdots, m, \tag{4}$$

$$\sum_{a_i \in \Lambda_j} f_j \leq x_i, \quad i = 1, 2, \cdots, n, \tag{5}$$

In other words, the maximal flow under X is $V(X) \equiv \max\{\sum_{j=1}^m f_j | F$ is feasible under $X\}$.

3.1 *Network Reliability Evaluation*

Let $\mathbf{T}_B$ be the set of state vectors under the components assignment B. The network reliability under B denoted by $SR_d(B)$ is defined as the probability that d units of homogenous commodity can be successively delivered from s to t under B, i.e., $SR_d(B) \equiv \Pr\{V(X) \geq d, X \in \mathbf{T}_B\}$. Let $\mathbf{X}_B = \{X | V(X) \geq d, X \in \mathbf{T}_B\}$, then $SR_d(B)$ can be obtained by summing up the probabilities of all $X \in \mathbf{X}_B$. Thus,

$$SR_d(B) = \sum_{X \in \mathbf{X}_B} \Pr(X). \tag{6}$$

Generally, it is not a wise way to enumerate all $X \in \mathbf{X}_B$ and then to sum up their probabilities in order to obtain $SR_d(B)$. Instead, this article evaluates $SR_d(B)$ based on the method proposed by [Lin *et al.* (1995)]. Any minimal state vector in the set $\mathbf{X}_B$ is called a lower boundary point for d or d-MP. A d-MP X means that $V(X) \geq d$ and $V(X) < d$ for any state vector Y such that $Y < X$ (where $Y \leq X$ if and only if $y_i < x_i$ for at least one i). Suppose $X_1, X_2, \cdots, X_q$ are q d-MPs, then $\mathbf{X}_B$ is equal to $\{\bigcup_{i=1}^q \{X | X \geq X_i\}\}$. That is,

$$R_d(B) = \sum_{X \in \mathbf{X}_B} \Pr(X) = \Pr\left\{\bigcup_{i=1}^q \{X | X \geq X_i\}\right\}. \tag{7}$$

Such a probability can be calculated by inclusion-exclusion principle [Lin (2001)] or state-space decomposition [Jane, Lin and Yuan (1993)]. Jane and Laih (2008) validated the modified version has better efficiency in computation and storage space. Thus, this article adopts Jane and Laih's state-space decomposition to evaluate the network reliability.

3.2   *Generate all d-MPs*

The flow vector F meets the exact demand if it satisfies constraints (2), (3) and (8),

$$\sum_{j=1}^{m} f_j = d, \tag{8}$$

where constraint (8) shows that the sum of flow from s to t needs to meet demand d. Let $\mathbf{F} = \{F | F$ satisfies constaints (2), (3) and (8)$\}$. If $X \in \mathbf{T}_B$ is a d-MP, then there exists an feasible $F \in \mathbf{F}$ under X such that

$$x_i = h_{b_i}(l) \text{ where a } l \in \{1, 2, \cdots, M_{b_i}\} \text{ such that}$$

$$h_{b_i}(l-1) < \sum_{a_i \in \Lambda_j} f_j \leq h_{b_i}(l), \text{ for } i = 1, 2, \cdots, n. \tag{9}$$

It is a necessary condition for a d-MP X. To explain this condition, consider a d-MP X and an $F \in \mathbf{F}$ feasible under X. Suppose to the contrary that there exists a x_u such that $x_u > h_{b_u}(l) \geq \sum_{a_i \in \Lambda_j} f_j$. Set $Y = (y_1, y_2, \cdots, y_n)$ where $y_u = h_{b_u}(l)$ and $y_i = x_i$ for all $i \neq u$. Then $Y < X$, and F is feasible under Y because of $f_j \leq \min_{a_i \in \Lambda_j}\{y_i\}$ for each j and $\sum_{a_i \in \Lambda_j} f_j \leq y_i$ for each i. That is, $V(Y) \geq d$ which contradicts X is a d-MP. Thus, $x_i = h_{b_i}(l)$.

Any $X = (x_1, x_2, \cdots, x_n)$ which is transformed from $F \in \mathbf{F}$ according to Eq. (11) must satisfy $V(X) \geq d$ and will be taken as a d-MP candidate. Then each d-MP candidate can be checked whether it is a d-MP or not by employing the following algorithm.

Algorithm I.

Step 1. $I = \emptyset$ (I is the stack that stores the index of non-d-MP X. Initially, I is empty.)

Step 2. For $i = 1$ to v and $i \notin I$. // Suppose $X_1, X_2, \cdots, X_v$ are all d-MP candidates.

Step 3. For $j = i + 1$ to v and $j \notin I$.

Step 4. If $X_j < X_i$, then X_i is not a d-MP, I $= I \bigcup \{i\}$, and go to step 7.
      Else if $X_j \geq X_i$, then X_j is not a d-MP, I $= I \bigcup \{j\}$.
Step 5. Next j.
Step 6. X_i is a d-MP.
Step 7. Next i.

All d-MPs generated from Algorithm I can be utilized to evaluate $SR_d(B)$.

4 Problem Formulation

The mathematical programming formulation for the MCAP with optimal network reliability subject to budget C is thus given by

$$Maximize \ SR_d(B) \tag{10}$$

$$Subject \ to$$

$$b_i = k, k \in \{1, 2, \cdots, z\} \quad for \quad i = 1, 2, \cdots, n, \tag{11}$$

$$b_i \neq b_r \quad for \quad i \neq r, \quad and \tag{12}$$

$$\sum_{i=1}^{n} c_{b_i} \leq C. \tag{13}$$

Both of constraints (11) and (12) satisfy the assumption that each component can be assigned to at most one arc and each arc must be given with exact one component. Constraint (13) says the total assignment cost that is the total cost of the assigned components can not exceed budget C. All feasible components assignments can be obtained through constraints (11), (12) and (13). Note that $SR_d(B) = 0$ as there is no $X \in \mathbf{X}_B$. Then, the optimal components assignment with maximal network reliability is obtained by maximizing objective function (10).

The MCAP can be solved by the IEM. However, this method is very time-consuming if the network is large. Therefore, an efficient algorithm to solve the MCAP with optimal network reliability subject to budget C should be developed.

5 Proposed Algorithm

This article proposes a GA based algorithm to solve the MCAP. Before executing the GA-based algorithm, the parameters - population size (P_{size}), number of generation g_{time}, crossover rate P_{cr} and mutation rate P_{mr}, must be defined and the chromosome should be encoded. The population size represents the number of chromosomes in the population. The number of generation is the terminal condition of the GA-based algorithm. The crossover rate controls the probability of executing crossover. The mutation rate controls the probability of executing mutation.

This article utilizes the integer encoding to represent a chromosome so that the chromosome fits in with the components assignment (i.e. chromosome G is equivalent to components assignment B). Hence, a chromosome is denoted as $G = (g_1, g_2, \cdots, g_n)$, where $g_i \in \{1, 2, \cdots, z\}$ signifies that arc a_i is given with component ω_k if $g_i = k$. Such an encoding satisfies constraints (11) and (12), and has less memory utilization than the binary encoding when the network is large.

The GA-based algorithm starts with a population, in which there are (P_{size}) chromosomes. Subsequently, all chromosomes must be evaluated by the fitness function. To be worthy of attention, the fitness value of a chromosome is equal to the network reliability. However, if the total assignment cost of a chromosome exceed the assignment budget or there is no $X \in \mathbf{X}_G$, then the network reliability of the chromosome is given with a random penalty value that is suggested not to exceed 10^{-4}. The network reliability of each chromosome is calculated by the following steps.

Algorithm II.

Step 1. Determine whether the chromosome satisfies the constraint (14) or not,

$$\sum_{i=1}^{n} c_{g_i} \leq C. \tag{14}$$

If the chromosome does not satisfy the constraint (14), give a random penalty value to be its network reliability, and then evaluate the next chromosome.

Step 2. Find all F satisfying the following constraints.

$$f_j \leq \min_{a_i \in \Lambda_j} \{h_{g_i}(M_{g_i})\}, \quad j = 1, 2, \cdots, m, \tag{15}$$

$$\sum_{a_i \in \Lambda_j} f_j \leq \{h_{g_i}(M_{g_i})\} \quad i = 1, 2, \cdots, n, \quad \text{and} \tag{16}$$

$$\sum_{j=1}^{m} f_j = d. \tag{17}$$

If no feasible F exists, give a random penalty value to be its network reliability, and then evaluate the next chromosome.

Step 3. Transform each F into X via the following equation.

$$x_i = h_{g_i}(l) \text{ where } l \in \{1, 2, \cdots, M_{g_i}\} \text{ such that}$$

$$h_{g_i}(l-1) < \sum_{a_i \in \Lambda_j} f_j \le h_{g_i}(l), \text{ for } i = 1, 2, \cdots, n. \quad (18)$$

Step 4. Apply Algorithm **I** to determine whether each X generated from step 3 is a d-MP or not.

Step 5. Suppose all d-MPs from step 4 are $X_1, X_2, \cdots, X_q$. Calculate the network reliability (i.e. fitness value) of the chromosome, $\Pr\{\bigcup_{i=1}^q \{X | X \ge X_i\}\}$, by state-space decomposition.

After evaluating all chromosomes in population, the next stage is to execute the evolution process consisting of selection, crossover and mutation. This article adopts roulette wheel selection, single-point crossover and uniform mutation in the evolution process. Since the single-point crossover and uniform mutation are mainly utilized in the binary-encoding cases, this article modifies them to be suitable for integer encoding. Suppose that two chromosomes, (2 5 1 3 6 4) and (1 3 4 2 5 6), are selected from the population. Through the modified single-point crossover, the two chromosomes exchange with each other to be (2 5 1 3 5̲ 6̲) and (1 3 4 2 6̲ 4̲) when the crossover point is the 5th gene. After exchanging, the 2nd gene and the 5th gene in the first chromosome are identical, as are the 3rd gene and the 4th gene in the second chromosome. For the first (resp. second) chromosome, the 2nd (resp.3rd) gene is randomly changed to be one of the genes on the left hand side of the crossover point from the second (resp. first) chromosome. However, it results in the duplicate genes in the first chromosome that the 2nd gene in the first chromosome is changed to be the 1st, 2nd or 4th gene in the second chromosome. That is, the 2nd gene in the first chromosome can be only changed to be the 3rd gene in the second chromosome. Similarly, it results in the duplicate genes in the second chromosome that the 3rd gene in the second chromosome is changed to be the 1st, 3rd or 4th gene in the first chromosome. Hence, the 3rd gene in the second chromosome is only changed to be the 2nd gene in the first chromosome. Then, the two chromosomes are (2 4̲ 1 3 5 6) and (1 3 5̲ 2 6 4).

In the modified uniform mutation, a mutation point is generated randomly and supposed to be the 2nd gene. For example, utilize the chromosome (2 4 1 3 5 6). If the value of the 2nd gene changes into 3, then the 2nd gene exchanges with the 4th gene. Thus, the chromosome becomes (2 3 1 4 5 6). If the value of the 2nd gene changes into 7, then the chromosome is (2 7 1 3 5 6) due to no duplicate gene in the chromosome. Through the modified single-point crossover and the modified uniform mutation, the duplicate genes can be avoided, and the assumption that each component can be assigned to at most one arc and each arc must be given with exact one component can be satisfied.

Algorithm III.

Step 1. Determine P_{size}, P_{cr}, P_{mr}, and g_{time}, and generate an initial random population.

Step 2. Evaluate network reliability for each chromosome in population using Algorithm **II**.

Step 3. If the terminal condition is satisfied, return the optimal solution in the current generation, else go to step 4.

Step 4. Execute the evolution process to produce P_{size} chromosomes.
   4.1 Utilize roulette wheel selection.
   4.2. Implement the modified single-point crossover based on P_{cr}.
   4.3. Execute the modified uniform mutation based on P_{mr}.

Step 5. Evaluate network reliability for the P_{size} chromosomes from step 4 using Algorithm **II**.

Step 6. Mix the original P_{size} chromosomes and the new P_{size} chromosomes. Subsequently, choose the amount of better P_{size} chromosomes from the $2 \times P_{size}$ chromosomes to be the new population according to the network reliability, and then go to step 3.

6   Numerical Experiments

Two illustrative examples are adopted to demonstrate the GA-based algorithm. The GA-based algorithm and the IEM are programmed in MATLAB and implemented on a personal computer with Intel Core 2 Quad CPU 3.0G and 2G RAM.

6.1   *Example 1*

In example 1, the GA-based algorithm is compared with the IEM based on the network reliability and CPU time through a simple network (as presented in Fig. 1). All MPs of the network are $\Lambda_1 = \{a_1, a_2\}$, $\Lambda_2 =$

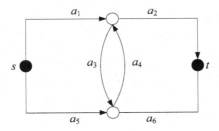

Fig. 1   A simple network.

Table 1   Probability distribution of capacities for 10 components.

Components (ω_k)	Cost	Capacity					
		0	5	10	15	20	25
1	60	0.05	0.1	0.25	0.6	0^a	0
2	10	0.1	0.3	0.6	0	0	0
3	20	0.1	0.9	0	0	0	0
4	50	0.1	0.9	0	0	0	0
5	60	0.1	0.9	0	0	0	0
6	20	0.05	0.25	0.7	0	0	0
7	50	0.3	0	0	0.7	0	0
8	80	0.15	0.05	0.2	0.1	0.5	0
9	100	0.3	0.05	0.05	0	0.05	0.55
10	70	0.05	0.25	0.2	0	0.1	0.4

^{a}The component does not provide this capacity.

$\{a_1, a_3, a_6\}$, $\Lambda_3 = \{a_2, a_4, a_5\}$ and $\Lambda_4 = \{a_5, a_6\}$, and there are 10 components ready to be assigned. Each component provides various capacities with a probability distribution (see Table 1). The possible capacity of each component is not formed with a consecutive range of integers. Several experiments are implemented by the GA-based algorithm with $P_{size} = 30$, $P_{cr} = 0.9$, $P_{mr} = 0.05$ and $g_{time} = 15$ to discuss the results of various demands and budgets. From the experimental results (see Table 2), the GA-based algorithm not only obtains the optimal solutions, but also has better computational efficiency than the IEM. The budget has obvious influence on the network reliability. Under the same demand level, the network reliability usually decreases when the budget decreases, especially under

Table 2 Probability distribution of capacities for 10 components.

(d, C)	Maximal reliability network	IEM Exact solution/ ♯ optimal solutions	CPU time[b]	Algorithm III[a] Optimal solution	CPU time
(5,240)	0.991848	(6,1,2,5,10,3)/216	18.45	(10,3,2,5,1,6)	0.17
(10,240)	0.935256	(1,10,5,3,2,6)/8	25.14	(1,6,3,5,2,10)	0.27
(15,240)	0.761084	(1,8,2,3,6,7)/2	18.94	(8,1,3,2,7,6)	0.25
(20,240)	0.549362	(1,10,2,3,6,7)/5	20.05	(1,10,3,2,6,7)	0.25
(25,240)	0.32487	(1,6,2,3,8,7)/8	16.19	(7,8,2,3,6,1)	0.28
(30,240)	0.1575	(1,10,3,2,7,6)/5	19.13	(1,6,2,3,7,10)	0.27
(35,240)	0	$\emptyset^c$	18.19	$\emptyset$	0.02
(5,210)	0.988515	(6,2,4,7,3,1)/24	14.16	(1,3,2,7,4,6)	0.13
(10,210)	0.8947	(2,7,4,3,6,1)/8	19.91	(6,1,3,4,2,7)	0.17
(15,210)	0.637245	(2,7,4,3,6,1)/8	13.98	(6,1,4,3,2,7)	0.16
(20,210)	0.465675	(7,1,4,3,2,6)/6	18.25	(7,1,4,3,2,6)	0.11
(25,210)	0.1764	(2,6,4,3,1,7)/8	14.83	(7,1,4,3,6,2)	0.16
(30,210)	0	$\emptyset$	17.03	$\emptyset$	0.03
(5,200)	0	$\emptyset$	10.24	$\emptyset$	0.02

[a] Algorithm III parameters: $P_{size} = 30$, $P_{cr} = 30$, $P_{mr} = 0.05$, $g_{time} = 15$.
[b] Unit: second.
[c] No solution.

the great demand level. In this case, there is no optimal solution under any demand level if budget C is less than 210.

6.2 *Example 2*

A large logistics service provider (LSP) deploys its logistics network as Fig. 2 in which there are six MPs. The LSP will deliver 600 units of homogenous commodity from s to t. However, the client can pay at most 2600 dollars. Hence, the LSP needs to plan the optimal utilization of its transportation resources to obtain the maximal network reliability under the expense which the client can pay. Tables 3-5 presents the transportation resources the LSP owns. Each transportation resource has an assignment cost. The capacity of each transportation resource represents load-carrying ability and may have multiple states.

The LSP utilizes Algorithm III with $P_{size} = 100$, $g_{time} = 500$, $P_{cr} = 0.6$ and $P_{mr} = 0.1$ to determine the best solution according to the LSP's transportation resources and the client's requirement. The LSP implements

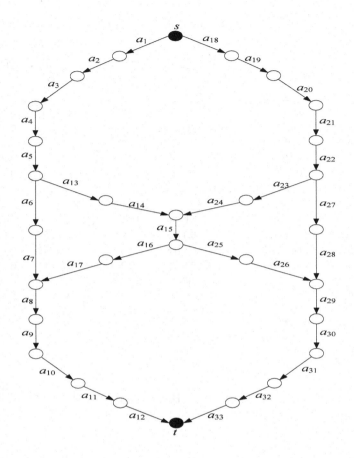

Fig. 2   A logistics network.

Algorithm III for 30 times and wants to obtain the largest and average maximal network reliability in the 30 experiments. The largest maximal network reliability is 0.936308 with the cost 2600 and the average maximal network reliability is 0.92718. Moreover in these 30 experiments, the average CPU time is 761 seconds. Obviously, Algorithm III obtains the best solution in a reasonable time. The progress of searching for the approximately optimal solution with the maximal network reliability 0.936308 is displayed in Fig. 3. The optimal assignment is (88, 71, 21, 55, 6, 41, 91, 5, 51, 1, 94, 44, 84, 34, 13, 92, 42, 82, 12, 38, 62, 29, 93, 3, 61, 11, 85, 63, 32, 60, 10, 35, 79).

Table 3 Probability distribution of capacities for 100 components.

ω_k	Cost	Capacity						
		0	200	400	600	800	1000	1200
1	150	0.01	0	0.01	0	0.01	0	0.97
2	80	0.05	0.05	0.05	0.15	0.2	0.5	0
3	65	0.07	0.08	0	0.85	0	0	0
4	30	0.7	0	0	0	0	0.3	0
5	110	0.01	0	0	0.05	0	0	0.94
6	180	0.01	0	0	0.01	0	0	0.98
7	40	0.5	0.5	0	0	0	0	0
8	35	0.25	0.25	0.5	0	0	0	0
9	55	0.15	0.25	0.1	0.1	0.1	0.1	0.2
10	70	0	0.05	0.05	0.9	0	0	0
11	25	0.01	0.99	0	0	0	0	0
12	90	0.02	0	0.05	0	0.05	0	0.88
13	50	0.07	0	0.28	0	0	0.65	0
14	95	0.05	0.05	0.9	0	0	0	0
15	50	0.6	0.4	0	0	0	0	0
16	80	0.15	0	0	0	0.85	0	0
17	65	0.1	0.1	0.1	0.7	0	0	0
18	70	0.7	0	0	0	0	0.3	0
19	50	0.07	0.18	0.75	0	0	0	0
20	15	0.4	0.4	0.2	0	0	0	0
21	120	0.01	0.01	0	0.03	0	0.95	0
22	160	0.02	0	0	0	0	0	0.98
23	100	0.05	0.05	0	0	0.1	0.1	0.7
24	35	0.7	0.3	0	0	0	0	0
25	40	0.3	0	0.7	0	0	0	0
26	80	0.1	0.1	0.8	0	0	0	0
27	10	0.8	0.2	0	0	0	0	0
28	5	0.9	0.1	0	0	0	0	0
29	115	0	0.05	0	0.05	0	0.1	0.8
30	55	0.03	0.27	0.3	0.4	0	0	0
31	130	0.01	0.01	0.02	0.02	0.02	0.02	0.9
32	60	0.05	0	0	0	0	0	0.95
33	30	0.1	0.9	0	0	0	0	0
34	20	0.08	0.37	0.55	0	0	0	0
35	75	0.02	0.03	0.05	0.9	0	0	0

Table 4   (cont. of Table 3) Probability distribution of capacities for 100 components.

ω_k	Cost	0	200	400	Capacity 600	800	1000	1200
36	140	0.05	0	0	0.05	0	0	0.9
37	80	0.07	0.23	0	0.2	0	0	0.5
38	35	0.02	0	0	0.08	0	0.9	0
39	90	0.5	0.5	0	0	0	0	0
40	100	0.7	0	0	0.3	0	0	0
41	80	0.01	0.02	0.03	0.04	0.05	0.06	0.79
42	40	0.07	0.07	0.07	0.07	0.07	0.65	0
43	15	0.15	0.15	0.15	0.15	0.4	0	0
44	130	0	0	0	0	0	0.1	0.9
45	60	0.02	0.48	0.5	0	0	0	0
46	155	0.01	0.02	0.03	0.04	0.05	0	0.85
47	95	0.05	0.3	0.65	0	0	0	0
48	90	0.4	0	0	0.2	0	0	0.4
49	85	0.5	0	0	0.5	0	0	0
50	50	0.1	0.2	0.7	0	0	0	0
51	150	0.01	0	0.01	0	0.01	0	0.97
52	80	0.05	0.05	0.05	0.15	0.2	0.5	0
53	65	0.07	0.08	0	0.85	0	0	0
54	30	0.7	0	0	0	0	0.3	0
55	110	0.01	0	0	0.05	0	0	0.94
56	180	0.01	0	0	0.01	0	0	0.98
57	40	0.5	0.5	0	0	0	0	0
58	35	0.25	0.25	0.5	0	0	0	0
59	55	0.15	0.25	0.1	0.1	0.1	0.1	0.2
60	70	0	0.05	0.05	0.9	0	0	0
61	25	0.01	0.99	0	0	0	0	0
62	90	0.02	0	0.05	0	0.05	0	0.88
63	50	0.07	0	0.28	0	0	0.65	0
64	95	0.05	0.05	0.9	0	0	0	0
65	50	0.6	0.4	0	0	0	0	0
66	80	0.15	0	0	0	0.85	0	0
67	65	0.1	0.1	0.1	0.7	0	0	0
68	70	0.7	0	0	0	0	0.3	0
69	50	0.07	0.18	0.75	0	0	0	0
70	15	0.4	0.4	0.2	0	0	0	0

Table 5   (cont. of Table 4) Probability distribution of capacities for 100 components.

ω_k	Cost	Capacity						
		0	200	400	600	800	1000	1200
71	120	0.01	0.01	0	0.03	0	0.95	0
72	160	0.02	0	0	0	0	0	0.98
73	100	0.05	0.05	0	0	0.1	0.1	0.7
74	5	0.7	0.3	0	0	0	0	0
75	40	0.3	0	0.7	0	0	0	0
76	80	0.1	0.1	0.8	0	0	0	0
77	10	0.8	0.2	0	0	0	0	0
78	5	0.9	0.1	0	0	0	0	0
79	115	0	0.05	0	0.05	0	0.1	0.8
80	55	0.03	0.27	0.3	0.4	0	0	0
81	130	0.01	0.01	0.02	0.02	0.02	0.02	0.9
82	60	0.05	0	0	0	0	0	0.95
83	30	0.1	0.9	0	0	0	0	0
84	20	0.08	0.37	0.55	0	0	0	0
85	75	0.02	0.03	0.05	0.9	0	0	0
86	140	0.05	0	0	0.05	0	0	0.9
87	80	0.07	0.23	0	0.2	0	0	0.5
88	35	0.02	0	0	0.08	0	0.9	0
89	90	0.5	0.5	0	0	0	0	0
90	100	0.7	0	0	0.3	0	0	0
91	80	0.01	0.02	0.03	0.04	0.05	0.06	0.79
92	40	0.07	0.07	0.07	0.07	0.07	0.65	0
93	15	0.15	0.15	0.15	0.15	0.4	0	0
94	130	0	0	0	0	0	0.1	0.9
95	60	0.02	0.48	0.5	0	0	0	0
96	155	0.01	0.02	0.03	0.04	0.05	0	0.85
97	95	0.05	0.3	0.65	0	0	0	0
98	90	0.4	0	0	0.2	0	0	0.4
99	85	0.5	0	0	0.5	0	0	0
100	50	0.1	0.2	0.7	0	0	0	0

7   Conclusions

The MCAP is different from the typical APs that focus on maximizing the total profit or minimizing the total cost. An efficient algorithm is developed based on GA, in which the fitness function evaluates the network reliability of a components assignment in terms of MPs and state-space

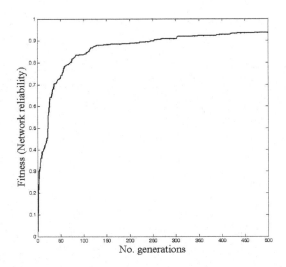

Fig. 3   The progress of searching for the optimal solution.

decomposition. For the larger network with 33 arcs and 100 components, the experimental results show that the approximately optimal solution can be found in no more than 1544. Hence, Algorithm III can be executed in a reasonable time.

Through the observation from example 1, there may be several optimal solutions with different total assignment costs under demand d and budget C. Generally, the decision maker focuses on determining the optimal solution with maximal network reliability and minimal assignment cost. Therefore, the addressed problem can be extended to the multi-objective optimization problem.

References

Chu, P. C. and Beasley, J. E. (1997). A genetic algorithm for the generalized assignment problem, *Computers and Operations Research* **24**, pp. 17–23.

Ford, L. R. and Fulkerson, D. R. (1962). *Flows in networks* (Princeton University, New Jersey).

Goldberg, D. (1989). *Genetic Algorithms in Search, Optimization and Machine Learning, Reading* (Addison-Wesley, Massachusetts).

Harper, P. R., De Senna, V., Vieira I. T. and Shahani, A. K. (2005). A genetic algorithm for the project assignment problem, *Computers and Operations Research* **32**, pp. 1255–1265.

Hsieh, C. C. and Chen, Y. T. (2005). Resource allocation decisions under various demands and cost requirements in an unreliable flow network, *Computer and Operations Research* **32**, pp. 2771–2784.

Jane, C. C. and Laih, Y. W. (2008). A practical algorithm for computing multi-state two-terminal reliability, *IEEE Transactions on Reliability* **57**, pp. 295–302.

Jane, C. C., Lin, J. S. and Yuan, J. (1993). Reliability evaluation of a limited-flow network in terms of minimal cutsets, *IEEE Transactions on Reliability* **42**, pp. 354–361.

Levitin, G. and Lisnianski, A. (2001). A new approach to solving problems of multi-state system reliability optimization, *Quality Reliability Engineering International* **17**, pp. 93–104.

Lin, Y. K. (2001). A simple algorithm for reliability evaluation of a stochastic-flow network with node failure, *Computer and Operations Research* **28**, pp. 1277–1285.

Lin, J. S., Jane, C. C. and Yuan, J. (1995). On reliability evaluation of a capacitated-flow network in terms of minimal pathsets, *Network* **25**, pp. 131–138.

Lisnianski, A. and Levitin, G. (2003). *Multi-state system reliability, assessment, optimization and application, Vol. 6* (World Scientific, Singapore).

Liu, Q., Zhang, H., Ma, X. and Zhao, Q. (2007). Genetic algorithm-based study on flow allocation in a multicommodity stochastic-flow network with unreliable nodes, in *Proc. the Eighth ACIS International Conference on Software Engineering, Artificial Intelligence, Networking, and Parallel/Distributed Computing*, pp. 576–581.

Majumdar, J. and Bhunia, A. K. (2007). Elitist genetic algorithm for assignment problem with imprecise goal, *European Journal of Operational Research* **177**, pp. 684–692.

Painton, L. and Campbell, J. (1995). Genetic algorithms in optimization of system reliability, *IEEE Transactions on Reliability* **44**, pp. 172–178.

Pentico, D. W. (2007). Assignment problems, a golden anniversary survey, *European Journal of Operational Research* **176**, pp. 774–793.

Wang, Y. Z. (2002). An application of genetic algorithm methods for teacher assignment problems, *Expert Systems with Applications* **22**, pp. 295–302.

Wilson, J. M. (1997). A genetic algorithm for the generalized assignment problem, *Journal of the Operational Research Society* **48**, pp. 804–809.

Winston, W. L. (1995). *Introduction to Mathematical Programming: Application and Algorithms* (Duxbury, California).

Chapter 9

Reliability Analysis of a Server System with Replication Schemes

Mitsutaka Kimura

Department of International Culture Studies,
Gifu City Woman's College,
7-1 Hitoichibakita-machi, Gifu 501-0192, Japan

1 Introduction

Recently, the server system with a backup site has been widely used to protect enterprise database. The backup site stands by the alert when a main site has broken down due to DoS attack, electricity failure, hurricanes, earthquakes, and so on. When a main site has broken down, the server system migrates the routine work from the main site to a backup site. The server in the main site transmits the database content from the main site to the backup site using a network link. This is called replication [Yamato, Kan and Kikuchi (2006); Imai, Araki, Sugiura, Fujita and Suemura (2004); Nakamura, Fujiyama, Kawai and Sunahara (2007); VERITAS Software corporation]. This chapter summarizes reliability analysis of various replication schemes.

In Section 2, we consider a server system with replication using journaling files. The server takes checkpoint for all database transactions in the main site and transmits the database content from the main site to the backup site. The replication should execute whenever the storage database in the main site is updated. But the replication generally executes at regular intervals because it has a prohibitive cost. However, it has a problem to compromise consistency of database content. Therefore, the server in the main site transmits journaling files (update files and update history) to the backup site as soon as the server updates the storage database when a client requests the data update. When the server in the backup site receives journaling files, they are not immediately reflected in the database to prevent delay. When the server in the main site breaks down, the server system migrates the routine work from the main site to a backup site and the state of a database is rolled back to the most recent checkpoint and restore a consistent state by reading back journaling files [Yamato, Kan and Kikuchi (2006); Fujita and Yata (2005); Watabe, Suzuki, Mizuno and Fujiwara (2007); Goda and Kitsuregawa (2007)].

When the database restores a consistent state by using some journaling files, the more journaling files the database has, it takes too much time to read back them, and the server system suffers losses caused by lower efficiency while it was down. They have been pointed out as a problem to be studied. In reference [Fujita and Yata (2005)] and [Watabe, Suzuki, Mizuno and Fujiwara (2007)], they have proposed the technique to speed up the database processing to restore a consistent state by being read back journaling files. In reference [Goda and Kitsuregawa (2007)], they have proposed the method to reduce operational cost in a disaster recovery system based on log forwarding remote copy. We formulate a stochastic model of a server system with asynchronous replication using journaling files and proposed an optimal policy to reduce waste of costs for replication and journaling files [Kimura, Imaizumi and Nakagawa (2010, 2011)].

In Section 3, we consider consider a server system with replication buffering relay method. The server system consists of a buffering relay unit as well as both main and backup sites. The server in a main site updates the storage database when a client requests the data update, and transfers the data and the address of the physical location updated data on the storage to a buffering relay unit. The server transmits all of the data in the buffering relay unit to a backup site at any time (replication). When a main site has broken down, the routine work is migrated from the main site to the backup site and is able to be executed immediately. When a client requests the data update or the date read, the server confirm the address table held in the buffering relay unit. If the requested data corresponds with the data held in the buffering relay unit, it is transmitted from the buffering relay unit to the backup site instantly. Then, the waste of cost for transmitting the updated data has been pointed out [Kan, Yamato, Kaneko and Kikuchi (2005)]. We consider the problem of reliability in a server system using the replication buffering relay method in order to reduce the cost of replication and transmitting the updated data in buffering relay unit. That is, we formulate the stochastic model of a server system with the replication buffering relay method and discuss an optimal replication interval to minimize it [Kimura, Imaizumi and Nakagawa (2011, 2012)].

2   Server System with Replication using Journaling Files

This section formulates a stochastic model of a server system with replication considering the number of transmitting journaling files. That is, the server updates the storage database and transmits journaling files when a client requests the data update. The server transmits the database content to a backup site either at a constant time or after a constant number of transmitting journaling files. We propose an optimal policy to reduce waste of the cost for execution of transmitting journaling files. We derive the expected numbers of the replication and of journaling files. Furthermore, we calculate the expected cost and discuss an optimal replication interval to minimize it. Finally, numerical examples are given.

2.1   *Reliability Quantities*

A server system consists of a monitor, a main site and a backup site as shown in Fig. 1.

Both main and backup sites consist of identical server and storage. The backup site stands by the alert when the main site breaks down. The server in the main site performs the routine work and updates the storage database when a client requests the data update. Then, the monitor transmits the update files and update history (journaling files) from the main site to the

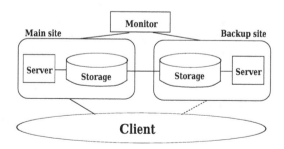

Fig. 1   Outline of a server system.

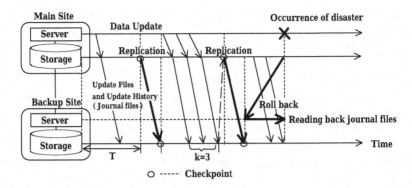

Fig. 2 Operation of an asynchronous replication scheme using journaling files.

backup site immediately. Furthermore, the monitor orders the server to transmit the database content to the backup site either at a constant time T or after a constant number $k(k = 1, 2, \cdots)$ of transmitting journaling files. The operation of a server system with asynchronous replication scheme using journaling files is shown in Fig. 2.

We formulate the stochastic model as follows:

(1) When the server in the main site breaks down by failures such as a disaster and crash failure and so on, the routine work is migrated from the main site to the backup site, and the state of a database is rolled back to the most recent checkpoint and can restore a consistent state by reading back journaling files (system migration). Both monitor and backup site do not breaks down, however, the server in the main site breaks down according to a general distribution $F(t)$.

(2) A client requests the data update to the storage. The request requires the time according to a general distribution $A(t)$, and the total of update and transmission of journaling files require the time according to a general distribution $B(t)$.

(3) The monitor orders the server to transmits the database content from the main site to the backup site, after the server in the main site updates the data by requesting from client and transmits journaling files at either $k(k = 1, 2, \cdots)$ times or time $T(0 \leq T < \infty)$, where $G(t) \equiv 0$ for $t < T$ and 1 for $t \geq T$.

   (a) The monitor orders the server to take checkpoints for all database transactions and to transmit the database content from the main site to the backup site (replication). The replication requires the time according to a general distribution $W(t)$.

(b) If the server in the main site breaks down while the server transmits the database content from the main site to the backup site, the monitor immediately orders to migrate the routine work from the main site to the backup site and the system migration is executed.

Under the above assumptions, we define the following states of the server system:

State 0: System begins to operate or restart.

State 1: When the monitor confirms the state of the main site at time T, the replication process begins to execute.

State 2: When the server in the main site has transmitted journaling files at $k(k = 1, 2, \cdots)$ times, the replication process begins to execute.

State 3: When the server in the main site breaks down, the process of a database is rolled back to the most recent checkpoint and can restore a consistent state by reading back journaling files.

The system states defined above form a Markov renewal process and represent time points of transition [Osaki (1992); Yasui, Nakagawa and Sandoh (2002)], where state 3 is an absorbing state and is a failure state. A transition diagram between system states is shown in Fig. 3.

Let Φ^* be the Laplace-Stieltjes (LS) transform of any function $\Phi(t)$, i.e., $\Phi^* \equiv \int_0^\infty e^{-st} d\Phi(t)$ for $Re(s) > 0$. The LS transforms of transition probabilities $Q_{i,j}(t)(i = 0, 1, 2; j = 0, 1, 2, 3)$ are given by the following

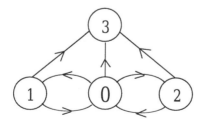

Fig. 3   Transition diagram between system states.

equations:

$$Q_{0,1}^*(s) = e^{-sT}\overline{F}(T)[1 - H^{(k)}(T)], \tag{1}$$

$$Q_{0,2}^*(s) = \int_0^T e^{-st}\overline{F}(t)dH^{(k)}(t), \tag{2}$$

$$Q_{0,3}^*(s) = \int_0^T e^{-st}[1 - H^{(k)}(t)]dF(t), \tag{3}$$

$$Q_{i,0}^*(s) = \int_0^\infty e^{-st}\overline{F}(t)dW(t) \qquad (i = 1, 2), \tag{4}$$

$$Q_{i,3}^*(s) = \int_0^\infty e^{-st}\overline{W}(t)dF(t) \qquad (i = 1, 2), \tag{5}$$

where $H(t) \equiv A(t) * B(t)$, $\overline{\Phi}(t) \equiv 1 - \Phi(t)$ represents a survival function of any function $\Phi(t)$, and $\Phi^{(i)}(t)$ is the i-fold convolution of $\Phi(t)$ and $\Phi^{(i)}(t) \equiv \Phi^{(i-1)}(t) * \Phi(t)(i = 1, 2, \cdots)$, $\Phi_1(t) * \Phi_2(t) \equiv \int_0^t \Phi_2(t-u)d\Phi_1(u)$, $\Phi^{(0)}(t) \equiv 1$ for $t \geq 0$.

We derive the expected number M_D of transmitting journaling files until state 3. The expected number $M_D(t)$ in $[0, t]$ is given by the following renewal equation:

$$
\begin{aligned}
M_D(t) = &\sum_{i=1}^k (i - 1) \int_0^t \overline{G}(t)H(t)^{(i-1)} * \overline{H}(t)dF(t) \\
&+ \left[\sum_{i=1}^k (i - 1) \int_0^t \overline{F}(t)H(t)^{(i-1)} * \overline{H}(t)dG(t)\right] * Q_{1,3}(t) \\
&+ k\left[\int_0^t \overline{G}(t)\overline{F}(t)dH(t)^{(k)}\right] * Q_{2,3}(t) \\
&+ \left[\sum_{i=1}^k \int_0^t \overline{F}(t)H(t)^{(i-1)} * \overline{H}(t)dG(t)\right] * Q_{1,0}(t) * [1 + M_D(t)] \\
&+ \left[\int_0^t \overline{G}(t)\overline{F}(t)dH(t)^{(k)}\right] * Q_{2,0}(t)[1 + M_D(t)]. \tag{6}
\end{aligned}
$$

By solving (6) for $M_D(t)$, its LS transform is

$$M_D^*(s) = \frac{\sum_{i=1}^{k} \int_0^T e^{-st}[H^{(i)}(t) - kH^{(k)}(t)]dF(t)}{1 - \left\{e^{-sT}\overline{F}(T)[1 - H^{(k)}(T)] + \int_0^T e^{-st}\overline{F}(t)dH^{(k)}(t)\right\}} \\ + \left\{e^{-sT}\overline{F}(T)\sum_{i=1}^{k}[H^{(i)}(T) - kH^{(k)}(T)]\right. \\ \left. + k\int_0^T e^{-st}\overline{F}(t)dH^{(k)}(t)\right\}\int_0^\infty e^{-st}\overline{W}(t)dF(t)}{} \\ \times \int_0^\infty e^{-st}\overline{F}(t)dW(t).$$ (7)

Hence, the expected number M_D is

$$M_D \equiv \lim_{t\to\infty} M_D(t) = \lim_{s\to 0} M_D^*(s) = \frac{Y(T,k)}{1 - X(T,k)\int_0^\infty W(t)dF(t)}, \quad (8)$$

where

$$X(T,k) \equiv \overline{F}(T) + \int_0^T H^{(k)}(t)dF(t),$$

$$Y(T,k) \equiv \sum_{i=1}^{k} \int_0^T \overline{F}(T)dH^{(i)}(t)$$

$$- \left[\overline{F}(T)\sum_{i=1}^{k} H^{(i)}(T) + k\int_0^T H^{(k)}(t)dF(t)\right]\int_0^\infty W(t)dF(t).$$

Similarly, The expected number $M_R(t)$ in $[0,t]$ is given by the following renewal equation:

$$M_R(t) = Q_{0,1}(t) + Q_{0,2}(t) + [Q_{0,1}(t) * Q_{1,0}(t) + Q_{0,2}(t) * Q_{2,0}(t)] * M_R(t). \quad (9)$$

The LS transform $M_R^*(s)$ of the expected number $M_R(t)$ in $[0,t]$ is

$$M_R^*(s) = \frac{Q_{0,1}^*(s) + Q_{0,2}^*(s)}{1 - Q_{0,1}^*(s)Q_{1,0}^*(s) - Q_{0,2}^*(s)Q_{2,0}^*(s)}. \quad (10)$$

Hence, the expected number M_R is

$$M_R \equiv \lim_{t\to\infty} M_R(t) = \lim_{s\to 0} M_R^*(s) = \frac{X(T,k)}{1 - X(T,k)\int_0^\infty W(t)dF(t)}. \quad (11)$$

2.2 *Optimal Policy*

We propose an optimal policy to reduce waste of costs for replication and journaling files. That is, we calculate the expected cost until state 3 and derive optimal replication intervals T^* and k^* to minimize it. Let c_D be the cost for transmitting journaling files and c_R be the cost for replication. We define the expected cost $C(T, k)$ as follows:

$$C(T, k) \equiv c_D M_D + c_R M_R. \qquad (12)$$

We seek an optimal replication interval T^* which minimizes $C(T, k)$ in (12) for $c_R \geq c_D$ and given $k(k \geq 1)$. Differentiating $C(T, k)$ with respect to T and setting it equal to zero,

$$Q_k(T)[1 - X(T, k)] \int_0^\infty W(t)dF(t) - Y(T, k) \int_0^\infty W(t)dF(t) = \frac{c_R}{c_D}, \qquad (13)$$

where

$$Q_k(T) \equiv -\frac{Y'(T, k)}{X'(T, k)},$$

and $\Phi'(t) \equiv d\Phi(t)/dt$. Denoting the left-hand side of (22) by $L_k(T)$,

$$L'_k(T) = Q_k'(T)[1 - X(T, k)] \int_0^\infty W(t)dF(t). \qquad (14)$$

Thus, if $Q_k'(T) > 0$ then $L_k'(T) > 0$ because

$$1 - X(T, k) \int_0^\infty W(t)dF(t) > 0.$$

Therefore, we have the following optimal policy:

(i) If $L_k(0) \geq c_R/c_D$, then $T^* = 0$.
(ii) If $L_k(\infty) > c_R/c_D > L_k(0)$, then there exists a finite and unique $T^*(0 < T^* < \infty)$ which satisfies (13).
(iii) If $L_k(\infty) \leq c_R/c_D$, then $T^* = \infty$.

Next, we consider the particular case where $A(t)$ is exponential and the time required for the data update can be neglected because it is much smaller than the other times, i.e., $A(t) \equiv 1 - e^{-\alpha t}$ and $B(t) \equiv 1$ for $t \geq 0$. In addition, $F(t) = 1 - e^{-\lambda t}$ and $W(t) = 1 - e^{-wt}$. Then,

$$Q_k(T) = \frac{\alpha}{\lambda + w} + \frac{w}{\lambda + w} \frac{\sum_{j=0}^{k-1} j[(\alpha T)^j / j!]}{\sum_{j=0}^{k-1} [(\alpha T)^j / j!]}.$$

Clearly,

$$Q_k(0) = \frac{\alpha + w}{\lambda + w}, \qquad\qquad Q_k(\infty) = \frac{\alpha + kw}{\lambda + w}$$

Suppose that $k = 1$. Then, $Q_k(T) \equiv (\alpha + w)/(\lambda + w)$, and hence, $T^* = \infty$, i.e., we should do no replication.

Next, for $k \geq 2$,

$$Q_k'(T) = \frac{w}{\lambda + w} \frac{\lambda}{\{\sum_{j=0}^{k-1}[(\alpha T)^j/j!]\}^2} \sum_{j=0}^{k-1} \frac{(\alpha T)^{j-1}}{j!} \sum_{i=0}^{j} \frac{(\alpha T)^i}{i!}(j - i)^2 > 0,$$

which follows that $Q_k(T)$ increases strictly with T from $(\alpha + w)/(\lambda + w)$ to $(\alpha + kw)/(\lambda + w)$. Thus, $L_k(T)$ increases strictly with T from $L_k(0) = \lambda(\alpha + w)/(\lambda + w)^2$ to $L_k(\infty)$.

Therefore, we have the following optimal policy for $k \geq 2$:

(i) If $L_k(0) \geq c_R/c_D$, then $T^* = 0$ and

$$C(0, k) = c_R\left(\frac{\lambda + w}{w}\right).$$

(ii) If $L_k(\infty) > c_R/c_D > L_k(0)$, then $0 < T^* < \infty$ and

$$C(T^*, k) = \frac{\lambda + w}{w}[c_D Q_k(T^*) - c_R].$$

(iii) If $L_k(\infty) \leq c_R/c_D$, then $T^* = \infty$, i.e., we should do no replication and

$$C(\infty, k) = \frac{c_D\left[\sum_{i=1}^{k}\left(\frac{\alpha}{\lambda+\alpha}\right)^i - \frac{\lambda}{\lambda+\alpha}\left(\frac{\alpha}{\lambda+\alpha}\right)^k\right] + c_R\left(\frac{\alpha}{\lambda+\alpha}\right)^k}{1 - \frac{w}{\lambda+w}\left(\frac{\alpha}{\lambda+\alpha}\right)^k}$$

Conversely, we seek an optimal replication interval k^* which minimizes $C(T, k)$ in (12) for $c_R \geq c_D$ and given $T(T > 0)$, and discuss analytically it. We can rewrite (12) as follows:

$$C(T, k) \equiv \frac{c_D Y(T, k) + c_R X(T, k)}{1 - \frac{w}{\lambda+w}X(T, k)} \qquad (k = 1, 2, \cdots). \qquad (15)$$

From the inequality $C(T, k + 1) - C(T, k) \geq 0$,

$$\frac{Y(T, k+1) - Y(T, k)}{Z(k+1) - Z(k)} + Y(T, k) \geq \frac{c_R}{c_D}\left(\frac{\lambda + w}{w}\right), \qquad (16)$$

where

$$Z(k) \equiv 1 - \frac{w}{\lambda + w}X(T, k).$$

Denoting the left-hand side of (16) by $L_T(k)$,

$$L_T(k) - L_T(k-1) = \left[\frac{Y(T, k+2) - Y(T, k+1)}{Z(k+2) - Z(k+1)} - \frac{Y(T, k+1) - Y(T, k)}{Z(k+1) - Z(k)} \right]$$
$$\times Z(k+1).$$

If $-[Y(T, k+1) - Y(T, k)]/[X(T, k+1) - X(T, k)]$ increases strictly with k and $L_T(\infty) > (c_R/c_D)[(\lambda + w)/w]$, then there exists a finite and unique $k^*(1 \leq k^* < \infty)$ which satisfies (16).

Next, we consider the particular case that $A(t) = 1 - e^{-\alpha t}$, $B(t) = 1$, $F(t) = 1 - e^{-\lambda t}$ and $W(t) = 1 - e^{-wt}$. Then,

$$-\frac{Y(T, k+1) - Y(T, k)}{X(T, k+1) - X(T, k)} = \frac{\alpha}{\lambda} \frac{w}{\lambda + w} + \frac{k\lambda}{\lambda + w}.$$

Therefore, there exists a finite and unique $k^*(1 \leq k^* < \infty)$ which satisfies (16).

Example 9.1.

We compute numerically an optimal replication interval T^* which minimizes $C(T, k)$ in (12). Suppose that the mean time $1/\beta$ required for the data update is a unit time. It is assumed that the mean generation interval of request the data update is $(1/\alpha)/(1/\beta) = 2 \sim 10$, the mean generation interval of a server down is $(1/\lambda)/(1/\beta) = 1000, 10000$, the mean time required for the replication is $(1/w)/(1/\beta) = 100 \sim 300$, the number of transmitting journaling files is $k = 5$. Further, we introduce the following costs: The cost for transmitting journaling files is $c_D = 1$, the cost for replication is $c_R/c_D = 5, 10$.

Table 1 gives the optimal T^* which minimizes the cost $C(T, k)$. For example, when $k = 5, c_D = 1, c_R = 5$ and $1/\beta = 1, 1/\alpha = 10, 1/w = 200, 1/\lambda = 1000$, the optimal interval is $T^* = 54.7$. This indicates that T^* increase with respect to $1/\alpha$ and $1/\lambda$. Further, T^* increase with respect to c_R/c_D. Thus, we should transmit more journaling files rather than execute replication. When c_R/c_D is large, T^* decrease with respect to $1/\lambda$. When $1/\lambda$ gets to a certain interval, we should execute replication rather than transmit journaling files. Moreover, T^* decrease with respect to $1/w$. In this case, we should execute replication rather than transmit journaling files.

Table 2 gives the optimal replication interval k^* which minimizes the cost $C(T, k)$. For example, when $c_D = 1, c_R = 5$ and $1/\beta = 1, 1/\alpha = 10, 1/w = 200, 1/\lambda = 1000, T = 10.0$, the optimal replication interval is $k^* = 5$. This indicates that k^* increase with respect to $1/\lambda$. Further, k^*

Table 1   Optimal replication time T^* to minimize $C(T, k)$ when $k = 5$

c_R/c_D	β/w	β/λ					
		1000			10000		
		β/α					
		2	5	10	2	5	10
5	100	10.9	27.8	57.2	11.1	28.1	57.7
	200	10.5	26.7	54.7	11.0	27.7	56.2
	300	10.2	25.9	52.9	10.9	27.4	55.5
10	100	11.2	29.5	63.3	11.3	29.6	62.9
	200	10.7	27.8	58.9	11.1	28.5	59.2
	300	10.4	26.8	56.4	11.0	28.0	57.7
50	100	13.6	43.1	108.7	13.3	40.9	101.6
	200	12.2	36.5	90.2	12.2	34.8	82.2
	300	11.6	33.9	82.4	11.7	32.5	74.3

Table 2   Optimal replication interval k^* to minimize $C(T, k)$

β/λ	c_R/c_D	β/w	β/α		
			2 ($T = 5.0$)	5 ($T = 10.0$)	10 ($T = 20.0$)
1000	5	100	17	6	6
		200	16	5	5
		300	16	5	5
	10	100	17	6	7
		200	17	6	6
		300	16	5	6
	50	100	20	9	7
		200	19	8	7
		300	18	7	7
10000	5	100	17	6	6
		200	17	6	6
		300	17	5	6
	10	100	18	6	7
		200	17	6	6
		300	17	6	6
	50	100	20	9	7
		200	19	7	7
		300	18	7	7

increase with c_R/c_D. Thus, we should transmit more journaling files rather than execute replication. When c_R/c_D is large, k^* do not depend on $1/\lambda$. This indicates that k^* decrease with respect to $1/w$. In this case, we should execute replication rather than transmit journaling files. Further, when $1/\lambda$ is large, k^* do not depend on $1/w$ and c_R/c_D and become constant.

3 Server System with Replication Buffering Relay Method

This section considers the problem of reliability in server system using the method in order to reduce the waste of costs for replication and transmitting the updated data in buffering relay unit. We formulate a stochastic model of a server system with the replication buffering relay method and derive the expected number of the replication and of updated data in buffering relay unit. Further, we calculate the expected cost and discuss an optimal replication interval to minimize it. Finally, a numerical example is given.

3.1 *Reliability Quantities*

A server system consists of a monitor, a buffering relay unit, a main site and a backup site as shown in Fig. 4.

Both main and backup sites consist of identical server and storage, and both backup site and buffering relay unit stand by the alert when a disaster has occurred. The server in the main site performs the routine work and updates the storage database when a client requests the data update. Then, the monitor transmits the updated data and the address of the physical location updated data on the storage to the buffering relay unit immediately. Furthermore, the monitor orders the buffering relay unit to transmit all of the data to the backup site after a constant number of transmitting data update. Then, we formulate the stochastic model as follows:

(1) A client requests the data update to the storage, and its time has an exponential distribution $(1 - e^{-\alpha t})$. The server in the main site updates

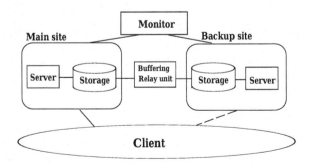

Fig. 4   Outline of a server system.

the storage database, and transmits the updated data and the address of the physical location updated data on the storage. The total of data update, the transmission of the data and the address of the location updated data requires the time according to an exponential distribution $(1 - e^{-\beta t})$.

(2) The monitor orders the buffering relay unit to transmit all of the data to the backup site, after the server in the main site updates the data by requesting from client and transmits the data at $n(n = 1, 2, \cdots)$ times (replication).

    (a) The replication time has an exponential distribution $(1 - e^{-wt})$.

    (b) If a disaster occurs in the main site while the replication is executed, the monitor immediately orders to migrate the routine work from the main site to the backup site.

(3) When a disaster has occurred in the main site, the routine work is migrated from the main site to the backup site, and the server performs the routine work in the backup site (system migration). The server system is shown in Fig. 5:

    (a) A disaster occurs according to an exponential distribution $(1 - e^{-\lambda t})$. A disaster does not occur in the monitor, the buffering relay unit and the backup site.

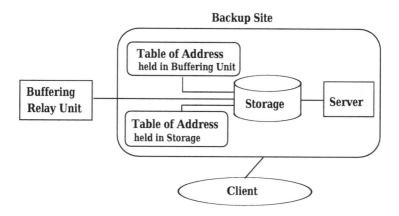

Fig. 5 Outline of a server system after system migration.

(b) When the routine work is migrated from the main site to the backup site, the monitor orders the buffering relay unit to transfer the address of updated data held in it to the backup site. The system migration executes according to an exponential distribution $(1 - e^{-gt})$.

(i) When a client requests the data update or the data read, the server confirms the table of address held in the buffering relay unit. If the address of the requested data corresponds with the address in the table, it is transmitted from the buffering relay unit to the backup site instantly and the table of the address held in the storage is updated.

(ii) The request time has an exponential distribution $(1 - e^{-\alpha_1 t})$. The confirmation of the address held in the buffering relay unit requires the time according to an exponential distribution $(1 - e^{-ht})$.

(iii) The probability that the address of the requested data corresponds with the address held in the buffering relay unit is $P(n)(n = 1, 2, \cdots)$. Where n represents the number of transmitting updated data in the buffering relay unit. The transmission of data and data update requires the time according to an exponential distribution $(1 - e^{-\beta_1 t})$.

(iv) The probability that the address of the requested data does not correspond with the address held in the buffering relay unit is $1 - P(n)(n = 1, 2, \cdots)$. The data update or the data read require the time according to an exponential distribution $(1 - e^{-\beta_2 t})$.

Under the above assumptions, we define the following states of the server system:

State 0: System begins to operate or restart.

States $a_1, a_2, \cdots, a_n$: When a client requests the data update to the storage, the data update begins.

States $b_1, b_2, \cdots, b_{n-1}$: Data update and the transmission of the data and the address of the location updated data is completed.

State b_n: When the server in the main site has transmitted the data and the address of the location updated data at $n(n = 1, 2, \cdots)$ times, replication begins.

States $F_0, F_1, \cdots, F_n$: Disaster occurs in the main site.

States $0_0, 0_1, \cdots, 0_n$: Routine work is migrated from the main site to the backup site, and the server performs the routine work in the backup site and the buffering relay unit.

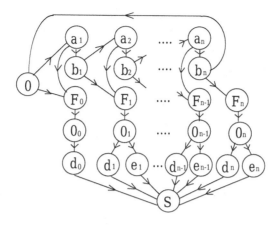

Fig. 6   Transition diagram between a system states.

States $e_1, e_2, \cdots e_n$: When a client requests the data update, the data update begins in the buffering relay unit.

States $d_1, d_2, \cdots d_n$: When a client requests the data update, the data update begins in the storage at the backup site.

State S: Data update is completed.

The system states defined above form a Markov renewal process [Osaki (1992); Yasui, Nakagawa and Sandoh (2002)], where S is an absorbing state. A transition diagram between system states is shown in Fig. 6.

The LS Transforms of transition probabilities $Q_{j,k}(t)(j = 0, a_i, b_i(i = 1, 2, \cdots, n); k = 0, a_i, b_i, F_0, F_i(i = 1, 2, \cdots, n))$ are given by the following equations:

$$Q^*_{0,a_1}(s) = Q^*_{b_{i-1},a_i}(s) = \frac{\alpha}{s+\alpha+\lambda}, \quad Q^*_{0,F_0}(s) = Q^*_{b_i,F_i}(s) = \frac{\lambda}{s+\alpha+\lambda},$$

$$Q^*_{a_i,b_i}(s) = \frac{\beta}{s+\beta+\lambda}, \quad Q^*_{a_i,F_{i-1}}(s) = \frac{\lambda}{s+\beta+\lambda}, \quad Q^*_{b_n,0}(s) = \frac{w}{s+w+\lambda},$$

$$Q^*_{b_n,F_n}(s) = \frac{\lambda}{s+w+\lambda}, \quad\quad\quad Q^*_{F_0,0_0}(s) = Q^*_{F_i,0_i}(s) = \frac{g}{s+g},$$

$$Q^*_{0_0,d_0}(s) = \frac{\alpha_1}{s+\alpha_1}\frac{h}{s+h}, \quad\quad Q^*_{0_i,d_i}(s) = \frac{\alpha_1}{s+\alpha_1}\frac{[1-P(i)]h}{s+h},$$

$$Q^*_{0_i,e_i}(s) = \frac{\alpha_1}{s+\alpha_1}\frac{P(i)h}{s+h}, \quad Q^*_{d_i,S}(s) = \frac{\beta_1}{s+\beta_1}, \quad Q^*_{e_i,S}(s) = \frac{\beta_2}{s+\beta_2},$$

$$(i = 1, 2, \cdots, n),$$

where $b_0 \equiv 0$. We derive the expected number $M_B(n)$ of the updated data in buffering relay unit until state S. The LS transforms $P^*_{0,F_i,e_i,S}(s)(i = 1, 2, \cdots, n)$ of the probability distributions $P_{0,F_i,e_i,S}(t)(i = 1, 2, \cdots, n)$ from state 0 to state S through state $F_i, e_i(i = 1, 2, \cdots, n)$ until time t are

$$P^*_{0,F_i,e_i,S}(s) = \frac{\Pi^i_{j=1}Q^*_{b_{j-1},a_j}(s)Q^*_{a_j,b_j}(s)[Q^*_{b_i,F_i}(s)+Q^*_{b_i,a_{i+1}}(s)Q^*_{a_{i+1},F_i}(s)]}{1-Q^*_{0,a_1}(s)\Pi^n_{j=1}Q^*_{a_j,b_j}(s)Q^*_{b_n,0}(s)} \times Q^*_{F_i,0_i}(s)Q^*_{0_i,e_i}(s)Q^*_{e_i,S}(s)$$

$$(i = 1, 2, \cdots n - 1), \qquad (17)$$

$$P^*_{0,F_n,e_n,S}(s) = \frac{Q^*_{0,a_1}(s)\Pi^n_{i=1}Q^*_{a_i,b_i}(s)Q^*_{b_n,F_n}(s)Q^*_{F_n,0_n}(s)Q^*_{0_n,e_n}(s)Q^*_{e_n,S}(s)}{1-Q^*_{0,a_1}(s)\Pi^n_{i=1}Q^*_{a_i,b_i}(s)Q^*_{b_n,0}(s)},$$

$$(18)$$

where $b_0 \equiv 0$.

Hence, the probability $P_{0,F_i,e_i,S}(i = 1, 2, \cdots, n)$ from state 0 to state $F_i(i = 1, 2, \cdots, n)$ through state $F_i, e_i(i = 1, 2, \cdots, n)$ is derived by $P_{0,F_i,e_i,S} \equiv \lim_{s \to 0}[P^*_{0,F_i,e_i,S}(s)](i = 1, 2, \cdots, n)$. Thereafter, the expected number $M_B(n)$ until state S is

$$M_B(n) \equiv \sum_{i=1}^{n} iP_{0,F_i,e_i,S} = \frac{Y(n)}{1 - \frac{w}{w+\lambda}X^n}, \qquad (19)$$

where

$$X \equiv \frac{\alpha\beta}{(\alpha + \lambda)(\beta + \lambda)},$$

$$Y(n) \equiv (1 - X)\sum_{i=1}^{n-1} iX^iP(i) + \frac{n\lambda}{\lambda + w}X^nP(n).$$

Similarly, we derive the expected number $M_R(n)$ of replication. The LS transform $M^*_R(s)$ of the expected number distribution $M_R(t)$ in $[0, t]$ are

$$M^*_R(s) = \frac{Q^*_{0,a_1}(s)\Pi^{n-1}_{i=1}Q^*_{a_i,b_i}(s)Q^*_{b_i,a_{i+1}}(s)Q^*_{a_n,b_n}(s)Q^*_{b_n,0}(s)}{1 - Q^*_{0,a_1}(s)\Pi^{n-1}_{i=1}Q^*_{a_i,b_i}(s)Q^*_{b_i,a_{i+1}}(s)Q^*_{a_n,b_n}(s)Q^*_{b_n,0}(s)},$$

$$= \frac{\left(\frac{w}{s+w+\lambda}\right)\left[\frac{\alpha\beta}{(s+\alpha+\lambda)(s+\beta+\lambda)}\right]^n}{1 - \left(\frac{w}{s+w+\lambda}\right)\left[\frac{\alpha\beta}{(s+\alpha+\lambda)(s+\beta+\lambda)}\right]^n}.$$

Hence, the expected number $M_R(n)$ is

$$M_R(n) \equiv \lim_{s \to 0}[M_R^*(s)] = \frac{\frac{w}{w+\lambda}X^n}{1 - \frac{w}{w+\lambda}X^n}, \tag{20}$$

3.2 *Optimal Policy*

We propose an optimal policy to reduce waste of costs for replication and transmitting the updated data in buffering relay unit. That is, we calculate the expected cost until state S and derive an optimal replication interval n^* to minimize it. Let c_B be the cost for transmitting the updated data in buffering relay unit and c_R the cost for replication. Then, we define the cost $C(n)$ as follows:

$$C(n) \equiv c_R M_R(n) + c_B M_B(n) \qquad (n = 1, 2, \cdots). \tag{21}$$

We seek an optimal replication interval $n^*(1 \leq n^* \leq \infty)$ which minimizes $C(n)$ in (21). From $C(n+1) - C(n) \geq 0$,

$$\frac{Y(n+1) - Y(n) - X^n \left(\frac{w}{\lambda+w}\right)[Y(n+1) - XY(n)]}{\left(\frac{w}{\lambda+w}\right)(1-X)X^n} \geq \frac{c_R}{c_B} \qquad (n = 1, 2, \cdots). \tag{22}$$

Denoting the left-hand side of (22) by $L(n)$,

$$L(n) - L(n-1)$$

$$= \frac{1 - \frac{w}{\lambda+w}X^n}{\frac{w}{\lambda+w}(1-X)X^n}\left\{Y(n+1) - Y(n) - X[Y(n) - Y(n-1)]\right\},$$

The bracket is

$$Y(n+1) - Y(n) - X[Y(n) - Y(n-1)]$$

$$= \frac{\lambda X^{n+1}}{\lambda+w}\left\{\frac{w(\alpha+\beta+\lambda) - \alpha\beta}{\alpha\beta}[nP(n) - (n-1)P(n-1)]\right.$$

$$\left. + (n+1)P(n+1) - nP(n)\right\},$$

Clearly,

$$L(\infty) = \infty.$$

Thus, if $w(\alpha+\beta+\lambda) - \alpha\beta > 0$ and $nP(n)$ is a increasing function of n then there exists a finite and unique minimum $n^*(1 \leq n^* < \infty)$ which satisfies (22).

In particular, when $P(n) = p$, $L(n)$ increases strictly with n to ∞, and hence, a finite n^* always exists.

Next, we consider the particular case that $P(n) \equiv 1 - e^{-\gamma n}(n = 1, 2, \cdots)$. Then,

$$nP(n) - (n-1)P(n-1) = (1 - e^{-\gamma})ne^{-\gamma(n-1)} + 1 - e^{-\gamma(n-1)} > 0,$$

Therefore, if $w(\alpha + \beta + \lambda) - \alpha\beta > 0$, $nP(n)$ is a increasing function of n and there exists a finite and unique n^* which satisfies (22).

Example 9.2.

We compute numerically an optimal replication interval n^* which minimizes $C(n)$ in (21). Suppose that the mean time $1/\beta$ required for the data update is a unit time. It is assumed that the mean generation interval of request the data update is $(1/\alpha)/(1/\beta) = 10 \sim 40$, the ratio that requested data corresponds with the data held in the buffering relay unit is $1/\gamma = 5, 50$, the mean generation interval of a disaster is $(1/\lambda)/(1/\beta) = 1000, 5000$, the mean time required for the replication is $(1/w)/(1/\beta) = 40, 80$, Further, we introduce the following costs: The cost for transmitting the data and the address of the location updated data is $c_B = 1$, the cost for replication is $c_R/c_B = 5, 10$.

Table 3 gives the optimal replication interval n^* which minimizes the cost $C(n)$. For example, when $c_B = 1, c_R = 5$ and $1/\beta = 1, 1/\gamma = 5, 1/\alpha = 10, 1/w = 40, 1/\lambda = 1000$, the optimal replication interval is $n^* = 27$. This indicates that n^* increase with $1/\gamma$, $1/\lambda$ and c_R/c_B. Thus, we should hold more amount of updated data in the buffering relay unit than execute replication. Moreover, n^* decrease with $1/\alpha$ and $1/w$. In this case, we should execute replication rather than hold large amounts of updated data in it.

Table 3   Optimal replication interval n^* to minimize $C(n)$

$1/\gamma$	c_R/c_B	β/w	β/λ					
			1000			5000		
			β/α					
			10	20	40	10	20	40
5	5	40	27	20	16	65	48	35
		80	23	18	15	61	46	34
	10	40	42	32	24	94	70	51
		80	38	30	23	91	68	50
50	5	40	35	30	25	68	53	42
		80	32	28	24	64	51	41
	10	40	48	40	34	94	73	57
		80	45	38	33	90	71	56

4   Conclusions

This chapter has studied analytically the two stochastic models of a server system by applying replication using journaling files and replication buffering relay method. Further, we have derived the reliability measures by using the theory of Markov renewal processes, and have discussed the optimal policy which minimizes the expected cost. Finally, we have given the numerical examples of each model in order to understand the results easily, and have evaluated them under some standard parameters.

References

Fujita, T. and Yata, K. (2005). *Asynchronous remote mirroring with journaling file systems*, Trans. IPS Japan, **46**, SIG 16, pp. 56–68.

Goda, K. and Kitsuregawa, M. (2007). *A study on power reduction method of disk storage for log forwarding based disaster recovery systems*, DBSJ Letters, **6**, 1, pp. 69–72.

Imai, T., Araki, S., Sugiura, T., Fujita, N. and Suemura, Y. (2004). *Session-uninterrupted Disaster Recovery across Distributed Data Centers IEICE Technical Report*, **NS2003-293**, IN2003-248, pp. 199–202.

Kan, M., Yamato, J., Kaneko, Y. and Kikuchi, Y. (2005). *An approach of shortening the recovery time in the replication buffering relay method for disaster recovery*, IEICE Technical Report, **CPSY2005-19**, pp. 25–30.

Kimura, M., Imaizumi, M. and Nakagawa, T. (2010). Optimal Replication Interval of an Asynchronous Replication using Journaling Files, in *The 4th Asia-Pacific International Symposium on Advanced Reliability and Maintenance Modeling* (Wellington, New Zealand), pp. 349–356.

Kimura, M., Imaizumi, M. and Nakagawa, T. (2011). Reliability Consideration of a Replication with Limited number of Journaling Files, in *The 7th International Conference on "Mathematical Methods in Reliability": Theory Methods Applications* (Beijing, China), pp. 948–953.

Kimura, M., Imaizumi, M. and Nakagawa, T. (2011). Reliability Consideration of a Server System with Replication Buffering Relay Method for Disaster Recovery, in *International Conferences ASEA/DRBC/EL, Communications in Computer and Information Science 257* (Jeju Island, Korea), pp. 392–398.

Kimura, M., Imaizumi, M. and Nakagawa, T. (2012). Reliability Modeling for a Server System with Buffering Relay Method, in *The 5th Asia-Pacific International Symposium on Advanced Reliability and Maintenance Modeling V* (Nanjing, China), pp. 255–262.

Nakamura, N., Fujiyama, K., Kawai, E. and Sunahara, H. (2007). *A Flexible Replication Mechanism with Extended Database Connection Layers for Disaster Recovery System*, Trans. IPS Japan, **48**, pp. 1562–1572.

Osaki, S., (1992). *Applied Stochastic System Modeling* (Springer-Verlag, Berlin).

VERITAS Software corporation. VERITAS volume replication for UNIX datasheet, http://eval.veritas.com/mktginfo/products/Datasheets/ High _Availability/vvr_datasheet_unix.pdf.

Watabe, S., Suzuki, Y., Mizuno, K. and Fujiwara, S. (2007). *Evaluation of speed-up methods of database redo processing for log-based disaster recovery systems*, DBSJ Letters, **6**, 1, pp. 133–136.

Yamato, J., Kan, M. and Kikuchi, Y. (2006). *Storage based data protection for disaster recovery*, J. IEICE, **89**, 9, pp. 801–805.

Yasui, K., Nakagawa,T. and Sandoh, H., (2002). *Reliability models in data communication systems, Stochastic Models in Reliability and Maintenance*, (edited by S.Osaki) (Springer-Verlag, Berlin), pp. 281–301.

PART 3
Computer Systems

Chapter 10

Two-Dimensional Software Reliability Growth Models

Shinji Inoue and Shigeru Yamada

Department of Social Management Engineering, Tottori University,
Minami 4-101, Koyama, Tottori-shi 680-8552, Japan

1 Introduction

Software reliability assessment in a testing-phase located in a final stage of the software development process is one of the important activities in developing a highly-reliable software system. In the testing-phase, an implemented software system is tested to detect and correct software faults latent in the software system. The software development manager has to assess the software reliability of the final software product especially in the final phase of the software development process for shipping a reliable software system to the user. A software reliability growth model (abbreviated as SRGM) [Musa (1972); Yamada and Osaki (1985); Pham (2003)] is known as one of the fundamental technologies for quantitative software reliability assessment, and plays an important role in software project management for producing a highly-reliable software system. The SRGM is a mathematical model, which describes a software reliability growth process in a testing phase of the software development process and the operational phase by regarding the number of software faults detected during a certain time-interval or the software failure-occurrence time interval as the random variables. After describing the software reliability growth process by using the SRGM, we assess software reliability quantitatively based on the software reliability assessment measures, which are derived from the SRGM applied for the software reliability analysis.

Most of SRGMs being classified with software failure-detection count models are developed under the basic assumption that a software reliability growth process depends only on the testing-time or the operational-time, such as calender time. However, it is difficult to say that the reliability of the software system depends only the testing-time duration since it is known that the software reliability growth process depends not only on the testing-time but also the several testing-effort factors which are related to the software reliability growth process, such as test-execution time (CPU hours) [Yamada *et al.* (1986)], testing-skill of test engineers [Fujiwara and Yamada (2001)], and testing-coverage [Inoue and Yamada (2004)]. Under the background mentioned above, two-dimensional (or bivariate) software reliability growth modeling approaches have been discussed in recent years. For examples, Ishii and Dohi [Ishii and Dohi (2006)] proposed software reliability modeling framework based on a two-dimensional nonhomogeneous Poisson process (abbreviated as NHPP), where they assumed that the software reliability growth process depends on simultaneous calender time and CPU hours. Ishii *et al.* [Ishii *et al.* (2008)] proposed a two-dimensional geometric software reliability model by taking account of calender time and the number of executed test cases.

We discuss other two-dimensional software reliability growth modeling approaches, in which the software reliability growth process depends on the testing-time and the testing-effort factors. Especially, our modeling approaches consider with the degree of the impact of the these two factors to the software reliability growth process and the effect of the program size to the reliability growth process, respectively. Therefore, we can expect more feasible software reliability measurement and assessment by our two-dimensional software reliability growth modeling approaches. We define some stochastic quantities for describing the two-dimensional software failure-occurrence or fault-detection phenomenon before discussing our two-dimensional software reliability growth modeling frameworks. After that, two-types of two-dimensional software reliability growth modeling approaches are discussed. Then, comparing with existing one-dimensional SRGMs, we check performance of our models in terms of a goodness-of-fit criterion, and show examples of the application of two-dimensional software reliability analysis based on our two-dimensional SRGM developed by our modeling framework by using actual fault count data.

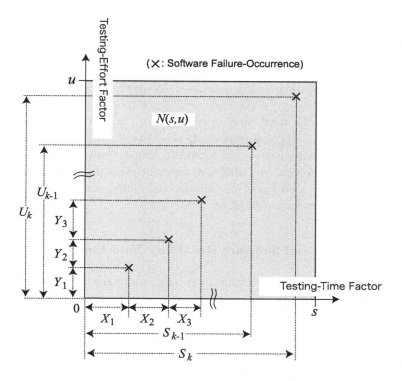

Fig. 1   Stochastic quantities for the two-dimensional software failure-occurrence or fault-detection phenomenon.

2   Stochastic Quantities for Two-Dimensional Modeling

Our two-dimensional SRGM describes a software reliability growth process depending on the following two-types of software reliability factors: Testing-time and testing-effort factors. Figure 1 illustrates the related stochastic quantities for the software failure-occurrence phenomenon on the two-dimensional space consisting of these two software reliability growth factors. According to Fig. 1, the stochastic quantities are defined as follows: $N(s, u)$ is a two-dimensional stochastic process representing the number of faults detected up to testing-time s and the amount of testing-effort expenditures u, S_k the k-th software failure-occurrence time $(k = 0, 1, 2, \cdots ; S_0 = 0)$, U_k the testing-effort expended up to the k-th software failure-occurrence $(k = 0, 1, 2, \cdots ; U_0 = 0)$, X_i the time-interval between the $(i-1)$-st and the i-th software failure-occurrences $(i = 1, 2, \cdots ; X_0 = 0)$, and Y_i the testing-effort expended during between the $(i-1)$-st and the i-th software failure-occurrences $(i = 1, 2, \cdots ; Y_0 = 0)$. In these stochastic quantities above, it is apparent that $S_k = \sum_{i=1}^{k} X_i$ and $U_k = \sum_{i=1}^{k} Y_i$, where $X_i = S_i - S_{i-1}$ and $Y_i = U_i - U_{i-1}$.

The fault count data which consists of N data pairs, $(s_i, u_i, y_i)(i = 0, 1, 2, \cdots, N; s_0 < s_1 < \cdots < s_N, u_0 \leq u_1 \leq u_2 \leq \cdots \leq u_N; s_0 = 0, u_0 = 0)$ with respect to the total number of faults, y_i, detected during a constant interval $(0, \Phi_i]$, where Φ_i represents elements of the space vector, i.e., $\Phi_i \equiv (s_i, u_i)$, are the realizations of S_i and U_i, respectively. Based on these quantities, several software reliability assessment measures, which are useful for quantitative assessment of software reliability, are derived by the probability mass function of $N(s, u)$.

3   Two-Dimensional Software Reliability Modeling

We discuss two-types of approaches for two-dimensional software reliability growth modeling.

3.1   *Weibull-Type Two-Dimensional SRGM*

We describe a software reliability growth process, which depends on the two software reliability growth factors, such as the testing-time and the testing-effort factors based on the one-dimensional Weibull-type software reliability growth model [Keiller and Miller (1991)]. To consider with the effect of such two reliability growth factors on the software reliability growth process simultaneously, we extend the testing-time factor of a one-dimensional SRGM as

$$t \equiv s^{\alpha} u^{1-\alpha} \quad (0 \leq \alpha \leq 1), \tag{1}$$

by using the Cobb-Douglas type function [Varian (1997); Ahn *et al.* (1998)]. In Eq. (1), s and u represent the testing-time and the testing-effort expenditure expended up to time s, respectively, and α the degree of the impact on the software reliability growth process.

Now we introduce a two-dimensional NHPP, which characterizes stochastic behavior of the two-dimensional software reliability growth process. Let $\{N(s, u), s \geq 0, u \geq 0\}$ be the two-dimensional stochastic process [Murthy *et al.* (1995, 2006)] representing the number of faults detected up to testing-time s and testing-effort expenditure u. The two-dimensional

NHPP is given as

$$
\left.
\begin{aligned}
\Pr\{N(s,u) = n\} &= \frac{\{H(s,u)\}^n}{n!} \exp[-H(s,u)] \\
&\qquad (n = 0, 1, 2, \cdots) \\
H(s,u) &= \int_0^s \int_0^u h(\tau, \upsilon) \mathrm{d}\tau \mathrm{d}\upsilon
\end{aligned}
\right\},
\tag{2}
$$

where $H(s,u)$ and $h(\tau,\upsilon)$ represent a mean value function and the intensity function of the two-dimensional NHPP, respectively.

As one of the examples, we apply such approach in Eqs. (1) and (2) to the following one-dimensional Weibull-type SRGM [Keiller and Miller (1991)]:

$$
\begin{aligned}
H(t) &\equiv \gamma(t) \\
&= \left(\frac{t}{\rho}\right)^\beta \quad (0 < \beta < 1; \rho > 0),
\end{aligned}
\tag{3}
$$

where $H(t)$ is a mean value function of a one-dimensional NHPP [Nakagawa (2005)], β the software reliability growth parameter, and ρ the scale parameter. If we apply our approach to the one-dimensional Weibull-type SRGM, then we can extend the mean value function in Eq. (3) as

$$
H(s,u) \equiv \gamma(s,u) = \left(\frac{s^\alpha u^{1-\alpha}}{\rho}\right)^\beta.
\tag{4}
$$

Equation (4) has the following properties: when $\alpha = 1$, Eq. (4) can be regarded as the conventional (or one-dimensional) Weibull-type SRGM, in which the software reliability growth process depends only on the testing-time. On the other hand, Eq. (4) becomes a testing-effort dependent SRGM when $\alpha = 0$. As we mentioned above, our approach keeps consistency with one-dimensional SRGMs. We call this two-dimensional SRGM in Eq. (4) "Our Model 1".

3.2 *Binomial Two-Dimensional SRGM*

We discuss another approach for two-dimensional software reliability growth modeling. Our another approach is based on the following basic assumptions [Ishii and Dohi (2006)]:

(A1) Whenever a software failure is observed, the fault is detected immediately, and no new faults are introduced in the fault-detection procedures.

(A2) Each software failure occurs at independently and identically distributed random time with the bivariate probability distribution function $F(s, u) \equiv \Pr\{S \le s, U \le u\} = \int_0^s \int_0^u f(x, y) \mathrm{d}x \mathrm{d}y$, where $f(x, y)$ and $\Pr\{A\}$ represent the joint density function and the probability of event A, respectively.

(A3) The initial number of faults in the software system, $N_0 (> 0)$, is a random variable, and is finite.

From the basic assumptions above, we develop a two-dimensional software reliability growth modeling framework with program size. The program size is one of the metrics of software complexity, which influences the software reliability growth process. From the basic assumptions, we have the probability mass function of $N(s, u)$ as

$$\Pr\{N(s, u) = m\} = \sum_n \binom{n}{m} \{F(s, u)\}^m \{1 - F(s, u)\}^{n-m}$$
$$\times \Pr\{N_0 = n\} \quad (m = 0, 1, 2, \cdots). \quad (5)$$

We note that the range of n cannot be fixed because the probability distribution function of the initial fault content N_0 is not still given in Eq. (5). In our modeling approach, we assume the following binomial distribution with parameter (K, λ) for the probability mass function of the initial fault content in Eq. (5) to incorporate the effect of the program size on the two-dimensional software reliability growth process:

$$\Pr\{N_0 = n\} = \binom{K}{n} \lambda^n (1 - \lambda)^{K-n}$$
$$(0 < \lambda < 1 \, ; \, n = 0, 1, \cdots, K). \quad (6)$$

Equation (6) has the following physical assumptions for the initial fault content [Kimura *et al.* (1993); Inoue and Yamada (2007)]:

(a) The software system consists of K lines of code (LOC) at the beginning of the testing-phase.
(b) Each code has a fault with the constant probability λ.
(c) Each software failure caused by a fault remaining in the software system occurs independently and randomly.

Substituting Eq. (6) into Eq. (5), we have

$$\Pr\{N(s, u) = m\} = \binom{K}{m} \{\lambda F(s, u)\}^m \{1 - \lambda F(s, u)\}^{K-m}$$
$$(0 < \lambda < 1 \, ; \, m = 0, 1, 2, \cdots K). \quad (7)$$

From Eq. (7), we can see that several types of two-dimensional SRGMs with the effect of the program size can be developed by assuming a suitable bivariate software failure-occurrence time distribution, $F(s, u)$.

As one of the examples, if we assume that each software failure-occurrence time follows the following bivariate probability distribution function proposed by Gumbel [Gumbel (1960)]:

$$F(s, u) = (1 - e^{-as})(1 - e^{-bu})(1 + ze^{-as-bu})$$
$$(a > 0, b > 0, -1 \le z \le 1), \tag{8}$$

then the expectation of $N(s, u)$ is derived as

$$E[N(s, u)] = K\lambda F(s, u)$$
$$= K\lambda(1 - e^{-as})(1 - e^{-bu})(1 + ze^{-as-bu}). \tag{9}$$

We call the two-dimensional SRGM in Eq. (9) "Our Model 2", which is developed by our modeling approach in Eq. (7).

4 Parameter Estimation

We discuss parameter estimation methods for two-dimensional SRGMs in Eqs. (4) and (9). Suppose that N data pairs have been observed. We now discuss a parameter estimation method for Our Model 1 in Eq. (4). In Our Model 1, we need to estimate parameters, α, β, and ρ, respectively. The parameter estimates of this two-dimensional SRGM can be easily obtained by using the multiple regression analysis since we can derive the following equation from Eq. (4) as

$$\log \gamma(s, u) = -\beta \log \rho + \alpha\beta \log s + (1 - \alpha)\beta \log u, \tag{10}$$

by taking the natural logarithm of Eq. (4). Then, we have the following multiple regression equation:

$$Y_i = a_0 + a_1 K_i + a_2 L_i + \epsilon_i, \tag{11}$$

where

$$\begin{cases} Y_i = \log y_i, \\ K_i = \log s_i, \\ L_i = \log u_i, \\ a_0 = -\beta \log \rho, \\ a_1 = \alpha\beta, \\ a_2 = (1 - \alpha)\beta. \end{cases} \tag{12}$$

In Eq. (11), ϵ_i is a standard normal error term with homoscedasticity, i.e., equality of variance. From Eqs. (11) and (12), the sum of the squared vertical distances from the actual data points to the presumed polynomial, $S(a_0, a_1, a_2)$, is derived as

$$S(a_0, a_1, a_2) = \sum_{i=1}^{N} \epsilon_i^2$$

$$= \sum_{i=1}^{N} \{Y_i - (a_0 + a_1 K_i + a_2 L_i)\}^2. \qquad (13)$$

Parameter estimates, $\widehat{a}_0$, $\widehat{a}_1$, and $\widehat{a}_2$, of the parameters a_0, a_1, and a_2 are estimated by minimizing Eq. (13). That is, solving the simultaneous equations, $\partial S / \partial a_0 = \partial S / \partial a_1 = \partial S / \partial a_2 = 0$, yields the parameter estimates of a_0, a_1, and a_2, respectively. Finally, we can obtain parameter estimates $\widehat{\alpha}$, $\widehat{\beta}$, and $\widehat{\rho}$ of the parameter α, β, and ρ as

$$\begin{cases} \widehat{\alpha} = \dfrac{\widehat{a}_1}{\widehat{a}_1 + \widehat{a}_2}, \\[2mm] \widehat{\beta} = \widehat{a}_1 + \widehat{a}_2, \\[2mm] \widehat{\rho} = \exp\left[-\dfrac{\widehat{a}_0}{\widehat{a}_1 + \widehat{a}_2}\right], \end{cases} \qquad (14)$$

by using the estimated partial regression coefficients, $\widehat{a}_0$, $\widehat{a}_1$, and $\widehat{a}_2$.

And we also discuss a parameter estimation method for Our Model 2 in Eq. (9). In Eq. (9), considering that information of the program size can be easily obtained from an actual software project, we need to estimate parameters, λ, a, b, and z, respectively. We can estimate the parameters of this two-dimensional SRGM by using the method of maximum-likelihood. Now, we derive the likelihood function, l, for the two-dimensional stochastic process, $N(s, u)$, in Eq. (5). Based on the basic assumptions in 3.2, we have the likelihood function as

$$l = \Pr\{N(s_1, u_1) = y_1, N(s_2, u_2) = y_2, \cdots, N(s_N, u_N) = y_N\}$$

$$= \prod_{i=1}^{N} \Pr\{N(s_i, u_i) = y_i \mid N(s_{i-1}, u_{i-1}) = y_{i-1}\}$$

$$\cdot \Pr\{N(s_1, u_1) = y_1\}. \qquad (15)$$

by using the Bayes' formula and Markov property. The conditional probability in Eq. (15) can be derived as

$$\Pr\{N(s_i, u_i) = y_i \mid N(s_{i-1}, u_{i-1}) = y_{i-1}\}$$

$$= \binom{K - y_{i-1}}{y_i - y_{i-1}} \{z(\Phi_{i-1}, \Phi_i)\}^{y_i - y_{i-1}} \{1 - z(\Phi_{i-1}, \Phi_i)\}^{K - y_i}, \quad (16)$$

where

$$z(\Phi_{i-1}, \Phi_i) = \frac{\lambda\{F(s_i, u_i) - F(s_{i-1}, u_{i-1})\}}{1 - \lambda F(s_{i-1}, u_{i-1})}. \quad (17)$$

By using Eqs. (16) and (17), we can rewrite Eq. (15) as

$$l = \prod_{i=1}^{N} \binom{K - y_{i-1}}{y_i - y_{i-1}} \{z(\Phi_{i-1}, \Phi_i)\}^{y_i - y_{i-1}} \{1 - z(\Phi_{i-1}, \Phi_i)\}^{K - y_i}, \quad (18)$$

where $F(s_0, u_0) = F(s_0, u) = F(s, u_0) = 0$. Based on the logarithmic likelihood function of Eq. (18), we can derive the simultaneous logarithmic likelihood equations with respect to each parameter of the two-dimensional SRGM in Eq. (9). Accordingly, we can obtain maximum-likelihood estimates of the parameters by solving the simultaneous logarithmic likelihood equations numerically.

5   Model Comparison

We conduct model comparison of our two-dimensional SRGMs with the following existing one-dimensional SRGMs: One-dimensional Weibull-type (called "ONE-D WEIBULL") in Eq. (3), a testing-coverage dependent SRGM ("TCD") [Inoue and Yamada (2004)], an exponential SRGM ("EXPO") [Goel and Okumoto (1979)], and a delayed S-shaped SRGM ("DELAYED-S") [Yamada *et al.* (1983)]. The following has more details on the testing-coverage dependent, exponential, and delayed S-shaped SRGMs:

- **Testing-Coverage Dependent SRGM** [Inoue and Yamada (2004)]: This SRGM is developed by considering the relationship between the testing-coverage attainment and the number of detected faults. Let $H_{TC}(t)$ be the expected number of the faults detected up to testing-time t and the mean value function of the one-dimensional NHPP.

Then, we have

$$H_{TC}(t) = a \left(1 - \exp\left[-s \int_0^t c(x)\mathrm{d}x \right] \right)$$

$$(a > 0, 0 < s < 1), \tag{19}$$

where a is the initial fault content, s the fault-detection rate per attained testing-coverage and $c(t)$ the ratio of testing-coverage attained at testing-time t, respectively. And letting $C(t)$ be the ratio of testing-coverage attained up to testing-time t, we have $c(t) \equiv \mathrm{d}C(t)/\mathrm{d}t$. The parameters a and s are estimated by the method of maximum-likelihood based on the NHPP after getting estimation of the testing-coverage function $C(t)$. For the testing-coverage function $C(t)$, they gives the following function with the testing-skill of test-case designers:

$$C(t) = \frac{\alpha(1 - e^{-b_{sta} \cdot t})}{1 + z \cdot e^{-b_{sta} \cdot t}}, \tag{20}$$

where α is the target value of testing-coverage to be attained, $z = (1 - r)/r$, $r = b_{ini}/b_{sta}(0 \leq r \leq 1)$, b_{ini} the initial testing-skill factor of the test-case designers, and b_{sta} the steady-state one. The parameters α, b_{sta}, and r are estimated by the least-square method based on the corresponding testing-coverage data

- **Exponential SRGM** [Goel and Okumoto (1979)]:
  This SRGM is one of the classical continuous-time NHPP models, but this model is still well-applied to practical software reliability assessment. Let $H_E(t)$ be the expected number of faults detected up to testing-time t and the mean value function of the one-dimensional NHPP. Then,

$$H_E(t) = a(1 - e^{-bt}) \qquad (a > 0, b > 0), \tag{21}$$

  where a represents the expected total number of potential faults detected in an infinitely long duration, and b the fault-detection rate. The parameters can be estimated by the method of maximum-likelihood.

- **Delayed S-shaped SRGM** [Yamada *et al.* (1983)]:
  This SRGM is also well-applied to practical software reliability assessment, and indicates an S-shaped software reliability growth curve. Let

Table 1   Result of model comparison based on MSE.

	DS1	DS2
OUR MODEL 1	**1.96356**	**3.09947**
OUR MODEL 2	54.7535	75.0829
ONE-D WEIBULL	1425.24	205.688
TCD	111.838	49.9692
EXPO	603.198	73.0885
DELAYED-S	150.005	36.6406

$H_{DS}(t)$ be the expected number of faults detected up to testing-time t. Then,

$$H_{DS}(t) = a\left[1 - (1 + bt)e^{-bt}\right] \qquad (a > 0, b > 0), \qquad (22)$$

where a represents the expected initial fault content and b the failure-occurrence rate or the fault-detection rate. The parameters are estimated by the method of maximum-likelihood.

We use two actual data sets, DS1 and DS2 [Fujiwara and Yamada (2002)]. DS1 consists of 24 data pairs: $(s_k, u_k, y_k)(k = 1, 2, \cdots, 24; s_{24} = 24$ (weeks), $u_{24} = 0.9095, y_{24} = 296, K = 1.972 \times 10^5(\text{LOC}))$ and DS2 22 data pairs: $(s_k, u_k, y_k)(k = 1, 2, \cdots, 22; s_{22} = 22$ (weeks), $u_{22} = 0.9198, y_{24} = 212, K = 1.630 \times 10^5(\text{LOC}))$, where u_i, the testing-effort factor, represents the testing-coverage attained up to testing-time s_i .

Now we conduct model comparison based on the well-known mean square errors (abbreviated as MSE). The MSE is calculated by dividing the sum of squared vertical distance between the observed and estimated cumulative number of faults, y_i and $\widehat{y}(\Psi_i)$, detected during the time-interval $(0, \Psi_i]$, respectively, by the number of observed data pairs. Ψ_i is the software reliability growth factor in each model. The MSE is calculated as

$$\text{MSE} = \frac{1}{N}\sum_{i=1}^{N}\{y_i - \widehat{y}(\Psi_i)\}^2. \qquad (23)$$

From Eq. (23), the smaller value of MSE represents fitting better to the actual data. Table 1 shows the results of model comparisons based on MSE. From Table 1, we can see that our two-dimensional SRGMs have better performance in terms of MSE among SRGMs cited in this model comparison. Especially, Our Model 1 has the best performance. And, we can say that two-dimensional software reliability measurement, in which it is assumed that the software reliability growth process depends not only on testing-time but also testing-effort factors, is useful and feasible for actual assessment of software reliability.

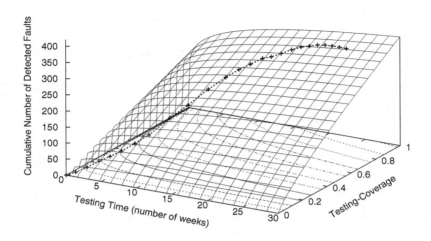

Fig. 2   Estimated two-dimensional mean value function, $\widehat{\gamma}(s, u)$ (DS1).

6   Numerical Examples

We show examples of application of Our Model 1 in Eq. (4), which has the best performance in our model comparison in terms of MSE, by DS1. From DS1, we can estimate the parameters, α, β, and ρ, as $\widehat{\alpha} = 0.17555$, $\widehat{\beta} = 0.95652$, and $\widehat{\rho} = 0.00415$, respectively, following to the parameter estimation method discussed in 4.

Figure 2 depicts the behavior of the estimated two-dimensional mean value function in Eq. (4), which depends on the two-dimensional space. In Fig. 2, the dotted lines and the curved surface represent the behavior of the actual data and the estimated behavior of the expected number of detected faults, respectively. From Fig. 2, we can see that the estimated two-dimensional SRGM can describe the actual phenomenon that the software reliability growth does not observed even if a lot of testing-time is expended under the situation that the amount of testing-effort expenditures is not increased. Such actual phenomenon can not be described by one-dimensional SRGMs.

We also show a numerical example of an operational software reliability based on the estimated two-dimensional mean value function in Fig. 2. The operational software reliability [Ishii and Dohi (2006)] is defined as the probability that a software failure does not occur in the operational time-interval $(s_e, s_e + \eta](s_e \geq 0, \eta \geq 0)$ given that the amount of the testing-effort

(✕ : Software Failure-Occurrence)

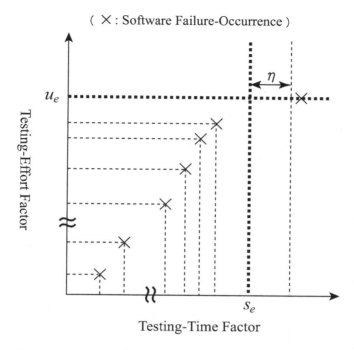

Fig. 3 Illustration for concept of operational software reliability.

expenditure has been taking up to u_e by testing-time s_e. Figure 3 shows an illustration for the concept of the operational software reliability. Therefore, from the properties of the two-dimensional NHPP, the operational software reliability is formulated as

$$R(\eta \mid s_e, u_e) = \exp\left[-\left\{H(s_e + \eta, u_e \mid \widehat{\Theta}) - H(s_e, u_e \mid \widehat{\Theta})\right\}\right], \qquad (24)$$

where $\widehat{\Theta}$ indicates a set of parameter estimates of a two-dimensional SRGM. Figure 4 shows the estimated operational software reliability function, $\widehat{R}(\eta \mid 24, 0.9095)$. From Fig. 4, we can estimate the operational software reliability, $\widehat{R}(1.0 \mid 24, 0.9095)$, to be about 0.127.

7   Concluding Remarks

From the view point of actual software failure-occurrence phenomenon, it is natural to consider that a software reliability growth process depends

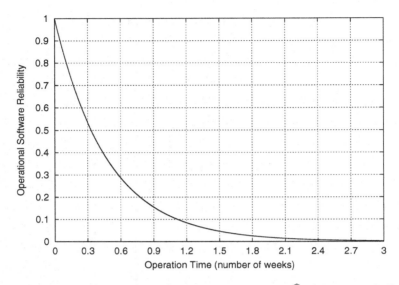

Fig. 4   Estimated operational software reliability function, $\widehat{R}(1.0 \mid 24, 0.9095)$ (DS1).

not only on the testing-time factor but also the testing-effort factor, such as test-execution time, testing-skill of test engineers, testing-coverage. We discussed two-types of two-dimensional software reliability growth modeling frameworks, which describe a reliability growth process depending on two-types of software reliability growth factors: testing-time and testing-effort factors. And we developed two-dimensional SRGMs in our two-types of the two-dimensional modeling frameworks. After that, we discussed parameter estimation methods of our each modeling framework, and compared performance of our SRGMs with existing one-dimensional SRGMs by using fault count data observed in actual testing phases. By our model comparison based on MSE and showing numerical examples for software reliability analysis based on our two-dimensional SRGM by using actual data, we can say that our two-dimensional software reliability measurement technologies enable us to conduct more feasible software reliability analysis than the existing software reliability measurement approach, in which it is assumed that the software reliability growth process depends only on testing-time.

In our future studies, we need to investigate usefulness and validity of our two-dimensional software reliability growth modeling frameworks by applying them to many actual software development projects and comparing with existing one- and two-dimensional SRGMs. And regarding our two-dimensional software reliability growth modeling framework in Eq. (7), we have to research the appropriate range of program size because the stochastic property on the initial fault content, a binomial distribution, can be regarded as a Poisson distribution as the parameter K and λ, respectively, tend to the infinite and zero. Further we also need to derive more useful software reliability assessment measures based on two-dimensional modeling framework.

References

Ahn, C. W., Chae, K. C. and Clark, G. M. (1998). Estimating parameters of the power law process with two measures of failure time, *J. Qual. Tech.* **30**, 2, pp. 127–132.

Fujiwara, T. and Yamada, S. (2001). Software reliability growth modeling based on testing-skill characteristics: Model and Application, *Elec. Commu. Japan — Part 3* **84**, 6, pp. 42–49.

Fujiwara, T. and Yamada, S. (2002). C0 coverage-measure and testing-domain metrics based on a software reliability growth model, *Intern. J. Rel. Quali. Safe. Eng.* **9**, 4, pp. 329–340.

Goel, A. L. and Okumoto, K. (1979). Time-dependent error-detection rate model for software reliability and other performance measures, *IEEE Trans. Reliab.* **R-28**, 3, pp. 206–211.

Gumbel, E. J. (1960). Bivariate exponential distributions, *J. ASA* **55**, pp. 698–707.

Inoue, S. and Yamada, S. (2004). Testing-coverage dependent software reliability growth modeling, *Intern. J. Reliab. Qual. Safe. Eng.* **11**, 4, pp. 303–312.

Inoue, S. and Yamada, S. (2007). Discrete program-size dependent software reliability assessment: Modeling, estimation, and goodness-of-fit comparisons, *IEICE Trans. Fundamentals,* **E90-A**, 12, pp. 2891–1609.

Ishii, T. and Dohi, T. (2006). Two-dimensional software reliability models and their application, *Proc. 12th Pacific Rim Intern. Symp. Depend. Comput.,* pp. 3-10.

Ishii, T., Fujiwara, T. and Dohi, T. (2006). Bivariate extension of software reliability modeling with number of test cases, *Intern. J. Reliab. Qual. Safe. Eng.* **15**, 1, pp. 1–17.

Keiller, P. A. and Miller, D. R. (1991). On the use and the performance of software reliability growth models, *Rel. Eng. Syst. Safe.* **32**, 1-2, pp. 95–117.

Kimura, M., Yamada, S., Tanaka, H. and Osaki, S. (1993). Software reliability measurement with prior-information on initial fault content, *Trans. IPS Japan* **34**, 7, pp. 1601–1609.

Murthy, D. N. P., Baik, J., Wilson, R. J. and Bulmer, M. R. (2006). *Two-dimensional failure modeling* in *Springer Handbook of Engineering Statistics* (Pham, H. Ed.), Springer-Verlag, Berlin, pp. 97–111.

Murthy, D. N. P., Iskandar, B. P. and Wilson, R. J. (1995). Two-dimensional failure-free warranty policies: Two-dimensional point process models, *Opera. Res.*, **43**, 2, pp. 356–366.

Musa, J. D., Iannio, D. and Okumoto, K. (1987). *Software ReliabilityFMeasurement, Prediction, Application*, McGraw-Hill, New York.

Nakagawa, T. (2005). Maintenance Theory of Reliability, Springer-Verlag, London.

Pham, H. (2000). *Software Reliability*. Springer-Verlag, Singapore.

Varian, H. R. (1991). *Intermediate Microeconomics —A Modern Approach*, (2nd Edition), W. W. Norton & Company, New York.

Yamada, S., Ohba, M. and Osaki, S. (1983). S-shaped reliability growth modeling for software error detection, *IEEE Trans. Reliab.* **R-32**, 5, pp. 475–478, 485.

Yamada, S., Ohtera, S. and Narihisa, H. (1986). Software reliability growth models with testing-effort, *IEEE Trans. Reliab.* **R-35**, 1, pp. 19–23.

Yamada, S. and Osaki, S. (1985). Software reliability growth modelingFModels and applications, *IEEE Trans. Soft. Eng.* **SE-11**, 12, pp. 1431–1437.

Chapter 11

Hybrid Coordinated Checkpointing Technique Using Incremental Snapshots

Mamoru Ohara, Masayuki Arai, Satoshi Fukumoto and Kazuhiko Iwasaki

Faculty of System Design, Tokyo Metoropolitan University,
6-6 Asahigaoka, Hino, Tokyo 191-0065, Japan

1 Introduction

In distributed systems, checkpointing/recovery must maintain consistent system states. In coordinated checkpointing, processes coordinate with each other in order to save a consistent global state. Because global states saved by coordinated checkpointing are always consistent, recovery operation is simple. However, overhead for coordination is sometimes unacceptably large when communication cost is high. In contrast, uncoordinated checkpointing techniques attempt to reduce checkpointing overhead by independently creating checkpoints in each process. As a result, recovery cost of uncoordinated techniques is usually higher than coordinated ones. We can see a trade-off relation between checkpointing overhead in normal operation and recovery cost [Tohma (1988)].

Elnozahy *et al.* concluded that recent great reduction of communication cost makes coordinated checkpointing most attractive [Elnozahy *et al.* (2002)]. Large recovery cost of uncoordinated checkpointing may be an obstacle in today's distributed systems. However, when processes create checkpoints with coordinated techniques, they must suspend message exchange and wait for all processes. Therefore, when there are rigorous requirements for rapid recovery after a failure, that is, a system needs to create checkpoints frequently in order to decrease recovery cost, the system performance may be significantly decreased. It is not acceptable in systems consisting of numerous processes such as grid computing systems [Camargo *et al.* (2004); Krishnan and Gannon (2004); Woo *et al.* (2004)].

In this chapter, we present a hybrid C/R technique, which reduces frequency of coordinated checkpointing and still realizes relatively rapid recovery. In the proposed technique, the system performs coordinated checkpointing only at a portion of periodic checkpointing opportunities. At the other checkpointing opportunities, processes independently save their state differentials and message histories, chaged or received after the last coordinated checkpoint. In this chapter, we call the saved information consisting of the differentials and the message histories as an incremental snapshot. By merging checkpoint data saved by coordinated checkpointing and the incremental snapshots, we can obtain the same image as that obtained by uncoordinated checkpointing. That is, the proposed technique is a hybrid technique of coordinated and uncoordinated C/R for reducing communication cost. Moreover, the proposed technique can also reduce the overhead for uncoordinated checkpointing by substituting the incremental snapshots.

In the proposed technique, the number of recovery lines is obviously decreased compared to that in traditional coordinated techniques. However, we can probabilistically obtain additional recovery lines which consist of coordinated checkpoints and the incremental snapshots. Thus, we may be able to create a more effective fault-tolerance mechanism than the traditional techniques in the gross. That is, we have possibilities of improving recovery performance without additional checkpointing overhead, or inversely, fulfilling restrictions in recovery cost with less state saving overhead. Moreover, processes can discard all existing coordinated checkpoints and the incremental snapshots when they create a new coordinated checkpoint, it is easy to estimate storage overhead priori. This is an advantage of the proposed technique against the communication-induced checkpointing [Elnozahy and Zwaenepoel (1992)].

In this chapter, we construct a simple analytical model for validating effectiveness of the proposed technique. We construct models estimating state saving overhead and recovery cost, respectively. In evaluation of the recovery cost, we speculated the probability distribution, which expresses magnitude of rollback, using simulations. We examine two techniques to generate the incremental snapshots: the divided snapshotting technique, in which a process saves differences in its state between two successive snapshots/checkpoints, and the cumulative snapshotting technique, where a process always saves the differences between the last coordinated checkpoint and the current state. Through the analyses and numerical examples, we found that with low message frequency, the proposed technique is more effective than the traditional coordinated technique. Comparing the two snapshot techniques, the cumulative one was more advantageous in many situations.

Note that the analytical models presented in this chapter evaluate state saving overhead and recovery cost independently. We describe the analytical models with setting checkpoint interval as a major parameter. In many traditional models, overheads in normal and recovery operations are often evaluated in an integrated manner by defining the failure frequency [Chandy *et al.* (1975); Nicola and Spanje (1990); Ozaki *et al.* (2004); Soliman and Elmaghraby (1998)]. However, such models had difficulties in that the failure frequency, which is very difficult to speculate in running systems, has a significant impact on the evaluation.

This chapter is organized as follows. In section 2, we propose a combinational use of coordinated and uncoordinated checkpointing with using the incremental snapshots. We construct analytical models evaluating state saving overhead and recovery cost of the proposed technique in section 3 and present numerical examples and discussions in section 4. Finally, section 5 conclude the chapter.

2   Proposed Technique: Hybrid Coordinated Checkpointing Using Incremental Snapshots

In this section, we propose a checkpointing technique which combines coordinated and uncoordinated checkpointing by substituting the incremental snapshots from most of uncoordinated checkpointing. We use the distributed snapshot algorithm proposed by Chandy and Lamport [Chandy and Lamport (1985)] for coordinated checkpointing. Since each process independently generates the incremental snapshots, we use the online algorithm of uncoordinated checkpointing introduced by Jefferson [Jefferson (1985)] for recovery.

Figure 1 illustrates the concept of the proposed checkpointing technique. Each process has a local clock maintaining its own local time. The local clock is reset to 0 at the system initiation and every completion of the coordinated checkpointing algorithm; it is incremented by one for every event. Events are generated internally or are delivered by messages from other processes. Unlike with the global clock indicating real time shared in the whole distributed system, the progress rates of the local clocks differ

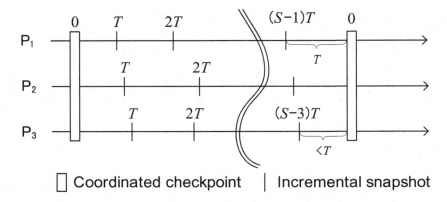

□ Coordinated checkpoint | Incremental snapshot

Fig. 1   Conceptual diagram of the proposed checkpointing technique

between processes. For every application message sent by a process, the process saves its destination, the message identifier, and the local time at which it is sent into the stable storage in order to enable the process to generate the corresponding anti-message in recovery.

Every time the local clock goes a constant number of ticks (T ticks), the process get a chance to save its state and generate an incremental snapshot. The process saves changes in its state and the history of messages received after the last coordinated checkpoint or the last snapshot. The local clock is not advanced by the snapshot creation.

Repeating the snapshotting cycle, a process reaches the Sth chance for snapshotting, and then, it invokes the coordinated checkpointing algorithm as an initiator and saves its whole state as a checkpoint. There may be two or more initiators in coordinated checkpointing. In addition, when another process reaches the Sth chance during the initiators are saving their states, it also becomes an initiator process. After completion of saving the internal state in an initiator, it sends a control message to all of the other processes. When a process receives the control message before it create S snapshots, it gives up the remaining snapshotting chances and participates the coordinated checkpointing: it also starts saving of its whole internal state. Therefore, each process initiates coordinated checkpointing when its local clock reaches $T \cdot S$ or when it receives the control message.

After completion of the coordinated checkpointing, all the old checkpoints, the incremental snapshots, and message histories are discarded. In addition, the local clocks of all processes are reset to 0.

Hereafter, we denote T as local checkpoint interval and S as prescribed checkpoint number. We also denote product of T and S, $W \equiv S \cdot T$ as coordinated checkpoint interval. Note that the proposed technique with $S = 1$, that is $W = T$, is the same as the traditional coordinated checkpointing technique.

In this chapter, we examine two incremental snapshot techniques: the cumulative snapshotting technique and the divided snapshotting technique. Usually, the cumulative technique has higher overhead for generating an incremental snapshot. However, amount of the overhead depends on data update pattern of applications. For example, in applications where data items are added or updated uniformly in whole data space, the overhead of the divided technique can be almost constant while that of the cumulative one linearly grows with time. On the other hand, in applications having hot spots, overheads of the two techniques are sometimes comparable because data items in the hot spots are intensively updated.

Recovery upon a process failure can be realized by using the incremental snapshots in the similar manner as one illustrated in [Jefferson (1985)]. Note that the incremental snapshots hold only differences in process states, changed after the last checkpoint or snapshot. Therefore, processes have to merge the checkpoint data with the snapshot data in order to roll back to the states saved in the snapshots. In the cumulative technique, a process only has to directly merge the last checkpoint data with the snapshot, to which it would return. In contrast, in the divided technique, the process must merge all snapshots generated after the last checkpoint repeatedly. Thus, recovery cost of the divided one is usually higher than that of the cumulative one.

Although the domino effect can occur in recovery of the proposed technique, a recovery line is always found at a coordinated checkpoint. Thus, we can explicitly estimate upper bound of recovery cost.

3 Evaluation Model

In this section, we construct stochastic models for validating effectiveness of the proposed technique. In the models, coordinated checkpoint interval W is used as one of major design parameters. First, we describe a model estimating the expected overhead for state saving in time, $H(W)$, added to every ordinary event execution. Next, we estimate recovery cost $R(W)$ by deriving the expected time taken for rollback and rerunning after a failure.

3.1 *State Saving Overhead*

We denote time interval between two successive coordinated checkpointing excluding time for generating snapshots as X. We first derive the expected value $E[X]$ for X.

Suppose that there are N processes, $P_1, P_2, \cdots, P_N$ in a system. We describe real time taken for the local clocks reaches W in the processes, excluding time for generating snapshots, by $X_1, X_2, \cdots, X_N$, respectively. These are discrete random variables, which are independent from each other. Definitely from the relation between logical time and real time, $W \leq X_k$ ($1 \leq k \leq N$). If we assume events occur at a constant rate p in a unit time, the random variables $X_1, X_2, \cdots, X_N$ obey negative binomial distributions (also called as Pascal distribution). The probability function is

$$f(x) = Pr\{X_k = x\}$$
$$= \binom{x-1}{W-1} p^W (1-p)^{x-W}. \tag{1}$$

Note that $x = W, W+1, W+2, \cdots$. Also, the cumulative distribution function is

$$F(x) = \sum_{x'=W}^{x} f(x'). \tag{2}$$

Because the coordinated checkpointing is initiated by a process which first reaches the Sth snapshot, $X = min\{X_1, X_2, \cdots X_N\}$. Defining $Pr\{X \leq W-1\} \equiv 0$ and $F(W-1) \equiv 0$, the expected value of X is obtained by

$$E[X] = \sum_{x=W}^{\infty} x \cdot Pr\{X = x\}$$

$$= \sum_{x=W}^{\infty} x \cdot [Pr\{X \leq x\} - Pr\{X \leq x-1\}]$$

$$= \sum_{x=W}^{\infty} x \cdot \left(\left[1 - \{1 - F(x)\}^N\right] - \left[1 - \{1 - F(x-1)\}^N\right] \right)$$

$$= W + \sum_{x=W}^{\infty} \{1 - F(x)\}^N. \tag{3}$$

Next, we estimate overhead for the incremental snapshotting. The overhead can be estimated by accumulating time taken for generating

$S - 1(= W/T - 1)$ times snapshotting in the initiator process. We can assume dispersion of time for generating an incremental snapshot is negligible against the coordinated checkpointing interval. That is, time for generating the sth incremental snapshot $C_{snap}(s)$ $(1 \leq s \leq S - 1)$ can be deterministically obtained. Concretely speaking, we express time for generating an incremental snapshot by the cumulative technique as

$$C_{snap}(s) = \gamma + \delta(s - 1). \tag{4}$$

γ denotes time for generating the first snapshot and δ is incremental time for saving added/updated data items in the second or later snapshotting.

On the other hand, time for the divided snapshotting can be expresses as a special case where δ is 0 in Eq. (4), i.e., $C_{snap}(s)$ is always γ.

We next derive overhead for a process to perform coordinated checkpointing. Today, message overhead is very small and it can be negligible compared to time for saving whole process states. Moreover, since time for generating a coordinated checkpoint in each process is much shorter than the coordinated checkpoint interval, we can assume the coordinated checkpointing overhead is constant. Therefore, we suppose that time for coordinated checkpointing can be estimated as time interval from start of state saving in the first initiator process till completion of state saving in the process which is lastly involved in the checkpointing.

We define $X_{min} = min\{X_1, X_2, \cdots, X_N\}$ and $X_{max} = max\{X_1, X_2, \cdots, X_N\}$ for the cumulative real time $X_1, X_2, \cdots, X_N$. We also define $Y = X_{max} - X_{min}$. Y means the time interval from the local clock in the first process reaches W to the clock in the last process does. We also denote ascending ordered random variables for the cumulative time in N processes by $X_{(1)}, X_{(2)}, \cdots, X_{(N)}$. That is, $X_{(1)} \leq X_{(2)} \leq \cdots \leq X_{(N)}$.

As mentioned earlier, we suppose the synchronized algorithm introduced by Chandy and Lamport [Chandy and Lamport (1985)] for coordinated checkpointing. In this algorithm, an initiator sends control messages after it saves checkpoint data, and the control messages invoke the algorithm in other processes. Therefore, if time needed to save checkpoint data in each process is C_0, the algorithm is invoked in all processes within C_0 from it is invoked in the first initiator. Thus, we can obtain overhead for total coordinated checkpointing by

$$v(Y) = \begin{cases} Y + C_0 \ (Y < C_0) \\ 2C_0 \ \ \ \ (Y \geq C_0). \end{cases} \tag{5}$$

With definitions above, the probability distribution function for Y can be derived as

$$
\begin{aligned}
G(y) &= Pr\{Y \le y\} \\
&= Pr\{X_{max} - X_{min} \le y\} \\
&= Pr\{X_{max} \le X_{min} + y\} \\
&= \sum_{x=W}^{\infty} Pr\{X_{min} = x \cap X_{max} \le x + y\} \\
&= \sum_{x=W}^{\infty} \sum_{k=1}^{N} Pr\{X_{(1)} = X_{(2)} = \cdots = X_{(k)} = x \cap \\
&\qquad\qquad x+1 \le X_{(k+1)}, X_{(k+2)}, \cdots, X_{(N)} \le x+y\} \\
&= \sum_{x=W}^{\infty} \sum_{k=1}^{N} \binom{N}{k} f^k(x) \cdot \{F(x+y) - F(x)\}^{N-k}.
\end{aligned}
\tag{6}
$$

Thus, expected overhead for coordinated checkpointing is

$$
E[v(Y)] = \sum_{y=0}^{C_0 - 1} \{1 - G(y)\} + C_0.
\tag{7}
$$

Combining overheads derived above, we obtain state saving overhead per unit time as

$$
H(W) = \frac{\sum_{s=1}^{S-1} C_{snap}(s) + E[v(Y)]}{E[X]}.
\tag{8}
$$

3.2 *Recovery Cost*

When a process rolls back by a failure, other processes may also be required to roll back because of anti-messages. Processes can resume processing of lost tasks after the cascading rollbacks reach a recovery line. Note that since events non-deterministically occur in processes [Elnozahy *et al.* (2002)], a process which reruns up to the point at which the failure is detected does not necessarily resume the same state the process had before the failure. In this study, we estimate recovery cost by time interval from a process detects a failure until it completes rerunning. We discuss the expected recovery cost $R(W)$ along the recovery procedure.

Each process holds the history of messages sent after the last coordinated checkpoint in addition to checkpoint/snapshot data in the stable storage. This enables a process to generate necessary anti-messages before rolling back. Number of anti-messages generated is at most $(N - 1)$

for N processes in the system [Matsumoto and Taki (1996); Elnozahy *et al.* (2002)]. Supposing each entry in the message history is very small and random-accessible, we can regard time needed to generate anti-messages negligible compared to that for reading checkpoint data from the stable storage. Similarly, time for sending anti-messages and message delay are also small, therefore we assume that recovery operations are invoked at the same time in all processes involved.

Time interval between two successive coordinated checkpoints can be divided into S local intervals by $(S-1)$ incremental snapshots. The local intervals represent for time points of failures. If a process fails in the sth local interval $(0 \le s \le S-1)$, it holds s incremental snapshots at the beginning of recovery. We denote time cost for the process reconstructing its state saved at the latest s^{th} incremental snapshot by $C_{load}(s)$. Regarding the last coordinated checkpoint as the 0th incremental snapshot,

$$C_{load}(s) = C_0 + C_{snap}(s) \tag{9}$$

for the cumulative snapshotting and

$$C_{load}(s) = C_0 + \sum_{i=1}^{s} C_{snap}(i) \tag{10}$$

for the divided snapshotting where we define $C_{snap}(0) \equiv 0$ and $\sum_{i=1}^{0} \cdot \equiv 0$. We also assume that time for reading checkpoint/snapshot data equals to time for generating them.

Time cost needed for the cascading rollback reaches a recovery line and time for rerunning the lost tasks depend on the process which has the greatest rollback scale in the system; we denote the maximum number of rollbacks in a process in a recovery by Z. We assume each process rolls back one by one for each incremental snapshot it holds. Time interval from the process with the maximum rollback scale Z returns to the state of the latest incremental snapshot till it reaches a recovery line can be estimated as sum of costs for reading the remaining $Z-1$ snapshots, $C_{snap}(s-1), C_{snap}(s-2), \cdots, C_{snap}(s-Z+1)$.

If we assume that failures occur at the middle point of the local intervals in average, logical time taken for the failed process to rerun the lost tasks can be estimated as $ZT - (1/2)T$. Thus, the expected real time is $(Z-1/2)T/p$. In addition, a process has time cost of $C_{snap}(s-Z+2), C_{snap}(s-Z+3), \cdots, C_{snap}(s-1), C_{snap}(s)$ in the rerunning in order to regenerate $(Z-1)$ snapshots. Note that the regeneration cost is 0 when $Z=1$.

Therefore, when the process with the maximum Z-time rollback has s incremental snapshots upon a failure, total recovery cost $r_s(Z)$ is obtained by

$$r_s(Z) = C_{load}(s) + \sum_{i=1}^{Z-1} C_{snap}(s-i) + \frac{(Z-1/2)T}{p} + \sum_{i=1}^{Z-1} C_{snap}(s-i+1),$$

$$(11)$$

where random variable Z obeys a probability distribution $b_s(z)$, that is, $b_s(z) = Pr\{Z = z\}$. If we assume process failures can occur uniformly in the S local intervals, the expected recovery cost $R(W)$ is given by

$$R(W) = \sum_{s=0}^{S-1} \frac{1}{S} \sum_{z=1}^{s+1} b_s(z) \cdot r_s(z).$$

$$(12)$$

Note that $S = W/T$.

The probability $b_s(z)$ depends on the local and coordinated checkpoint intervals, number of processes, message frequency, communication patterns, and so on. It is quite difficult to analytically derive $b_s(z)$ except in special case where there are a few processes. In this study, we evaluate the distribution through Monte Carlo simulations. In the simulations, we randomly generate events of message exchange and internal processing and advance the local clocks. Message events are generated at probability q.

4　Numerical Examples and Discussions

In this section we present numerical examples suggesting trade-off relations between state saving overhead $H(W)$ and recovery cost $R(W)$ based on the proposed analytical models. Through the numerical examples, we discuss effectiveness of the proposed analytical models.

Figure 2 shows the state saving overhead for the cumulative snapshotting, the divided snapshotting, and a traditional coordinated technique (the algorithm proposed by Chandy and Lamport). The time for generating a checkpoint in coordinated checkpointing, C_0, was set to 25, and the parameter for the incremental snapshotting γ was 5, respectively. In the cumulative snapshotting, we set $\delta = 2$. We fixed $T = 100$ and changed the prescribed checkpoint number S so that $W = 100S$ in the proposed technique. Because S is always 1 in the traditional technique, T definitely equals to W. All examples presented in the section use the same parameters.

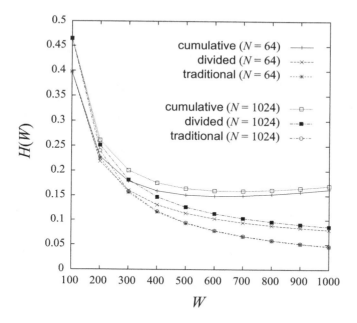

Fig. 2 State saving overhead $H(W)$ of the proposed and a traditional coordinated checkpointing techniques ($p = 0.9, C_0 = 25, \gamma = 5,$ and $\delta = 2$)

In the traditional technique, $H(W)$ monotonically decreases with W because overhead for coordinated checkpointing imposed per unit time is lower with larger W. In the proposed techniques, the incremental snapshot is combined with checkpointing. In the divided snapshotting, $H(W)$ also monotonically decreases because $C_{snap}(s)$ is always γ and overhead for generating snapshots is also constant. On the other hand, in the cumulative technique, overhead for generating snapshots monotonically increases because $C_{snap}(s)$ increases with s ($1 \leq s \leq W/T - 1$). This leads that $H(W)$ has the local minimum value in the cumulative technique.

In Fig. 2, the number of processes N is 64 or 1024. The larger N is, the greater expected time for coordinated checkpointing $E[v(Y)]$. Thus, $H(W)$ is greater with larger N. However, as we can see in Eq. (5), $E[v(Y)]$ is limited so that it is at most $2C_0$, therefore, $H(W)$ with $N = 1024$ is at most 1.2 times that with $N = 64$.

Next, we discuss the probability distribution of the maximum rollback distance Z obtained through simulations. When message probability is relatively low, i.e., $q = 0.01$, $b_s(1)$ is almost 1 for any combinations of

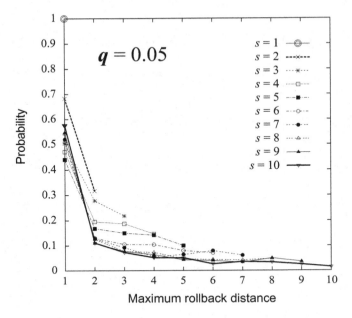

Fig. 3   Probability distribution of the maximum rollback distance with $N = 64$ and $q = 0.05$ ($S = 10, T = 100, p = 0.9, C_0 = 25, \gamma = 5$, and $\delta = 2$)

$N, S, T, C_0, \gamma, \delta$ and time location of failure points. This means a recovery line consisting of the current state or the latest snapshot of each process can be obtained in almost all cases.

With higher message probability $q = 0.05$, we can obtain the probability distribution as shown in Fig. 3. The figure shows numerical examples with the same parameters to Fig. 2, i.e., $N = 64, C_0 = 25$, and $C_{snap}(s) = 5 + 2s$. We fixed the coordinated checkpoint interval $S = 10, T = 100$, i.e., $W = 1000$. Each line denotes a rollback distance distribution when a failure occurs in a process which holds s snapshots. The probability $b_s(1)$ is the largest regardless of the failure points. In many cases, several times of rollback occurred due to the domino effect, however, probability of that a process rolls back up to the coordinated checkpoint, $b_s(s)$, is relatively low. Therefore, we can expect better recovery performance for the proposed technique when costs for merging the incremental snapshots are smaller than that for reading checkpoint data.

Figure 4 shows a probabilistic distributions for much higher message probability $q = 0.1$. Values of parameters other than q are the same as

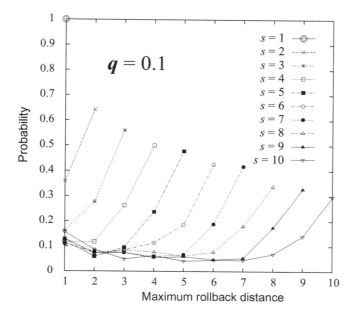

Fig. 4 Probability distribution of the maximum rollback distance with $N = 64$ and $q = 0.1$ ($S = 10, T = 100, p = 0.9, C_0 = 25, \gamma = 5$, and $\delta = 2$)

those in Fig. 3. In this example, $b_s(s)$, which is the probability where a process rolls back up to the coordinated checkpoint, is the highest. In such situations, the incremental snapshots cannot well contribute to reduction of recovery cost, they rather may increase the recovery cost of the proposed technique compared to the traditional technique.

We performed similar simulations with $N = 1024$ and found that growth of number of processes averagely increased the maximum rollback distance Z. We can read from these results that the larger message probability p is and the more processes a system has, the larger the maximum rollback distance Z is.

We calculated recovery cost $R(W)$ from Eq. (12) and $b_s(s)$ obtained by the simulation described above. Figure 5 shows numerical examples of $R(W)$ for message probability $q = 0.05$. Other parameters were set as $p = 0.9, C_0 = 25$, and $\gamma = 5$, respectively. The increment in cost for snapshotting, δ, was 2 for the cumulative technique and always 0 in the divided technique. We fix the local checkpoint interval $T = 100$, that is, $W = 100S$ in the two proposed snapshotting techniques and $T = W$ in the traditional

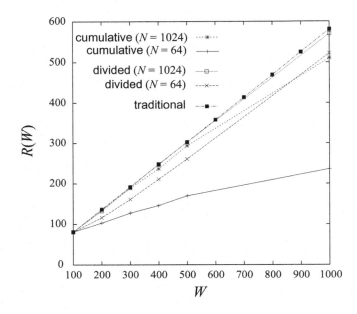

Fig. 5   Recovery cost $R(W)$ of the proposed and a traditional coordinated checkpointing techniques ($p = 0.9, q = 0.05, C_0 = 25, \gamma = 5$, and $\delta = 2$)

technique. $R(W)$ monotonically increased with increase of W. In the divided technique ($\delta = 0$), $R(W)$ increased with W almost linearly; this is the similar trend to that of the traditional technique. This is because in the divided technique we have to repeatedly merge all incremental snapshots from the one next to the coordinated checkpoint up to the one to which a process would like to roll back. On the other hand, recovery cost increased at a slower pace in the cumulative technique; it is always lower than that of the traditional one.

We should decide checkpointing/recovery policy based on requirements both in performance and reliability of the system, considering trade-off between the state saving overhead and the recovery cost. We now discuss validity of the proposed technique, gathering discussions above.

Figures 6–8 show trade-off relations between state saving overhead $H(W)$ and recovery cost $R(W)$ with setting message probability $q = 0.01, 0.05$ or 0.1. In all figures, number of processes $N = 64$. Similarly to Figs. 2 and 5, we set the parameters as $p = 0.9, C_0 = 25, \gamma = 5$, and $\delta = 2$. The local checkpoint interval T was set to 100 and $W = 100S$ in the proposed techniques; $T = W$ in the traditional coordinated technique.

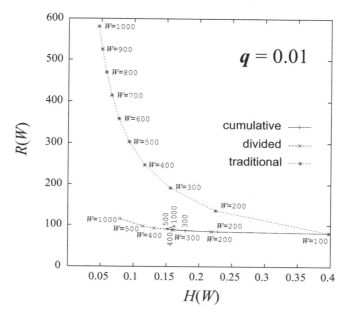

Fig. 6   State saving overhead $H(W)$ and recovery cost $R(W)$ with $N = 64$ and $q = 0.01$ $(p = 0.9, C_0 = 25, \gamma = 5, \text{ and } \delta = 2)$

Figure 6 shows numerical examples for $q = 0.01$. With such a low message probability, rollbacks usually stop at the latest snapshots or checkpoints. Therefore, the recovery cost of the proposed technique was not affected by W because the expected time for rerunning the lost tasks is $T/2p$ regardless of W. Comparing to the traditional technique in which the coordinated checkpoint interval has substantial impact on the recovery cost, the proposed techniques can achieve faster recovery with greatly lower state saving overhead. Uncoordinated checkpointing can also reduce the expected recovery cost when message frequency is low, however, there is no upper bound of the recovery cost because the worst-case rollback distance reaches infinity [Agbaria *et al.* (2001)]. The proposed techniques are better for systems having restrictions on the recovery cost. Especially, the divided technique can reduce the state saving overhead without increasing the recovery cost. Therefore it is suitable for applications where messages are infrequently exchanged.

Figure 7 present numerical examples for message probability $q = 0.05$. The proposed techniques have lower recovery cost than that of the tradi-

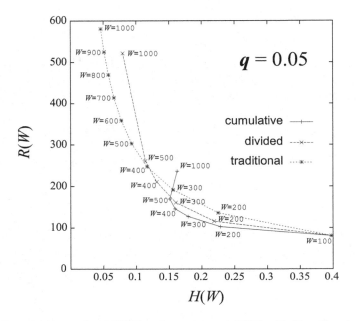

Fig. 7   State saving overhead $H(W)$ and recovery cost $R(W)$ with $N = 64$ and $q = 0.05$
($p = 0.9, C_0 = 25, \gamma = 5$, and $\delta = 2$)

tional technique when we can have more than 0.15 for state saving overhead
$H(W)$ in the cumulative technique and 0.1 in the divided technique, respectively. Moreover, under situations with a hard restriction in the recovery
cost, for example $R(W) \leq 200$, the proposed techniques can fulfill the restriction with lower $H(W)$. For applications with stronger restriction in the
recovery cost, the cumulative technique is very effective. In the cumulative
technique, since there is coordinated checkpoint interval W^* corresponding
to the local minimum value of state saving overhead ($W^* = 500$ in Fig. 7),
letting W more than W^* is useless in practice. The recovery cost is also
bound at most $R(W^*)$. However, with strong restrictions in state saving
overhead, we have to balance $H(W)$ and $R(W)$ by using the traditional
technique.

In Fig. 8, we set q to 0.1. With such a high message probability, rollbacks
to the coordinated checkpoint frequently occur due to the domino effect,
therefore, we cannot expect for the proposed technique to improve the
recovery performance. In such situations, cost for merging the incremental
snapshots may rather have a negative impact on recovery cost. Thus, the

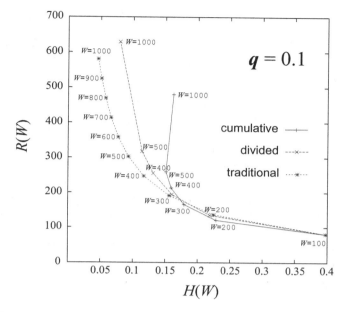

Fig. 8   State saving overhead $H(W)$ and recovery cost $R(W)$ with $N = 64$ and $q = 0.1$
($p = 0.9, C_0 = 25, \gamma = 5,$ and $\delta = 2$)

recovery cost of the proposed techniques is larger than that of the traditional one in many cases. We can obtain lower recovery cost than that of the traditional one when we set $H(W)$ more than 0.17, however, advantage of the proposed technique is relatively small.

We also evaluate the proposed technique for $N = 1024$. Similarly, the proposed technique can achieve better trade-off with small message probability q and they weak their advantages with growth of q. Moreover, the expected rollback distance is larger for $N = 1024$ as mentioned above, therefore, the proposed technique loses their edge with smaller q compared to numerical examples for $N = 64$.

5   Concluding Remarks

In this chapter, we proposed a coordinated C/R technique using the incremental snapshot and evaluated its state saving overhead and recovery cost by constructing simple stochastic models. In evaluation of the recovery cost, we specified the probabilistic distribution of the maximum rollback distance through simulations.

We compared the state saving overhead $H(W)$ and the recovery cost $R(W)$ of the proposed and the traditional coordinated C/R techniques. Using the proposed technique, we can achieve lower $R(W)$ with comparable low $H(W)$ to the traditional one. Even in environments with relatively high message frequency, the proposed techniques can achieve faster recovery if we can allow some growth in the state saving overhead.

There are different applications suitable for the two proposed snapshotting techniques, the divided technique and the cumulative technique. The divided technique, which has no lower bound in $H(W)$, is better for environments where the message frequency is low enough and the domino effects seldom occur. The cumulative technique is suitable for applications in which messages are exchanged more frequently.

The proposed technique utilizes Chandy and Lamport's checkpointing algorithm where all processes in a system synchronously save their state. It is possible that using more efficient coordinated algorithms such as one proposed by Koo and Toueg [Koo and Toueg (1987)] improves performance and reliability of distributed systems. We will tackle developing such more efficient techniques.

References

Agbaria, A., Attiya, H., Friedman, R. and Vitenberg, R. (2001). *Quantifying Rollback Propagation in Distributed Checkpointing, Proc. 20th IEEE Sympo. Reliable Distributed Systems (SRDS'01)*, pp. 36–45.

Camargo, R., Goldchleger, A., Kon, F. and Goldman, A. (2004). *Checkpointing-Based Rollback Recovery for Parallel Applications on the InteGrade Grid Middleware, Proc. 2nd Workshop on Middleware for Grid Computing*, pp. 35–40.

Chandy, K. M., Brown, J. C., Dissly, C. W. and Uhrig, W. R. (1975). *Analytic Models for Rollback and Recovery Strategies in Data Base Systems, IEEE Trans. Softw. Eng.*, Vol. SE-1, pp. 100–110.

Chandy, K. M., Lamport, L. (1985). *Distributed Snapshots: Determining Global States of Distributed Systems, ACM Trans. Comput. Syst.*, Vol. 3, No. 1, pp. 63–75.

Elnozahy, E. N. and Zwaenepoel, W. (1992). *Manetho: Transparent Rollback-Recovery wit Low Overhead, Limited Rollback and Fast Output Commit, IEEE Trans. Comput.*, Vol. 41, No. 5, pp. 526–531.

Elnozahy, E. N., Alvisi, L., Wang, Y.-M. and Johntson, D. B. (2002). *A Survey of Rollback-Recovery Protocols in Message-Passing Systems, ACM Comput. Surv.*, Vol. 34, No. 3, pp. 375–408.

Koo, R. and Toueg, S. (1987). *Checkpointing and Rollback-Recovery for Distributed Systems, IEEE Trans. Softw. Eng.*, Vol. SE-13, No. 1, pp. 23–31.

Jefferson, D. (1985). *Virtual Time, ACM Trans. Prog. Lang. Syst.*, Vol. 7, No. 3, pp. 404–425.

Krishnan, S. and Gannon, D. (2004). *Checkpoint and Restart for Distributed Componetns in XCAT3, Proc. 5th IEEE/ACM Int'l Workshop on Grid Computing (GRID'04)*, pp. 281–288.

Matsumoto, Y. and Taki, K. (1996). *Efficient Technique for Realizing Time Warp Mechanism for Parallel Logic Simulation, Trans. IPSJ*, Vol. 37, No. 4, pp. 654–665 (in Japanese).

Nicola, V. F. and Van Spanje, J. M. (1990). *Comparative Analysis of Different Models of Checkpointing and Recovery, IEEE Trans. Softw. Eng.*, Vol. SE-16, No. 8, pp. 807–821.

Ozaki, T., Dohi, T., Okamura, H. and Kaio, N. (2004). *Min-Max Checkpoint Placement under Incomplete Failure Information, Proc. IEEE Int'l Conf. Dependable Systems and Networks (DSN2004)*, pp. 721–730.

Soliman, H. M. and Elmaghraby, A. S. (1998). *An Analytical Model for Hybrid Checkpointing in Time Warp Distributed Simulation, IEEE Trans. Parallel Distrib. Syst.*, Vol. 9, No. 10, pp. 947–951.

Tohma, Y. and Mukaidono, M. ed. (1988). *Introduction to Reliability Techniques for Computer Systems*, (Japan Standards Association) (in Japanese).

Woo, N., Jung, H., Yeom, H., Park, T. and Park, H. (2004). *MPICH-GF: Transparent Checkpointing and Rollback-Recovery for Grid-Enabled MPI Processes, IEICE Trans. Inf. & Syst.*, Vol. E87-D, No. 7, pp. 1820–1828.

Chapter 12

Predicting Effort and Errors for Embedded Software Development Projects by Using an Artificial Neural Network

Kazunori Iwata[1], Sayori Maeji[2] and Toshio Nakagawa[3]

[1] *Department of Business Administration, Aichi University,*
4-60-6 Hiraike-cho, Nakamura, Nagoya 453-8777, Japan
[2] *Institute for Consumer Science and Human Life,*
Kinjo Gakuin University,
1723 Omori 2, Moriyama, Nagoya 463-8521, Japan
[3] *Department of Business Administration, Aichi Institute of Technology,*
1247 Yachigusa, Yakusa-cho, Toyota 470-0392, Japan

1 Introduction

Recently, growth in the information industry has caused a wide range of uses for information devices, and the associated need for more complex embedded software, that provides these devices with the latest performance and function enhancements [Hirayama (2004); Nakamoto *et al.* (1997)]. Consequently, it is increasingly important for embedded software-development corporations to ascertain how to develop software efficiently, whilst guaranteeing delivery time and quality, and keeping development low costs [Boehm (1976); Watanabe (2004); Tamaru (2004)]. Hence, companies and divisions involved in the development of such software are focusing on various types of improvement, particularly process improvement. Predicting effort requirements of new projects and guaranteeing quality of software are especially important, because the prediction relates directly to costs, while the quality reflects on the reliability of the corporation [Ogasawara and Kojima

215

(2003); Komiyama (2003); Takagi (2003); Nakashima (2004); N. (2004)]. In the field of embedded software, development techniques, management techniques, tools, testing techniques, reuse techniques, real-time operating systems and so on, have already been studied. However, there is little research on the relationship between the scale of the development and the number of errors, based of data accumulated from past projects. As a result, previously we described the prediction of the total scale using multiple regression analysis [Iwata *et al.* (2006); Nakashima *et al.* (2006)]. In this chapter we therefore, propose a method for creating effort and errors prediction model using an Artificial Neural Network (ANN). However, the ANN has a large margin of errors for some projects. We therefore, propose a method to reduce the margin of errors model. Finally, we also compare the accuracy of the proposed ANN model with that of a multiple regression analysis model using Welch's t-test [Student (1908); Welch (1947)].

The rest of the chapter is organized as follows. In Section 2, we explain software development management and discuss current problems and the objectives which this study is trying to achieve, and illustrate software development process and selection of data to establish the model in Section 3. In Section 4 we expound models to predict effort and errors. In Section 5 describes evaluation experiment. Section 6 concludes.

2 Software Project Management and Issues

The embedded software for financial institutions developed by "OMRON software Co." is based on the basic software customized for an individual customer's need to install it at various sites. To minimize the customization needed during the development of basic software, parameters are embedded to control the system. This engineering technique assures productivity and quality. This type of approach is actively taken during the process of software development. While using this type of technique, the pressures related to delivery time and quality are more and more intense. This requires improving further the quality and cost during software development. Hence, we have already studied costs of the processes by using analysis [Iwata *et al.* (2006); Nakashima *et al.* (2006)]. To cope with this situation, the tools that can manage the progress status or results in the database are used to improve the quality and productivity. However, the more the volume of software development project increases the more the errors increase. By analyzing the database, we determine the relationship between the volume of software development project and the errors.

3 Software Development Processes and Selection of Data

In software development division of "OMRON software Co.", the waterfall model [Boehm (1976)] is used as the basic development-process model. A general description of this model is given in Table 1.

Table 1   Software Development Process.

	Process	Contents of work
1	Conceptual design(CD)	This is so-called "system engineering work". They analyze customer requirements and detail the areas to be addressed as development factors.
2	Design	According to the development factors defined in CD process, designing of software functionality, combining of software modules, and writing of source code are performed.
3	Debugging	Verify the outcome of the design process with the actual machine to see if it is designed according to the design. The same designer in design process is assigned to debug.
4	Test	After finishing debugging, double-check the software to confirm that it satisfies customer's requirements. A different person (not those assigned to design and debug) is assigned to do this.

The software development of each individual piece of equipment is called "a project", and defined as follows.

Definition 12.1. Project
The software development for each piece of equipment that constitutes the system is called "the development project".

Each development project needs cost of human, which is called "effort" and makes some faults, which is called "errors" defined as follows.

Definition 12.2. Effort
Man-days cost for software development projects.

Definition 12.3. Errors
Faults in a machine, especially, in a system or program.

Regarding the data related to project productivity or quality, data for such things as internal effort and size information etc. are recorded as shown in Table 2 and Table 3, for 3 points in time:

Table 2   Classifications and Selection of Data 1.

Items	Data	Selection	Reason
Internal effort	Effort for each planned process and actual performance	Selected	The data are acquired by effort management system. These data are quantitative and accuracy is good.
Project-scale information	Number of lines in new, modified, original and reused software	Selected	These data are quantitative and accuracy is good.
Effort information	Actual effort in each process	Rejected	The data are acquired at the end of the project. Hence, they can not use to predict effort and errors at the beginning of the project.
	Effort for redo in each process	Rejected	The definition of redo is not consistent. The accuracy is poor.

1) At the beginning of the project.
2) During the project.
3) At the end of the project.

Before analyzing data, we examined the data and decided which data should be selected to make the model. Table 2 and Table 3 show the latter data and the results of the selection.

3.1   *Data Sets for Creating Models*

Using the following data, we create models to predict both the planning effort (*Eff*) and errors (*Err*).

Eff: "The amount of effort", which needs be predicted.

Err: "The number of errors" in a project.

V_{new}: "Volume of newly added", which denotes the number of steps in the newly generated functions of the target project.

V_{modify}: "Volume of modification", which denotes the number of steps modifying and adding to existing functions to use the target project.

V_{survey}: "Volume of original project", which denotes the original number of steps in the modified functions, and the number of steps deleted from the functions.

V_{reuse}: "Volume of reuse", which denotes the number of steps in functions of which only an external method of has been confirmed and which

Table 3   Classifications and Selection of Data 2.

Items	Data	Selection	Reason
Products information	Product classification and product models	Selected	It is necessary to make characteristics of the products and development process be reflected in the model.
	Customer name and sub project name	Rejected	The data is qualitative and it is difficult to obtain accurate data.
	Development type(new or modification)	Rejected	Because there are only two types, it is not appropriate as parameter for the model.
	Delivery time	Rejected	Because the delivery time is seldom changed, this is not selected.
Outsourcing	The estimation of outsourcing amount and actual situation	Rejected	Because outsourcing amount includes sales aspects, this is not appropriate for actual project error status.
	The estimation of outsourcing effort and actual situation	Rejected	Because this is estimated by outsourcing amount and includes sales aspects, this is not selected.
Quality information	The number of problems in each process	Selected	It is necessary to find relationship between a project and errors. These data consist of "Total Error", "Error in CD and Design", "Error in Debugging" and "Error in Test".

are applied to the target project design without confirming the internal contents.

4   Effort and Error Prediction Models

4.1 *An Artificial Neural Network Model*

Artificial Neural Networks (ANNs) are essentially simple mathematical models defining function.

$$f : X \to Y \tag{1}$$

where $X = \{x_i | 0 \le x_i \le 1, i \ge 1\}$ and $Y = \{y_i | 0 \le y_i \le 1, i \ge 1\}$.

ANNs are non-linear statistical data modeling tools and that can be used to model complex relationships between inputs and outputs. The

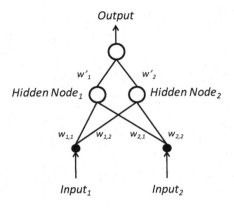

Fig. 1   Basic Artificial Neural Network

basic model is illustrated in Fig. 1, in which the output is calculated as follows.

1) Calculating values for hidden nodes. The value of *Hidden Node$_j$* is calculated using the following equation:

$$Hidden\ Node_j = f\left(\sum_i (w_{i,j} \times Input_i)\right) \qquad (2)$$

where $f(x)$ equals $\frac{1}{1+\exp(-x)}$ and the $w_{i,j}$ is weight calculated by the learning algorithm.

2) Calculating *Output* using *Hidden Node$_j$* as follows:

$$Output = f\left(\sum_k (w'_k \times Hidden\ Node_k)\right) \qquad (3)$$

where $f(x)$ equals $\frac{1}{1+\exp(-x)}$ and the w'_k is weight calculated by the learning algorithm.

We can use an ANN to create effort and error prediction models.

4.1.1   *Normalization of data*

In an ANN, a range of input values or output values is usually less than or equal to 1 and greater than or equal to 0. However, most selected data are grater than 1. Each data range is, therefore, converted to [0, 1] by normalization. The normalized value for t_{kind} is expressed as $f_n(t_{kind})$

(where *kind* denotes *Eff*, *Err*, V_{new}, V_{modify}, V_{survey} and V_{reuse}). The normalized value $f_n(t_{kind})$ is calculated using Eq. (4).

$$f_n(t_{kind}) = \frac{t_{kind} - min(T_{kind})}{max(T_{kind}) - min(T_{kind})} \tag{4}$$

where T_{kind} denotes the set of t_{kind}, and $max(T_{kind})$ and $min(T_{kind})$ denote the maximum and minimum values, respectively, of T_{kind}.

The normalization is flat and smooth, then, a small change in a normalized value influences a small-scale project to a greater degree than a large scale project.

For example, let $min(T_{Eff})$ equal 10, $max(T_{Eff})$ equal 300, t_{Eff1} equal 15, t_{Eff2} equal 250, predicted value for t_{Eff1} be $\widehat{t}_{Eff1}$ and t_{Eff2} be $\widehat{t}_{Eff2}$. If the prediction model has $+0.01$ error, then $f_{n_l}^{-1}(0.01) = 2.90$. The predicted values result in $\widehat{t}_{Eff1} = 17.90$ and $\widehat{t}_{Eff2} = 252.90$. Both cases has same errors, but their absolute of the relative errors (ARE) are follows:

$$ARE_{Eff1} = \left| \frac{\widehat{t}_{Eff1} - t_{Eff1}}{t_{Eff1}} \right| = \left| \frac{17.90 - 15}{15} \right| = 0.1933$$

$$ARE_{Eff2} = \left| \frac{\widehat{t}_{Eff2} - t_{Eff2}}{t_{Eff2}} \right| = \left| \frac{252.90 - 250}{250} \right| = 0.0116$$

The results indicate the absolute of the relative errors of former is greater than that of the latter.

These distributions for the amount of effort and the number of errors indicate the small-scale projects are major and more than the large scale projects. Therefore, in order to improve prediction accuracy, it is important to reconstruct the normalizing way.

4.2 New Normalization of Data

In order to solve the problem, we adopt new normalizing way in the following equation:

$$f_{n_c}(t) = \sqrt{1 - (f_{n_l}(t) - 1)^2} \tag{5}$$

The comparison between Eq. (4) and Eq. (5) is shown in Figs. 2 and 3. The Eq. (5) has a sharp inclination at the lower original data, then a small change at the lower original data get magnified.

Using the same assumption, the predicted values result in $\widehat{t}_{Eff1} = 15.56$ and $\widehat{t}_{Eff2} = 271.11$. Their absolute of the relative errors are in Eq. (6) and

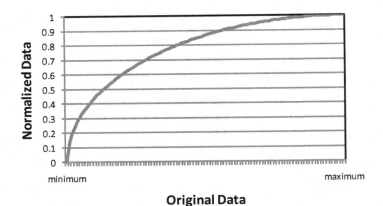

Fig. 2   Normalizing Results using Eq. (4)

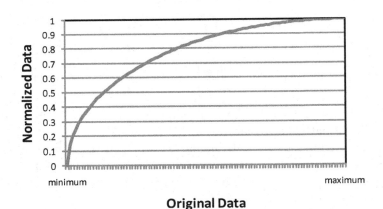

Fig. 3   Normalizing Results using Eq. (5)

Eq. (7).

$$ARE_{Eff1} = \left| \frac{15.56 - 15}{15} \right| = 0.0373 \tag{6}$$

$$ARE_{Eff2} = \left| \frac{271.11 - 250}{250} \right| = 0.0844 \tag{7}$$

The results show the absolute of the relative errors for the small-scale project is smaller than that of old normalization method and, in contrast,

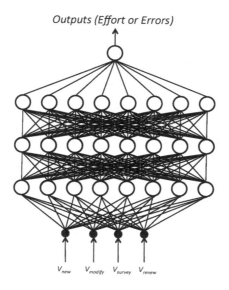

Fig. 4   Structure of Model

that for the large scale project is slightly larger than that of old normalization method. The more detailed comparison analyses are in Section 5.

4.2.1 *Structure of model*

In a feed-forward ANN, the information is moved from input nodes, through the hidden nodes to the output nodes. The number of hidden nodes is important, because if the number is too large, the network will over-training. The number of hidden nodes is, generally 2/3 of the number of input nodes or twice the number of input nodes. In this chapter, we use 8 hidden nodes in our model which is illustrated in Fig. 4.

4.3 *Multiple Regression Analysis Model*

The multiple regression analysis (MRA) model is derived from Eq. (8), in which the notation adheres to the meanings defined in Subsection 3.1. The model selects explanatory variables by using a best subset selection procedure based on Akaike's Information Criterion (AIC) [Akaike (1973)]. The smaller the AIC value, the higher the accuracy or fit of the model.

$$D = \alpha_1 \times S^2 + \alpha_2 \times S + \beta \tag{8}$$

where, D indicates *Eff* or *Err*, and S is calculated by Eq. (9).

$$S = V_{new} + V_{modify} + \theta_1 \times V_{survey} + \theta_2 \times V_{reuse} \qquad (9)$$

where, θ_1 and θ_2 are less than 1, thus Eq. (9) emphasizes V_{new} and V_{modify}.

5 Evaluation Experiment

5.1 *Evaluation Criteria*

Equations (10) to (13) are used as evaluation criteria for the effort and errors prediction models. The smaller the value of each evaluation criterion, the higher is the relative accuracy in Eqs. (10) to (13). The accuracy value is expressed as X, and the predicted value as $\widehat{X}$. Also, the number of data is expressed as n.

1) Mean of Absolute Errors (*MAE*).
2) Standard Deviation of Absolute Errors (*SDAE*).
3) Mean of Relative Errors (*MRE*).
4) Standard Deviation of Relative Errors (*SDRE*).

$$MAE = \frac{1}{n} \sum |\widehat{X} - X| \qquad (10)$$

$$SDAE = \sqrt{\frac{1}{n-1} \sum \left(|\widehat{X} - X| - MAE\right)^2} \qquad (11)$$

$$MRE = \frac{1}{n} \sum \left|\frac{\widehat{X} - X}{X}\right| \qquad (12)$$

$$SDRE = \sqrt{\frac{1}{n-1} \sum \left(\left|\frac{\widehat{X} - X}{X}\right| - MRE\right)^2} \qquad (13)$$

5.2 *Data Used in Evaluation Experiment*

The evaluation experiment uses the data for real projects. The project data is divided into two random sets. One of the two sets is used as training data, while the other is test data. The training data is used the generation of the effort (or errors) prediction model generation, which is used to predict the effort (or errors) of the projects in the test data. The prediction criteria presented in Subsection 5.1 are then used to confirm whether the effort were accurately predicted or not by. Both data sets, that is, the training data and test data, are divided into 10 sections and these are used to repeat the experiment 10 times.

5.3 *Results and Discussion*

A total of 1828 projects are used in the experiment, then 914 projects are used as test data for each experiment. For each method, averages of the experiments results for the 10 experiments are shown in Table 4.

Table 4   Experimental Results for Efforts Prediction

	MAE	$SDAE$	MRE	$SDRE$	
ANN Model	22.223	99.403	0.17782	0.96512	
MRA Model	107.61	213.00	2.0401	4.2476	

Table 5   Experimental Results for Errors Prediction

	MAE	$SDAE$	MRE	$SDRE$
ANN Model	21.738	93.016	0.20926	1.1854
MRA Model	110.62	237.37	2.6373	8.1617

5.3.1   *Validation analysis of the accuracy of the models*

We compare the accuracy of the ANN model with that of the regression analysis model using Welch's t-test [Welch (1947)]. The t-test (called Student's t-test) [Student (1908)] is used as a test of the null hypothesis that the means of two normally distributed populations are equal. Welch's t-test is used when the variances of two samples are assumed to be different to test the null hypothesis that the means of non two normally distributed populations are equal if the two sample sizes are equal [Aoki (2007)]. The t statistic to test whether the means are different is calculated as follows:

$$t_0 = \frac{|\overline{X} - \overline{Y}|}{\sqrt{\frac{s_x}{n_x} + \frac{s_y}{n_y}}} \tag{14}$$

where $\overline{X}$ and $\overline{Y}$ are the sample means, s_x and s_y are the sample standard deviations and n_x and n_y are the sample sizes. For use in significance testing, the distribution of the test statistic is approximated as an ordinary Student's t-distribution with the following degrees of freedom:

$$\nu = \frac{\left(\frac{s_x}{n_x} + \frac{s_y}{n_y}\right)^2}{\frac{s_x^2}{n_x^2(n_x-1)} + \frac{s_y^2}{n_y^2(n_y-1)}} \tag{15}$$

Thus once the a t-value and degrees of freedom have been determined, a p-value can be found using a table of values from the Student's t-distribution. If the p-value is smaller than or equal to the significance level, then the null hypothesis is rejected. The significance levels are usually 0.05 and 0.01, are represented by the Greek symbol, α.

The null hypothesis, in these cases, is "there is no difference between the means of the prediction errors for the ANN model and the MRA model". The results of the t-test for absolute errors and relative errors to predict the amount of effort are given in Tables 6 and 7, respectively. And the results of the t-test for absolute errors and relative errors to predict the number of errors are given in Tables 8 and 9, respectively.

The results indicate that the means of the absolute (or relative) errors between ANN models and MRA model shows a statistically significant difference, because the p-values are less than 0.01.

Table 6   Results of t-test for *MAE* for Effort

	ANN Model	MRA Model
Mean $(\overline{X})$	22.223	107.61
Standard deviation(s)	99.403	213.00
Sample size (n)	9140	9140
Degrees of freedom (ν)	12939.51	
t value (t_0)	34.7295	
p value	$< 2.2 \times 10^{-16}$	

Table 7   Results of t-test for *MRE* for Effort

	ANN Model	MRA Model
Mean $(\overline{X})$	0.17782	2.0401
Standard deviation(s)	0.96512	4.2476
Sample size (n)	9140	9140
Degrees of freedom (ν)	10080.13	
t value (t_0)	40.9737	
p value	$< 2.2 \times 10^{-16}$	

Table 8   Results of t-test for MAE for Errors

	ANN Model	MRA Model
Mean ($\overline{X}$)	21.738	110.62
Standard deviation(s)	93.016	237.37
Sample size (n)	9140	9140
Degrees of freedom (ν)	11881.02	
t value (t_0)	33.3305	
p value	$< 2.2 \times 10^{-16}$	

Table 9   Results of t-test for MRE for Errors

	ANN Model	MRA Model
Mean ($\overline{X}$)	0.20926	2.6373
Standard deviation(s)	1.1854	8.1617
Sample size (n)	9140	9140
Degrees of freedom (ν)	9524.393	
t value (t_0)	7.7881	
p value	$< 2.2 \times 10^{-16}$	

6   Conclusion

In this chapter, we have established effort and errors prediction models using artificial neural networks. In addition, we carried out an evaluation experiment that compared the accuracy of the ANN model with that of the MRA model using Welch's t-test. The results of the comparison indicate that the ANN model is more accurate than the MRA model, because the mean errors of the ANN are statistically significantly lower.

Our future works are the following:

1) In this study, we used a basic artificial neural network. More complex models need to be considered to improve the accuracy by avoiding over-training.
2) We implemented a model to predict the final amount of effort and number of errors in new projects. It is also important to predict effort and errors mid-way in the development process of a project.
3) We used all the data in implementing the model. However, the data include exceptions and there are harmful to the model. Data needs to be clustered in order to to identify these exceptions.

Acknowledgements

My deepest appreciation goes to Prof. Toyoshiro Nakashima whose comments and suggestions were innumerably valuable throughout the course of my study. I would also like to thank Mr. Yoshiyuki Anan and Prof. Naohiro Ishii whose comments made enormous contribution to my work.

References

Akaike, H. (1973). Information theory and an extention of the maximum likelihood principle, *2nd International Symposium on Information Theory, Petrov, B. N., and Csaki, F. (eds.)*, pp. 267–281.

Aoki, S. (2007). In testing whether the means of two populations are different(in Japanese), `http://aoki2.si.gunma-u.ac.jp/lecture/BF/index.html`.

Boehm, B. (1976). Software engineering, *IEEE Trans. Software Eng.* **C-25**, 12, pp. 1226–1241.

Hirayama, M. (2004). Current state of embedded software(in japanese), *Journal of Information Processing Society of Japan(IPSJ)* **45**, 7, pp. 677–681.

Iwata, K., Anan, Y., Nakashima, T. and Ishii, N. (2006). Improving accuracy of multiple regression analysis for effort prediction model, *Proceedings of 5th IEEE/ACIS International Conference on Computer and Information Science – ICIS 2006*, pp. 48–55.

Komiyama, T. (2003). Development of foundation for effective and efficient software process improvement(in japanese), *Journal of Information Processing Society of Japan(IPSJ)* **44**, 4, pp. 341–347.

N., U. (2004). Modeling techniques for designing embedded software (in japanese), *Journal of Information Processing Society of Japan(IPSJ)* **45**, 7, pp. 682–692.

Nakamoto, Y., Takada, H. and Tamaru, K. (1997). Current state and trend in embedded systems(in japanese), *Journal of Information Processing Society of Japan(IPSJ)* **38**, 10, pp. 871–878.

Nakashima, S. (2004). Introduction to model-checking of embedded software(in japanese), *Journal of Information Processing Society of Japan(IPSJ)* **45**, 7, pp. 690–693.

Nakashima, T., Iwata, K., Anan, Y. and Ishii, N. (2006). Studies on project management models for embedded software development projects, *Proceedings of 4th International Conference on Software Engineering Research, Management Applications – SERA 2006*, pp. 363–370.

Ogasawara, H. and Kojima, S. (2003). Process improvement activities that put importance on stay power(in japanese), *Journal of Information Processing Society of Japan(IPSJ)* **44**, 4, pp. 334–340.

Student (1908). The probable error of a mean, *Biometrika* **6**, 1, pp. 1–25.

Takagi, Y. (2003). A case study of the success factor in large-scale software system development project(in japanese), *Journal of Information Processing Society of Japan(IPSJ)* **44**, 4, pp. 348–356.

Tamaru, K. (2004). Trends in software development platform for embedded systems(in japanese), *Journal of Information Processing Society of Japan(IPSJ)* **45**, 7, pp. 699–703.

Watanabe, H. (2004). Product line technology for software development(in japanese), *Journal of Information Processing Society of Japan(IPSJ)* **45**, 7, pp. 694–698.

Welch, B. L. (1947). The generalization of student's problem when several different population variances are involved, *Biometrika* **34**, 28.

Chapter 13

Optimal Checkpoint Times for Database Systems

Kenichiro Naruse[1] and Sayori Maeji[2]

[1]*Nagoya Sangyo University,*
3255-5 Arai-cho, Owariasahi 488-8711, Japan
[2]*Institute of Consumer Science and Human Life,*
Kinjo Gakuin University,
1723 Oomori 2, Moriyama, Nagoya 463-8521, Japan

1 Introduction

Most computer systems in offices and industries execute successively tasks each of which has random processing times. In such systems, some errors often occur due to noises, human errors and hardware faults. To detect and mask errors, some useful fault tolerant computing techniques have been adopted [Lee and Anderson (1990); Siewiorek and Swarz (eds)]. Several studies of deciding optimal checkpoint frequencies have been made: The performance and reliability of redundant modular systems were evaluated [Pradhan and Vaidya (1992); Nakagawa (2008)], and the performance of checkpoint schemes with task duplication was evaluated [Ziv and Bruck (1997, 1998)]. The optimal instruction-retry period that minimizes the probability of the dynamic failure by a triple modular controller was derived [Kim and Shin (1996)]. The evaluation models with finite checkpoints and bounded rollback were discussed [Ohara *et al.* (2006-09-01)]. Furthermore, checkpointing scheme for a set of multiple tasks in real-time systems were investigated [Zhang and Chakrabarty (2004)].

In recent years, internet shopping is greatly increasing in the world. Many companies have to process big data which have been used for iden-tify users, shopping item, shopping history, and so on. These computers need to process big data with hi-speed response, and sometimes, may be slowed down by processing big data. To prevent such processing, we have to store the data on memory. Because the memory can read and write over 100 times faster than HDD, the database-nodes can process big data with hi-speed.

This chapter applies the checkpoint operation to the system which executes successively each task with random working times [Chen *et al.* (2010); Nakagawa *et al.* (2009); Naruse *et al.* (2006)] and a clustering database system. Section 2 considers three schemes of checkpoint in which two independent modules is executed and compare two states at checkpoint times. If two states of each module do not match with each other, we go back to the newest checkpoint and make their retrials in three types of checkpoint schemes. When failures occur in the system, it executes the recovery operation until the latest checkpoint and repeats such processes until the next checkpoint. Then, introducing checking costs and a loss cost from a failure to the latest checkpoint, the total expected cost between checkpoints is obtained, using an inspection policy. Section 3 considers a clustering database-node system [ORACLE (2012)] using an extended Scheme 3. The system works for several computers to process big data with hi-speed response. The big data are divided and stored in n database-nodes on memory which can read and write over 100 times faster than HDD. In this case, the database-nodes can process big data with hi-speed. We derive optimal checkpoint times and number of database-nodes, and compute both of them in numerical examples. Finally, we consider two types of database-node models such as single node and majority node [Nakagawa (2008)]. Finally, Sect. 4 summarizes the results in this chapter.

2   Random Checkpoint Models

We consider three checkpoint schemes which have different checkpoint types and present numerical examples for three schemes.

Suppose that we have to execute the successive tasks with a processing time $Y_k(k=1, 2, \cdots)$ (Fig. 1). A double database-nodes system of error detection for the processing of each task is adopted. Then, introducing two types of checkpoints; compare-and-store checkpoint (CSCP) and compare-checkpoint (CCP) [Nakagawa *et al.* (2003)], we consider the following three checkpoint schemes:

1) CSCP is placed at each end of tasks (Fig. 1),

2) CSCP is placed at the N-th ($N = 1, 2, \cdots$) end of tasks (Fig. 2),
3) CCP is placed at each end of tasks and CSCP is placed at the N-th end of tasks (Fig. 3).

Fig. 1   Task execution for Scheme 1

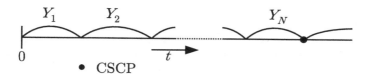

Fig. 2   Task execution for Scheme 2

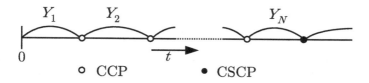

Fig. 3   Task execution for Scheme 3

The mean execution times per one task for each scheme are obtained, and optimal numbers N^* that minimize them for Schemes 2 and 3 are derived analytically and are compared numerically. This is one of applied models with random maintenance times [Nakagawa (2005); Sugiura *et al.* (2004)] to checkpoint models. Such schemes would be useful when it is better to place checkpoints at the end of tasks than those on one's way.

2.1 *Performance Analysis*

Suppose that task k has a processing time $Y_k(k = 1, 2, \cdots)$ with an identical distribution $G(t) \equiv \Pr\{Y_k \leq t\}$ and a finite mean $\mu = \int_0^\infty [1 - G(t)]\,dt < \infty$, and is executed successively. To detect errors, we provide two independent database-nodes where they compare two states at checkpoint times. Further, it is assumed that some errors occur at a constant rate $\lambda(\lambda > 0)$, *i.e.* , the probability that two database-nodes have no error during $(0, t]$ is $e^{-2\lambda t}$.

(1) Scheme 1
CSCP is placed at each end of task k: When two states of database-nodes match with each other at the end of task k, the process of task k is correct and its state is stored (Fig. 1). In this case, two database-nodes go forward and execute task $k + 1$. However, when two states do not match, it is judged that some errors have occurred. Then, two database-nodes go back and make the retry of task k again.

Let C be the overhead for the comparison of two states and Cs be the overhead for their store. Then, the mean execution time of the process of task k is given by a renewal equation:

$$\widetilde{L}_1(1) = \int_0^\infty \left\{ e^{-2\lambda t}(C + Cs + t) + \left(1 - e^{-2\lambda t}\right)\left[C + t + \widetilde{L}_1(1)\right]\right\}dG(t). \tag{1}$$

Solving (1) for $\widetilde{L}_1(1)$,

$$\widetilde{L}_1(1) = \frac{C + \mu + CsG^*(2\lambda)}{G^*(2\lambda)},$$

where $G^*(s)$ is the Laplace-Stieltjes (LS) transform of $G(t)$, *i.e.*, $G^*(s) \equiv \int_0^\infty e^{-st}dG(t)$ for $s > 0$. Therefore, the mean execution time per one task is

$$L_1(1) \equiv \widetilde{L}_1(1) = \frac{C + \mu}{G^*(2\lambda)} + Cs. \tag{2}$$

(2) Scheme 2
CSCP is placed only at the end of task N (Fig. 2): When two states of all task $k(k = 1, 2, \cdots, N)$ match at the end of task N, its state is stored and two database-nodes execute task $N + 1$. When two states do not match, two database-nodes go back to the first task 1 and make their retries. By the method similar to obtaining (1), the mean execution time of the process

of all task $k(k = 1, 2, \cdots, N)$ is

$$
\tilde{L}_2(N) = \int_0^\infty \{ e^{-2\lambda t} (NC + Cs + t)
$$
$$
+ (1 - e^{-2\lambda t}) \left[NC + t + \tilde{L}_2(N) \right] \} dG^{(N)}(t), \tag{3}
$$

where $G^{(N)}(t)$ is the N-fold Stieltjes convolution of $G(t)$ with itself, *i.e.*, $G^{(N)}(t) \equiv \int_0^t G^{(N-1)}(t - u) dG(u)$ $(N = 1, 2, \cdots)$, and $G^{(0)}(t) \equiv 1$ for $t \geq 0$ and $G^{(1)}(t) = G(t)$. Solving (3) for $\tilde{L}_2(N)$,

$$
\tilde{L}_2(N) = \frac{NC + N\mu + Cs\, [G^*(2\lambda)]^N}{[G^*(2\lambda)]^N}.
$$

Therefore, the mean execution time per one task is

$$
L_2(N) \equiv \frac{\tilde{L}_2(N)}{N} = \frac{C + \mu}{[G^*(2\lambda)]^N} + \frac{Cs}{N} \qquad (N = 1, 2, \cdots). \tag{4}
$$

When $N = 1$, $L_2(1)$ agrees with $L_1(1)$ in (2).

We find an optimal number N_2^* that minimizes $L_2(N)$. There exists a finite $N_2^*(1 \leq N_2^* < \infty)$ because $\lim_{N \to \infty} L_2(N) = \infty$. From the inequality $L_2(N + 1) - L_2(N) \geq 0$,

$$
\frac{N(N + 1)\left[1 - G^*(2\lambda)\right]}{[G^*(2\lambda)]^{N+1}} \geq \frac{Cs}{C + \mu} \qquad (N = 1, 2, \cdots). \tag{5}
$$

The left-hand side of (5) is strictly increasing to ∞ in N. Thus, there exists a finite and unique minimum $N_2^*(1 \leq N_2^* < \infty)$ which satisfies (5). If

$$
\frac{[1 - G^*(2\lambda)]}{[G^*(2\lambda)]^2} \geq \frac{Cs}{2(C + \mu)},
$$

then $N_2^* = 1$.

When $G(t) = 1 - e^{-t/\mu}$, (5) is rewritten as

$$
N(N + 1)\, 2\lambda\mu\, (2\lambda\mu + 1)^N \geq \frac{Cs}{C + \mu}. \tag{6}
$$

Example 13.1.
Table 1 presents the optimal number N_2^* and the resulting execution time $L_2(N_2^*)/\mu$ and $L_1(1)/\mu$ in (2) for $\lambda\mu$ and C/μ when $Cs/\mu = 0.1$. This indicates that optimal N_2^* decrease with $\lambda\mu$ and increase with C/μ. For example, when $\lambda\mu = 0.005$ and $C/\mu = 0.1$, $N_2^* = 3$ and $L_2(N_2^*)/\mu$ is 1.167 that is about 4% shorter than $L_1(1)/\mu = 1.211$ for Scheme 1.

Table 1 Optimal number N_2^* and the resulting execution time $L_2(N_2^*)/\mu$ for Scheme 2 and $L_1(1)/\mu$ for Scheme 1 when $Cs/\mu = 0.1$.

	$C/\mu = 0.5$			$C/\mu = 0.1$		
$\lambda\mu$	N_2^*	$L_2(N_2^*)/\mu$	$L_1(1)/\mu$	N_2^*	$L_2(N_2^*)/\mu$	$L_1(1)/\mu$
0.1	1	1.900	1.900	1	1.420	1.420
0.05	1	1.750	1.750	1	1.310	1.310
0.01	2	1.611	1.630	2	1.194	1.222
0.005	3	1.579	1.615	3	1.167	1.211
0.001	6	1.535	1.603	7	1.130	1.202
0.0005	8	1.525	1.602	10	1.121	1.201
0.0001	18	1.511	1.600	21	1.109	1.200

(3) Scheme 3

CSCP is placed at the end of task N and CCP is placed only at the end of task $k(k = 1, 2, \cdots, N-1)$ between CSCPs (Fig. 3): When two states of task $k(k = 1, 2, \cdots, N-1)$ match at the end of task k, two database-nodes execute task $k+1$. When two states of task $k(k = 1, 2, \cdots, N)$ do not match, two database-nodes go back to the first task 1. When two states of task N match, the process of all tasks N is completed, and its state is stored. Two database-nodes execute task $N + 1$.

Let $\widetilde{L}_4(k)$ be the mean execution time from task k to the completion of task N. Then, by the method similar to obtaining (3),

$$\widetilde{L}_4(k) = \int_0^\infty \left\{ e^{-2\lambda t} \left[C + t + \widetilde{L}_4(k+1) \right] + \left(1 - e^{-2\lambda t}\right) \left[C + t + \widetilde{L}_4(k+1) \right] \right\} dG(t)$$
$$(k = 1, 2, \cdots, N-1),$$

$$\widetilde{L}_4(N) = \int_0^\infty \left\{ e^{-2\lambda t} (C + t + Cs) + \left(1 - e^{-2\lambda t}\right) \left[C + t + \widetilde{L}_4(1) \right] \right\} dG(t).$$

$$(7)$$

Solving (7) for $\widetilde{L}_4(1)$,

$$\widetilde{L}_4(1) = \frac{(C + \mu) \left\{ 1 - [G^*(2\lambda)]^N \right\}}{[1 - G^*(2\lambda)] [G^*(2\lambda)]^N} + Cs.$$

Therefore, the mean execution time per one task is

$$L_4(N) \equiv \frac{\widetilde{L}_4(1)}{N} = \frac{(C + \mu) \left\{ 1 - [G^*(2\lambda)]^N \right\}}{N [1 - G^*(2\lambda)] [G^*(2\lambda)]^N} + \frac{Cs}{N} \qquad (N = 1, 2, \cdots).$$

$$(8)$$

When $N = 1$, $L_4(1)$ agrees with (2). By comparing (4) with (8), Scheme 3 is better than Scheme 2.

It can be clearly seen that a finite $N_4^*(1 \leq N_4^* < \infty)$ that minimizes $L_4(N)$ exists. From the inequality $L_4(N + 1) - L_4(N) \geq 0$,

$$\frac{1}{[G^*(2\lambda)]^{N+1}} \sum_{j=1}^{N} \left\{ 1 - [G^*(2\lambda)]^j \right\} \geq \frac{Cs}{C + \mu} \quad (N = 1, 2, \cdots), \quad (9)$$

whose left-hand side is strictly increasing to ∞ in N. Thus, there exists a finite and unique minimum $N_4^*(1 \leq N_4^* < \infty)$ which satisfies (9). If $(C + \mu)[1 - G^*(2\lambda)] \geq Cs[G^*(2\lambda)]^2$, then $N_4^* = 1$. By comparing (9) with (5), it can be easily seen that $N_4^* \geq N_3^*$.

When $G(t) = 1 - e^{-t/\mu}$, (9) is rewritten as

$$(2\lambda\mu + 1)^{N+1} \sum_{j=1}^{N} \left[1 - \left(\frac{1}{2\lambda\mu + 1} \right)^j \right] \geq \frac{Cs/\mu}{C/\mu + 1}. \quad (10)$$

Example 13.2.
Table 2 presents the optimal number N_4^* and the resulting execution time $L_4(N_4^*)/\mu$ in (8) for $\lambda\mu$ and C/μ when $Cs/\mu = 0.1$. Clearly, Scheme 3 is better than Scheme 2 and $N_4^* \geq N_3^*$. However, in general, the overhead C for Scheme 2 would be less than that for Scheme 3. In such case, Scheme 2 might be better than Scheme 3.

Table 2   Optimal number N_4^* and the resulting execution time $L_4(N_4^*)/\mu$ for Scheme 3 when $Cs/\mu = 0.1$.

$\lambda\mu$	$C/\mu = 0.5$		$C/\mu = 0.1$	
	N_4^*	$L_4(N_4^*)/\mu$	N_4^*	$L_4(N_4^*)/\mu$
0.1	1	1.900	1	1.420
0.05	1	1.750	1	1.310
0.01	3	1.594	3	1.178
0.005	4	1.563	4	1.153
0.001	8	1.526	9	1.122
0.0005	12	1.518	13	1.115
0.0001	26	1.508	30	1.107

3   Database-node Models

We consider two types of database-node models such as a single database-node model (Fig. 4) and a majority database-node model (Fig. 5), and present numerical examples for both of them.

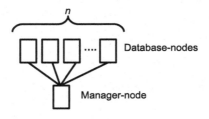

Fig. 4   Single node model.

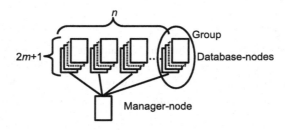

Fig. 5   Majority node model.

(1) Single node

We consider one manager-node and n database-nodes in a database system. The database-nodes process each task, and the manager-node watches the database-nodes in Fig. 4. First of all, the database-nodes store database data from HDD to memory by dividing them into the number of database-nodes, and store a part of data on memory. So that, the database-nodes can process each task with hi-speed. However, if some errors occur in the database-nodes, the data is lost on memory. To prevent such data loss, we suppose the following checkpoint schemes such as Journal Checkpoint (JC) and Flush Checkpoint (FC) schemes [ORACLE (2012)]: JC is placed at each end of tasks and FC is placed at the Nth end of tasks. JC stores command and result of a task to HDD, and FC updates the HDD database according to JC data. The management-node watches n database-nodes. If some errors occur in the database-nodes, the management-node restarts them and transports a part of corresponding data from HDD to memory. This checkpoint scheme is the same as Scheme 3 of Sect. 2 except that Scheme 3 compare two database-node results to detect failures, but this scheme can detect failures by manager-node.

(2) Majority node

We extend a single node model to a redundant database-node model which consists of n groups of $(2m + 1)$ database-nodes and one manager-node in Fig. 5. The redundant model is a majority decision system with $2m + 1$ database-nodes as an error masking system, *i.e.*, $(m + 1)$-out-of-$(2m + 1)$ system [Nakagawa (2008)]. If more than $m + 1$ results of $2m + 1$ database-nodes agree, the manager-node judges the database-node is correct, and the database-node can execute the next task. Thus, if less than m errors occur in the group of $2m + 1$ database-nodes, the database-nodes can mask the error using one of a majority decision policy. If more than $m + 1$ errors occur in the one database-nodes, the management-node restarts the database-node and transports a part of corresponding data from HDD to memory.

3.1 *Performance Analysis*

Let C be the overhead for JC and Cs be the overhead for FC with $C < Cs$. Suppose that we have to execute the successive tasks with processing times $Y_k(k = 1, 2, \cdots)$ (Fig. 3). Let W be the time for transferring all database data to the database-node, and R be its restart time. The mean execution times per one task are obtained, and optimal numbers N^* and n^* that minimize them are derived analytically. This is one of applied models with random maintenance times [Nakagawa (2005)] to checkpoint models. Such schemes would be useful when it is better to place checkpoints at the end of tasks than those at periodic times.

FC is placed at the end of task N $(N = 1, 2, \cdots)$ and JC is placed only at the end of task $k(k = 1, 2, \cdots, N - 1)$ between FCs (Fig. 3): When there is no error in the process of task $k(k = 1, 2, \cdots, N - 1)$, the database-nodes execute task $k + 1$. When errors occur in task $k(k = 1, 2, \cdots, N)$, the manager-node restarts the database-nodes and restores both FC and JC data, and transfers them to database, irrespective of manager-node failures.

When the management-node does not detect any error until task N, the process of all tasks N is completed, and its state is stored. The database-node executes task $N + 1$. It is assumed $F(t)$ is a failure probability of database-node. Let $\widetilde{L}(n, k)$ be the mean execution time from task k to the completion of task N. Then, by the similar method of obtaining (2) [Maeji *et al.* (2010)],

$$
\widetilde{L}(n, k) = \int_0^\infty \left\{ \overline{F}(t) \left[C + t + \widetilde{L}(n, k+1) \right] \right.
$$
$$
\left. + F(t) \left[C + t + \widetilde{L}(n, 1) + \frac{W}{n} + R \right] \right\} \, dG(t) \quad (k = 1, 2, \cdots, N-1),
$$

$$
\widetilde{L}(n, N) = \int_0^\infty \left\{ \overline{F}(t) \left[C + t + Cs \right] \right.
$$
$$
\left. + F(t) \left[C + t + \widetilde{L}(n, 1) + \frac{W}{n} + R \right] \right\} \, dG(t), \tag{11}
$$

where $\overline{F}(t) \equiv 1 - F(t)$.

(1) Single node

It is assumed that some errors occur at a constant rate $\lambda (\lambda > 0)$, *i.e.*, the probability that n database-nodes have no error during $(0, t]$ is $\overline{F}(t) = e^{-n\lambda t}$ $(n = 1, 2, \cdots)$.

Solving (11) for $\widetilde{L}_s(n, 1)$,

$$
\widetilde{L}_s(n, 1) = \frac{\left(C + \mu + \frac{W}{n} + R \right) \left\{ 1 - [G^*(n\lambda)]^N \right\}}{[1 - G^*(n\lambda)][G^*(n\lambda)]^N} + Cs. \tag{12}
$$

Therefore, the mean execution time per one task is

$$
L_s(n, N) \equiv \frac{\widetilde{L}_s(n, 1)}{N} = \frac{\left(C + \mu + \frac{W}{n} + R \right) \left\{ 1 - [G^*(n\lambda)]^N \right\}}{N [1 - G^*(n\lambda)][G^*(n\lambda)]^N} + \frac{Cs}{N}
$$
$$
(n, N = 1, 2, \cdots). \tag{13}
$$

When $G(t) = 1 - e^{-t/\mu}$, (13) is rewritten as

$$
L_s(n, N) = \frac{\left(C + \mu + \frac{W}{n} + R \right)}{N} \sum_{j=1}^N (n\lambda\mu + 1)^j + \frac{Cs}{N}. \tag{14}
$$

We find an optimal number N^* that minimizes $L_s(n, N)$ in (14) for $n \geq 1$. From the inequality $L_s(n, N+1) - L_s(n, N) \geq 0$,

$$
(n\lambda\mu + 1)^{N+1} \sum_{j=1}^N \left[1 - \left(\frac{1}{n\lambda\mu + 1} \right)^j \right] \geq \frac{\frac{Cs}{\mu}}{\frac{C}{\mu} + 1 + \frac{W/n+R}{\mu}}
$$
$$
(N = 1, 2, \cdots). \tag{15}
$$

Clearly, if

$$n\lambda\mu(n\lambda\mu + 1) \geq \frac{\dfrac{Cs}{\mu}}{\dfrac{C}{\mu} + 1 + \dfrac{W/n + R}{\mu}},$$

then $N^* = 1$. Clearly, N^* decreases with n.

Example 13.3.

Table 3 presents an optimal number N^* in (15) and the resulting execution time $L_s(n, N^*)/\mu$ in (14) for $\lambda\mu$ and n when $Cs/\mu = 0.8$, $C/\mu = 0.2$, $W/\mu = 5$ and $R/\mu = 10$. This indicates that N^* decrease with both n and $\lambda\mu$, and its resulting mean times $L_s(n, N^*)$ decrease with n and increase with $\lambda\mu$ for $N^* \geq 2$. The case of $N^* = 1$ means that we should provide no JC between FC. For example, when $\lambda\mu = 0.005$ and $n = 2$, we make 3 JC checkpoints between FC checkpoints. In this case, the execution time is 14.242.

Table 3   Optimal number N^* and resulting execution time $L_s(n, N^*)/\mu$ for $\lambda\mu$ and n when $Cs/\mu = 0.8, C/\mu = 0.2, W/\mu = 5$ and $R/\mu = 10$.

$\lambda\mu$	$n = 1$		$n = 2$		$n = 5$		$n = 10$	
	N^*	$\dfrac{L_s(n,N^*)}{\mu}$	N^*	$\dfrac{L_s(n,N^*)}{\mu}$	N^*	$\dfrac{L_s(n,N^*)}{\mu}$	N^*	$\dfrac{L_s(n,N^*)}{\mu}$
0.1	1	18.620	1	17.240	1	19.100	1	24.200
0.05	1	17.810	1	15.870	1	16.050	1	18.350
0.01	3	16.793	2	14.514	2	13.530	1	13.670
0.005	4	16.604	3	14.242	2	13.061	2	12.992
0.001	10	16.369	8	13.924	5	12.544	4	12.195
0.0005	14	16.318	11	13.855	7	12.437	5	12.037
0.0001	31	16.252	24	13.768	16	12.302	12	11.843
0.00005	44	16.236	34	13.748	23	12.271	17	11.800

Next, we find an optimal n^* that minimizes $L_s(n, N)$ in (14) for $N \geq 1$. From the inequality $L_s(n + 1, N) - L_s(n, N) \geq 0$,

$$\left(C + \mu + \frac{W}{n+1} + R\right) \sum_{j=1}^{N} [(n+1)\lambda\mu + 1]^j$$

$$\geq \left(C + \mu + \frac{W}{n} + R\right) \sum_{j=1}^{N} (n\lambda\mu + 1)^j \qquad (n = 1, 2, \cdots). \qquad (16)$$

Especially, when $N = 1$, (16) is

$$n(n+1) \geq \frac{W}{C + \mu + R}, \qquad (17)$$

and when $N = 2$,

$$(C + \mu + R)\lambda\mu\left[(2n + 1)\lambda\mu + 3\right] + W(\lambda\mu)^2 \geq \frac{W}{n(n + 1)}. \qquad (18)$$

The left-hand side of (18) is strictly increasing to ∞, and the right-hand side is decreasing from $W/2$ to 0. Thus, an optimal number $n^*(1 \leq n^* < \infty)$ which satisfy (18) exists uniquely. If

$$3\lambda\mu\left(\lambda\mu + 1\right)\left(C + \mu + R\right) + W\left(\lambda\mu\right)^2 \geq \frac{W}{2},$$

then $n^* = 1$.

Example 13.4.
Table 4 presents an optimal number n^* in (16) and the resulting execution time $L_s(n^*, N)/\mu$ in (14) for $\lambda\mu$ and N when $Cs/\mu = 0.8$, $C/\mu = 0.2$, $W/\mu = 5$ and $R/\mu = 10$. The resulting execution times $L_s(n^*, N)$ increase with N and $\lambda\mu$. This shows a similar tendency to Table 1.

Table 4   Optimal number n^* and resulting execution time $L_s(n^*, N)/\mu$ for $\lambda\mu$ and N when $Cs/\mu = 0.8, C/\mu = 0.2, W/\mu = 5$ and $R/\mu = 10$.

$\lambda\mu$	$N = 1$		$N = 2$		$N = 5$		$N = 10$	
	n^*	$\frac{L_s(n^*,N)}{\mu}$	n^*	$\frac{L_s(n^*,N)}{\mu}$	n^*	$\frac{L_s(n^*,N)}{\mu}$	n^*	$\frac{L_s(n^*,N)}{\mu}$
0.1	2	17.240	2	18.484	1	21.919	1	28.480
0.05	3	15.597	2	16.224	2	18.561	1	21.475
0.01	7	13.548	5	13.530	4	14.186	3	15.273
0.005	9	13.085	8	12.944	5	13.306	4	13.985
0.001	21	12.478	17	12.189	12	12.202	9	12.433
0.0005	30	12.337	24	12.015	17	11.951	13	12.087
0.0001	67	12.150	54	11.784	38	11.621	28	11.635
0.00005	94	12.106	77	11.730	54	11.544	40	11.530

Finally, we find both optimal n^* and N^* that minimize $L_s(n, N)$ in (13). To compute both optimal n^* and N^*, we substitute $n = 1$ into (15), and compute n_1^* for $N = N_1^*$ in (16), and compute N_2^* for $n = n_1^*$ into (15). If $N_I^* = N_{I+1}^*$ and $n_I^* = n_{I+1}^*$ then $N^* = N_I^*$ and $n^* = n_I^*$. Both n^* and N^* decrease with $\lambda\mu$. This indicates an optimal combination of the numbers of database-nodes and JC between FC.

Table 5 presents optimal number n^*, N^* and the resulting execution time $L_s(n^*, N^*)/\mu$ when $Cs/\mu = 0.8$, $C/\mu = 0.2$, $W/\mu = 5$ and $R/\mu = 10$. This indicates n^*/N^* decrease gradually with $\lambda\mu$, and $L_s(n^*, N^*)$ increase

Table 5   Optimal number n^*, N^* and the resulting execution time $L_s(n^*, N^*)/\mu$ for $\lambda\mu$ when $Cs/\mu = 0.8$, $C/\mu = 0.2$, $W/\mu = 5$ and $R/\mu = 10$.

$\lambda\mu$	n^*	N^*	$\dfrac{L_s(n^*,N^*)}{\mu}$
0.1	2	1	17.240
0.05	3	1	15.597
0.01	5	2	13.530
0.005	8	2	12.944
0.001	15	3	12.149
0.0005	19	4	11.938
0.0001	33	7	11.617
0.00005	44	8	11.526

with $\lambda\mu$. It can be easily seen that the pair of (n^*, N^*) is better than any (n, N^*) and (n^*, N) for the same $\lambda\mu$ in Tables 3 and 4.

(2) Majority node model

We consider a 2-out-of-3 database-node system. Then, the probability that one database-node is correct during (0,T] is [Nakagawa (2008); Naruse *et al.* (2006)]

$$\overline{F}(t) = 3e^{-2\lambda t} - 2e^{-3\lambda t}. \tag{19}$$

Thus, the probability that n database-node systems are correct is

$$\left[\overline{F}(t)\right]^n = \left(3e^{-2\lambda t} - 2e^{-3\lambda t}\right)^n. \tag{20}$$

From (13), the mean execution time per one task is

$$L_m(n,N) = \frac{(C + \mu + \frac{W}{n} + R)\left\{1 - \left[\int_0^\infty \left(3e^{-2\lambda t} - 2e^{-3\lambda t}\right)^n dG(t)\right]^N\right\}}{N\left[1 - \int_0^\infty (3e^{-2\lambda t} - 2e^{-3\lambda t})^n dG(t)\right]\left[\int_0^\infty (3e^{-2\lambda t} - 2e^{-3\lambda t})^n dG(t)\right]^N} + \frac{Cs}{N}$$
$$(n, N = 1, 2, \cdots). \tag{21}$$

When $G(t) = 1 - e^{-t/\mu}$, (21) is

$$L_m(n, N) = \frac{(C + \mu + \frac{W}{n} + R)\sum_{j=1}^N A^{-j}}{N} + \frac{Cs}{N}, \tag{22}$$

where

$$A \equiv \sum_{i=0}^n \frac{\binom{n}{i}(-1)^i 3^{n-i} 2^i}{(2n + i)\lambda\mu + 1}.$$

We find an optimal number N^* for 2-out-of-3 node model that minimizes $L_m(n, N)$ in (22) for $n \geq 1$.

Example 13.5.
Table 6 presents optimal number N^* and the resulting execution time $L_m(n, N^*)/\mu$ for $\lambda\mu$ and n when $Cs/\mu = 0.8$, $C/\mu = 0.2$, $W/\mu = 5$, $R/\mu = 10$ and $\mu = 1$. This shows a similar tendency to Tables 3 and 4, and the mean execution time is a little shorter than a single node model.

Table 6   Optimal number N^* and resulting execution time $L_m(n, N^*)/\mu$ for $\lambda\mu$ and n when $Cs/\mu = 0.8, C/\mu = 0.2, W/\mu = 5$, $R/\mu = 10$ and $\mu = 1$.

$\lambda\mu$	$n = 1$		$n = 2$		$n = 5$		$n = 10$	
	N^*	$\frac{L_m(n,N^*)}{\mu}$	N^*	$\frac{L_m(n,N^*)}{\mu}$	N^*	$\frac{L_m(n,N^*)}{\mu}$	N^*	$\frac{L_m(n,N^*)}{\mu}$
0.1	2	17.585	1	15.554	1	15.131	1	16.128
0.05	3	16.859	2	14.589	2	13.654	1	13.745
0.01	13	16.326	10	13.866	7	12.454	5	12.060
0.005	26	16.263	20	13.782	13	12.324	10	11.874
0.001	129	16.212	99	13.716	66	12.224	48	11.734
0.0005	257	16.206	198	13.708	132	12.212	96	11.717
0.0001	1283	16.201	987	13.702	661	12.202	477	11.703
0.00005	2566	16.201	1973	13.701	1322	12.201	981	11.702

Next, we find an optimal number n^* for 2-out-of-3 node model that minimizes $L_m(n, N)$ in (22) for $N \geq 1$.

Example 13.6.
Table 7 presents an optimal number n^* and the resulting execution time $L_m(n^*, N)/\mu$ for $\lambda\mu$ and n when $Cs/\mu = 0.8$, $C/\mu = 0.2$, $W/\mu = 5$, $R/\mu = 10$ and $\mu = 1$. The resulting execution times $L_m(n^*, N)$ increase with N and $\lambda\mu$. This shows a similar tendency to Tables 1, 2 and 4.

Finally, we find both optimal n^* and N^* that minimize $L_m(n, N)$ in (22). We compute both optimal n^* and N^* by the similar method used in Scheme 3 of Sect. 2. This indicates an optimal combination of numbers of database-nodes and JC between FC for a 2-out-of-3 node model.

Example 13.7.
Table 8 presents optimal number n^*, N^* and the resulting execution time $L_m(n^*, N^*)/\mu$ when $Cs/\mu = 0.8$, $C/\mu = 0.2$, $W/\mu = 5$, $R/\mu = 10$ and

Table 7   Optimal number n^* and resulting execution time $L_m(n^*, N)/\mu$ for $\lambda\mu$ and n when $Cs/\mu = 0.8, C/\mu = 0.2, W/\mu = 5, R/\mu = 10$ and $\mu = 1$.

$\lambda\mu$	$N = 1$		$N = 2$		$N = 5$		$N = 10$	
	n^*	$\frac{L_m(n^*,N)}{\mu}$	n^*	$\frac{L_m(n^*,N)}{\mu}$	n^*	$\frac{L_m(n^*,N)}{\mu}$	n^*	$\frac{L_m(n^*,N)}{\mu}$
0.1	4	15.041	3	15.496	2	17.365	1	20.308
0.05	7	13.633	5	13.654	4	14.432	2	15.692
0.01	29	12.356	23	12.039	16	11.987	12	12.139
0.005	56	12.180	46	11.822	32	11.675	23	11.710
0.001	274	12.037	224	11.645	158	11.423	117	11.366
0.0005	547	12.018	446	11.622	315	11.392	233	11.323
0.0001	2729	12.004	2228	11.604	1575	11.366	1163	11.289
0.00005	5457	12.002	4455	11.602	3150	11.363	2326	11.284

Table 8   Optimal number n^*, N^* and resulting execution time $L_m(n^*, N^*)/\mu$ for $\lambda\mu$ when $Cs/\mu = 0.8, C/\mu = 0.2, W/\mu = 5, R/\mu = 10$ and $\mu = 1$.

$\lambda\mu$	n^*	N^*	$\frac{L_m(n^*,N^*)}{\mu}$
0.1	4	1	15.041
0.05	5	2	13.654
0.01	18	4	11.971
0.005	30	6	11.675
0.001	94	16	11.357
0.0005	151	25	11.298
0.0001	448	73	11.233
0.00005	713	116	11.221

$\mu = 1$. This indicates n^*/N^* decrease gradually with $\lambda\mu$, and $L_m(n^*, N^*)$ increase with $\lambda\mu$ and are less than those in Tables 6 and 7.

4   Conclusion

This chapter have considered three types of checkpoint schemes and two types of database-nodes systems, and have solved the problems in what places we should place suitable checkpoints, provide optimal number of database-nodes and compare which database-node model shows good performance. We have derived the optimal JC number between FC for the number of database-nodes, the optimal database-nodes for the number of tasks for FC, the optimal database-nodes and JC number between FC, and the optimal execution time for the same number of database-nodes. It has been shown that the optimal mean execution times decrease when the rates

$\mu/(1/\lambda)$ of the mean processing time with the mean error time decrease, and that the mean time of execution time in a 2-out-of-3 node model is smaller than a single node model in some range. This would be applied to cloud systems and other database systems requested for high reliability and high speed processing.

References

Chen, M., Nakamura, S. and Nakagawa, T. (2010). Replacement and preventive maintenance models with random working times, *IEICE TRANSACTIONS on Fundamentals of Electronics, Communications and Computer Sciences* **93-A**, 2, pp. 500–507, doi:10.1587/transfun.E93.A.500, http://ci.nii.ac.jp/naid/10026863281/.

Kim, H. and Shin, K. G. (1996). Design and analysis of an optimal instruction-retry policy for tmr controller computers, *IEEE Transactions on Computers* **45**, 11, pp. 1217–1225.

Lee, P. A. and Anderson, T. (1990). *Fault Tolerance Principles and Practice* (Springer, Wien).

Maeji, S., Naruse, K. and Nakagawa, T. (2010). Optimal checking models with random working times, in *Advanced Reliability and Modelling IV*, pp. 488–495.

Nakagawa, S., Fukumoto, S. and Ishii, N. (2003). Optimal checkpointing intervals of three error detection schemes by a double modular redundancy, *Mathematical and Computing Modeling* **38**, 11–13, pp. 1357–1363.

Nakagawa, T. (2005). *Maintenance Theory of Reliability* (Springer, London).

Nakagawa, T. (2008). *Advanced Reliability Models and Maintenance Policies* (Springer, London).

Nakagawa, T., Naruse, K., and Maeji, S. (2009). Random checkpoint models with N tandem tasks, *IEICE TRANSACTIONS on Fundamentals of Electronics, Communications and Computer Sciences* **E92-A**, 2, pp. 1572–1577.

Naruse, K., Nakagawa, T. and Maeji, S. (2006). Optimal checpoint intervals for error detection by multiple modular redundancies, in *Advanced Reliability and Modelling II*, pp. 293–300.

Ohara, M., Suzuki, R., Arai, M., Fukumoto, S. and Iwasaki, K. (2006-09-01). Analytical model on hybrid state saving with a limited number of checkpoints and bound rollbacks(reliability, maintainability and safety analysis), *IEICE TRANSACTIONS on Fundamentals of Electronics, Communications and Computer Sciences* **89**, 9, pp. 2386–2395, http://ci.nii.ac.jp/naid/110007537954/.

ORACLE (2012). My sql :: Mysql cluster cge, URL http://www-jp.mysql.com/products/cluster/.

Pradhan, D. and Vaidya, N. (1992). Roll-forward checkpointing scheme: Concurrent retry with non-dedicated spares, in *Proceedings of the IEEE Workshop on Fault-Tolerant Parallel and Distributed Systems*, pp. 166–174.

Siewiorek, D. P. and Swarz(eds), R. S. (1982). *The Theory and Practice of Reliable System Design* (Digital Press, Bedford, Massachusetts).

Sugiura, T., Mizutani, S. and Nakagawa, T. (2004). Optimal random replacement polices, in *Tenth ISSAT International Conference on Reliability and Quality in Design*, pp. 99–103.

Zhang, Y. and Chakrabarty, K. (2004). Dynamic adaptation for fault tolerance and power management in embedded real-time systems, *ACM Transactions on Embedded Computing* **3**, 2, pp. 336–360.

Ziv, A. and Bruck, J. (1997). Performance optimization of checkpointing schemes with task duplication, *IEEE Transactions on Computers* **46**, 12, pp. 1381–1386.

Ziv, A. and Bruck, J. (1998). Analysis of checkpointing schemes with task duplication, *IEEE Transactions on Computers* **47**, 2, pp. 222–227.

Chapter 14

Periodic and Random Inspections for a Computer System

Mingchih Chen[1], Xufeng Zhao[2] and Syouji Nakamura[3]

[1] *Graduate Institute of Business Administration,*
Fu Jen Catholic University,
No. 510, Zhongzheng Rd., Xinzhuang, New Taipei City 24205, Taiwan
[2] *Graduate School of Management and Information Sciences,*
Aichi Institute of Technology,
1247 Yachigusa, Yakusa-cho, Toyota 470-0392, Japan
[3] *Department of Human Life and Information,*
Kinjo Gakuin University,
1723 Omori 2, Moriyama, Nagoya 463-8521, Japan

1 Introduction

It has been well-known that faults in computer systems sometimes occur intermittently [Malaiya and Su (1981); Castillo *et al.* (1982); Nakagawa (2005)]: Faults are hidden and become permanent failure when the duration of hidden faults exceeds a threshold level [Nakagawa *et al.* (1993); Nakagawa (2008)]. To prevent such faults, some inspection policies for computer systems were considered [Malaiya and Su (1981); Su *et al.* (1978); Malaiya (1982)], data transmission strategies for communication systems were considered [Yasui *et al.* (2002)], and some properties for security measures of software in computer systems were observed [Liu and Traore (2007)].

The reliability models and a variety of maintenance models play an important role in manufacturing or computer systems. Some applications of reliability models in computer systems, such as communications, backup policies, checkpoint intervals, were summarized [Nakamura and Nakagawa (2010)]. The latest work proposed new reliability and fault-tolerant methods which were applied in manufacturing modules [Savsar and Aldai-hani (2012)], system reliability allocation based on Bayesian network [Qian *et al.* (2012)], supporting real-time data services [Xiao *et al.* (2012)], and software reliability modeling based on gene expression [Zhang and Xiao (2012)].

Some systems in offices and industries execute successive jobs and com-puter processes. For such systems, it would be impossible or impractical to maintain them in a strictly periodic fashion [Barlow and Proschan (1965)]. For example, when a job has a variable working cycle and processing time, it would be better to do some maintenance after it has completed its work and process [Nakagawa (2005); Zhao and Nakagawa (2012)]. The opti-mal periodic policy for such systems when the working time is exponential was derived [Nakagawa (2005)] and periodic and random inspection policies were summarized [Nakagawa (2008)]. Furthermore, several backup policies for database systems with random working times were discussed [Naruse *et al.* (2009); Maeji *et al.* (2010)], by applying the inspection policy to the backup policy.

We firstly apply a standard inspection policy [Nakagawa (2005)] with imperfect checks to a computer system in this chapter: The system has to be operated for an infinite time span. To detect fault, the system is checked at periodic times and random times according to computer processes, which are called periodic inspection and random inspection, respectively. System fault is detected at the next checking time with a certain probability and undetected fault is detected at the second checking time with the same probability.

Secondly, we apply periodic and random inspections into fault detection, rollback operation, and backup process [Reuter (1984); Fukumoto (1992)] for a computer system, whose whole processes is called backup operation in this chapter. Two cases when the fault could be detected immediately and at the following checking time are discussed.

For the above two models, expected costs are obtained and their opti-mal policies which minimize them are derived when fault and process times are exponential. Comparisons between optimal times for periodic and ran-dom inspections when both fault and processing times are exponential are given. It is shown that periodic inspection is better than the random pol-icy; however, if a modified random inspection cost is lower than that for periodic policy, then two inspections are almost the same. Furthermore, we consider random inspection policies in which the system is checked at the Nth interval of random process times for respective models.

2 Model I

2.1 *Periodic Inspection*

Consider a standard inspection policy [Nakagawa (2005)] with imperfect checks: A computer system should operate for an infinite time span and is checked at periodic times kT $(k = 1, 2, \cdots)$. System fault is detected at the following checking time with probability q $(0 < q \le 1)$ and maintenance is done immediately. The undetected fault that occurs with probability $p \equiv 1 - q$ could be detected at the second checking time with the same probability q. Such a procedure is continued by the computer system until all faults are detected.

It is assumed that system fault occurs according to a distribution $F(t)$ with finite mean $1/\lambda$ and density function $f(t)$, i.e., $F(t) \equiv \int_0^t f(u)\mathrm{d}u$. All time intervals spent for checks and maintenances are negligible. Let c_T be the cost of one check and c_D be the loss cost per unit of time for interval elapsed between a fault and its detection at some checking time. Then, the probability that the system fault occurs between the jth and $(j+1)$th $(j = 0, 1, 2, \cdots)$ checks, and it is detected after the $(k+1)$th $(k = 0, 1, 2, \cdots)$ check, i.e., at the $(j + k + 1)$th check, is

$$p^k q[F((j+1)T) - F(jT)].$$

Clearly,

$$\sum_{k=0}^{\infty} p^k q \sum_{j=0}^{\infty} [F((j+1)T) - F(jT)] = 1.$$

Then, the expected number of checks until fault is detected is

$$\sum_{k=0}^{\infty} (k+1)p^k q + \sum_{j=0}^{\infty} j[F((j+1)T) - F(jT)]$$

$$= \sum_{j=1}^{\infty} \overline{F}(jT) + \frac{1}{q}, \tag{1}$$

and the mean time from fault occurrence to its detection is

$$\sum_{j=0}^{\infty} \sum_{k=0}^{\infty} p^k q \int_{jT}^{(j+1)T} [(j + k + 1)T - t]\mathrm{d}F(t)$$

$$= T\left[\sum_{j=1}^{\infty} \overline{F}(jT) + \frac{1}{q}\right] - \frac{1}{\lambda}. \tag{2}$$

Therefore, the total expected cost until maintenance is, from (1) and (2),

$$C_P(T) = (c_T + c_D T)\left[\sum_{j=1}^{\infty} \overline{F}(jT) + \frac{1}{q}\right] - \frac{c_D}{\lambda}. \tag{3}$$

In particular, when $F(t) = 1 - e^{-\lambda t}$ $(0 < 1/\lambda < \infty)$,

$$C_P(T) = (c_T + c_D T)\left(\frac{1}{1 - e^{-\lambda T}} + \frac{p}{q}\right) - \frac{c_D}{\lambda}. \tag{4}$$

Differentiating $C_P(T)$ with respect to T and setting it equal to zero,

$$\frac{p}{q}(1 - e^{-\lambda T})^2 e^{\lambda T} + e^{\lambda T} - 1 - \lambda T = \frac{\lambda c_T}{c_D}, \tag{5}$$

whose left-hand side increases from 0 to ∞. Thus, there exists an optimal T^* $(0 < T^* < \infty)$ which satisfies (5), and the resulting cost rate is

$$\frac{C_P(T^*)}{c_D/\lambda} = (e^{\lambda T^*} - 1)\left[\frac{p^2}{q^2}(1 - e^{-\lambda T^*}) + \frac{p+1}{q}\right]. \tag{6}$$

2.2   *Random Inspection*

Consider a random inspection policy (Nakagawa, 2005) with imperfect checks: The computer system is checked at random successive times such as working and processing times for a job which are denoted by Y_j $(j = 1, 2, \cdots)$, where Y_j is independent and has an identical distribution $G(t)$ with finite mean $1/\mu$, and $S_j \equiv \sum_{i=1}^{j} Y_i$ $(j = 1, 2, \cdots)$ and $S_0 \equiv 0$. Then, the probability that the computer system process exactly j times in $[0, t]$ is $\Pr\{S_j < t \le S_{j+1}\} = G^{(j)}(t) - G^{(j+1)}(t)$, where $G^{(j)}(t)$ $(j = 1, 2, \cdots)$ denote the j-fold Stieltjes convolution of $G(t)$ with itself and $\Phi^{(0)}(t) \equiv 1$ for $t \ge 0$. Suppose that every checking cost for such a random policy is c_R and other assumptions are the same as those in Section 2.1.

The probability that system fault occurs between the jth and $(j+1)$th $(j = 0, 1, 2, \cdots)$ checking times and its fault is detected after the $(k+1)$th check is

$$p^k q \int_0^{\infty} \left[\int_{t_1}^{\infty} [F(t_2) - F(t_1)] dG(t_2 - t_1)\right] dG^{(j)}(t_1). \tag{7}$$

Clearly,

$$\sum_{k=0}^{\infty} p^k q \sum_{j=0}^{\infty} \int_0^{\infty} \left[\int_{t_1}^{\infty} [F(t_2) - F(t_1)] dG(t_2 - t_1)\right] dG^{(j)}(t_1)$$

$$= \sum_{j=0}^{\infty} \int_0^{\infty} \left[\int_0^{\infty} [\overline{F}(t_1) - \overline{F}(t_1 + t_2)] dG(t_2) \right] dG^{(j)}(t_1)$$

$$= \sum_{j=0}^{\infty} \int_0^{\infty} \overline{F}(t) dG^{(j)}(t) - \sum_{j=1}^{\infty} \int_0^{\infty} \overline{F}(t) dG(t) = 1.$$

Then, the expected number of checks until maintenance is

$$\sum_{k=0}^{\infty} (k+1) p^k q + \sum_{j=0}^{\infty} j \int_0^{\infty} \left[\int_{t_1}^{\infty} [F(t_2) - F(t_1)] dG(t_2 - t_1) \right] dG^{(j)}(t_1)$$

$$= \int_0^{\infty} M(t) dF(t) + \frac{1}{q}, \tag{8}$$

where $M(t) \equiv \sum_{j=1}^{\infty} G^{(j)}(t)$ denotes the expected number of checks in $[0, t]$, which is called a renewal function in stochastic processes (Nakagawa, 2011). The mean time from fault to its detection is

$$\sum_{j=0}^{\infty} \int_0^{\infty} dG^{(j)}(t_1) \int_{t_1}^{\infty} dG(t_2 - t_1) \int_{t_1}^{t_2} dF(t)$$

$$\times \sum_{k=0}^{\infty} p^k q \int_{t_2}^{\infty} (t_3 - t) dG^{(k)}(t_3 - t_2)$$

$$= \sum_{j=0}^{\infty} \int_0^{\infty} dG^{(j)}(t_1) \int_{t_1}^{\infty} dG(t_2 - t_1) \int_{t_1}^{t_2} \left(\frac{p}{q\mu} + t_2 - t \right) dF(t)$$

$$= \frac{p}{q\mu} + \sum_{j=0}^{\infty} \int_0^{\infty} dG^{(j)}(t_1) \int_0^{\infty} dG(t_2) \int_{t_1}^{t_1+t_2} [\overline{F}(t_1) - \overline{F}(t)] dt$$

$$= \sum_{j=0}^{\infty} \int_0^{\infty} dG^{(j)}(t_1) \int_0^{\infty} \overline{G}(t) [\overline{F}(t_1) - \overline{F}(t + t_1)] dt$$

$$+ \frac{p}{q\mu} = \frac{1}{q\mu} + \frac{1}{\mu} \int_0^{\infty} M(t) dF(t) - \frac{1}{\lambda}. \tag{9}$$

Therefore, the total expected cost until maintenance is, from (8) and (9),

$$C_R(G) = \left(c_R + \frac{c_D}{\mu} \right) \left[\int_0^{\infty} M(t) dF(t) + \frac{1}{q} \right] - \frac{c_D}{\lambda}.$$

In particular, when $G(t) = 1 - e^{-\mu t}$ $(0 < 1/\mu < \infty)$, i.e., $M(t) = \mu t$, the expected cost is a function of μ which is given by

$$C_R(\mu) = \left(c_R + \frac{c_D}{\mu} \right) \left(\frac{\mu}{\lambda} + \frac{1}{q} \right) - \frac{c_D}{\lambda}. \tag{10}$$

Differentiating $C_R(\mu)$ with respect to μ and setting it equal to zero,

$$\left(\frac{\lambda}{\mu}\right)^2 = \frac{q\lambda c_R}{c_D}, \tag{11}$$

and the resulting cost is

$$\frac{C_R(\mu^*)}{c_D/\lambda} = \frac{\lambda}{q\mu}\left(\frac{\lambda}{q\mu} + 2\right). \tag{12}$$

2.3 *Comparison of Periodic and Random Inspections*

We compare periodic and random inspection policies theoretically and numerically when $F(t) = 1 - e^{-\lambda t}$ and $G(t) = 1 - e^{-\mu t}$. For the simplicity of notations, it is assumed that $q = 1$, $\lambda = 1$, $c_T = c_R$, and $c \equiv \lambda c_T/c_D \leq 1$ because expected loss cost for the mean fault delayed time $1/\lambda$ would be much higher than cost c_T of one check in most inspection models.

Under the above assumptions, the total expected cost for the periodic inspection is, from (4),

$$\frac{C_P(T)}{c_D} = \frac{c + T}{1 - e^{-T}} - 1. \tag{13}$$

An optimal T^* which minimizes $C_P(T)$ is given by a solution of equation

$$e^T - 1 - T = c, \tag{14}$$

and the resulting cost is

$$\frac{C_P(T^*)}{c_D} = c + T^* = e^{T^*} - 1. \tag{15}$$

The total expected cost for random inspection is, from (10),

$$\frac{C_R(\mu)}{c_D} = c(1 + \mu) + \frac{1}{\mu}. \tag{16}$$

An optimal μ^* which minimizes $C_R(\mu)$ is given by

$$\frac{1}{\mu^*} = \sqrt{c}, \tag{17}$$

and the resulting cost is

$$\frac{C_R(\mu^*)}{c_D} = c(1 + \mu^*) + \frac{1}{\mu^*} = \left(\frac{1}{\mu^*}\right)^2 + \frac{2}{\mu^*}. \tag{18}$$

When $c = 1$, $T^* = 1.1462$ from (14), and $1/\mu^* = 1$ from (17). Thus, it is easily proved that $0 < T^* < 1.1462$ and $0 < 1/\mu^* \leq 1$.

From (14) and (17), compute a solution of equation

$$Q(T) \equiv e^T - (1 + T + T^2) = 0. \tag{19}$$

Clearly, a solution of (19) is about 1.79. Thus, $Q(T) < 0$ for $0 < T < 1.79$, and hence,

$$0 < \frac{1}{\mu^*} < T^* \leq 1.1462.$$

Next, prove that $2/\mu^* > T^*$. From (14),

$$c = e^T - (1 + T) > \frac{T^2}{2},$$

which follows that $T^* < \sqrt{2c}$. In addition, from (17),

$$\frac{2}{\mu^*} = 2\sqrt{c} > \sqrt{2c} > T^*.$$

Thus,

$$\frac{1}{\mu^*} < T^* < \frac{2}{\mu^*}.$$

Furthermore, from (15) and (18),

$$\frac{C_R(\mu^*) - C_T(T^*)}{c_D} = c(1 + \mu^*) + \frac{1}{\mu^*} - c - T^*$$

$$= c\mu^* + \frac{1}{\mu^*} - T^* > c\mu^* - \frac{1}{\mu^*} = 0.$$

From the above results, $T^* > 1/\mu^*$ and $C_P(T^*) < C_R(\mu^*)$ when $c_T = c_R$, i.e., periodic inspection is better than the random policy and optimal interval T^* is greater than $1/\mu^*$.

Suppose that $F(t) = 1 - e^{-\lambda t}$, $G(t) = 1 - e^{-\mu t}$, $c_T = c_R$, and $q = 0.9$ and $\lambda = 1$. Then, Table 1 presents optimal T^*, $1/\mu^*$ and their resulting costs for c_T/c_D. This indicates as estimated previously that $T^* > 1/\mu^*$ and $C_P(T^*) < C_R(\mu^*)$, i.e., periodic inspection time is larger than the random one, and hence, when $c_T = c_R$, periodic inspection is better than the random policy.

It has been assumed in Table 1 that two checking costs c_T and c_R are the same. In general, cost c_R would be lower than c_T because the system is checked at random times according to random processes. Such random inspections may not break off the random and successive procedures in computers. We compute a modified random inspection cost $\hat{c}_R$ when the

Table 1 Optimal T^*, $1/\mu^*$, and $C_P(T^*)/c_D$, $C_R(\mu^*)/c_D$ when $q = 0.9$ and $\lambda = 1$.

c_T/c_D	T^*	$C_P(T^*)/c_D$	$1/\mu^*$	$C_R(\mu^*)/c_D$
0.001	0.0403	0.0502	0.0300	0.0678
0.002	0.0566	0.0714	0.0424	0.0965
0.005	0.0893	0.1143	0.0671	0.1546
0.010	0.1256	0.1639	0.0949	0.2219
0.020	0.1764	0.2363	0.1342	0.3204
0.050	0.2746	0.3879	0.2121	0.5270
0.100	0.3824	0.5716	0.3000	0.7778
0.200	0.5282	0.8556	0.4243	1.1650
0.500	0.7994	1.5052	0.6708	2.0463
1.000	1.0757	2.3807	0.9487	3.2193

expected costs of two inspection policies are the same. From (6) and (12), we compute $\widehat{\mu}$ which satisfies

$$(e^{\lambda T^*} - 1)\left[\frac{p^2}{q}(1 - e^{-\lambda T^*}) + p + 1\right] = \frac{1}{q}\left(\frac{\lambda}{\widehat{\mu}}\right)^2 + \frac{2\lambda}{\widehat{\mu}},$$

and compute

$$\frac{\widehat{c}_R}{c_D/\lambda} = \frac{1}{q}\left(\frac{\lambda}{\widehat{\mu}}\right)^2.$$

Table 2 presents $1/\widehat{\mu}$, $\widehat{c}_R/c_D$ and $\widehat{c}_R/c_T$ for c_T/c_D when $q = 0.9$ and $\lambda = 1$. This indicates that $\widehat{c}_R$ is a little more than the half of c_T. In other words, when $c_R \approx c_T/2$, two expected costs for periodic and random inspections are almost the same.

Table 2 Values of $1/\widehat{\mu}$, $\widehat{c}_R/c_D$ and $\widehat{c}_R/c_T$ when $q = 0.9$ and $\lambda = 1$.

c_T/c_D	$1/\widehat{\mu}$	$\widehat{c}_R/c_D$	$\widehat{c}_R/c_T$
0.001	0.0224	0.0005	0.5039
0.002	0.0318	0.0010	0.5054
0.005	0.0504	0.0025	0.5086
0.010	0.0715	0.0051	0.5118
0.020	0.1016	0.0103	0.5160
0.050	0.1621	0.0263	0.5253
0.100	0.2313	0.0535	0.5352
0.200	0.3312	0.1097	0.5485
0.500	0.5355	0.2868	0.5735
1.000	0.7738	0.5987	0.5987

It is noted from (14) and (17) that $T^* \to 0$ and $1/\mu^* \to 0$ as $c \to 0$. Thus, from (15) and (18),

$$\lim_{T \to 0} \frac{e^T - 1}{T^2 + 2T} = \frac{1}{2}.$$

This shows that if $c \to 0$, then $C_T(T^*) \to C_R(\mu^*)/2$. So that, it would be estimated that if $c \to 0$ and $c_R/c_T = 0.5$, then two expected costs of the periodic and random policies would be the same as shown in Table 2.

2.4 Nth Random Inspection

Suppose that the computer system is checked at times S_{jN} ($j = 1, 2, \cdots, N = 1, 2, \cdots$), i.e., at times $S_{1N}, S_{2N}, \cdots$. When $N = 1$, the system is checked at every S_j in Section 2.2. Then, from $C_R(G)$, replacing $G(t)$ with $G^{(N)}(t)$, $1/\mu$ with N/μ, and $M(t)$ with $M^{(N)}(t) = \sum_{j=1}^{\infty} G^{(jN)}(t)$ ($N = 1, 2, \cdots$), the total expected cost until fault detection is

$$C_R(N) = \left(c_R + \frac{Nc_D}{\mu}\right)\left[\int_0^\infty M^{(N)}(t)\mathrm{d}F(t) + \frac{1}{q}\right] - \frac{c_D}{\lambda}$$
$$(N = 1, 2, \cdots). \qquad (20)$$

In particular, when $F(t) = 1 - \mathrm{e}^{-\lambda t}$,

$$\int_0^\infty \mathrm{e}^{-\lambda t}\mathrm{d}M^{(N)}(t) = \frac{[G^*(\lambda)]^N}{1 - [G^*(\lambda)]^N},$$

where $G^*(\lambda) \equiv \int_0^\infty \mathrm{e}^{-\lambda t}\mathrm{d}G(t)$, and the expected cost in (20) is

$$C_R(N) = \left(c_R + \frac{Nc_D}{\mu}\right)\left(\frac{1}{1 - A^N} + \frac{p}{q}\right) - \frac{c_D}{\lambda}. \qquad (21)$$

where $A = G^*(\lambda) < 1$. From the inequality $C_R(N+1) - C_R(N) \geq 0$,

$$\frac{1 - A^N}{(1 - A)A^N}\left[1 + \frac{p}{q}(1 - A^{N+1})\right] - N \geq \frac{c_R}{c_D/\mu}, \qquad (22)$$

which increases strictly with N to ∞. Thus, there exists a unique minimum N^* ($1 \leq N^* < \infty$) which satisfies (22). Clearly, N^* increases strictly with q from 1 to a solution of the equation

$$\frac{1 - A^N}{(1 - A)A^N} - N \geq \frac{c_R}{c_D/\mu}.$$

Note that when $G(t) = 1 - \mathrm{e}^{-\mu t}$, $A = \mu/(\lambda + \mu)$.

Table 3 presents optimal N^* and the resulting cost $C_R(N^*)/c_D$ for $1/\mu$ and c_R/c_D when $q = 0.9$ and $1/\lambda = 1$. This indicates that optimal N^* decreases with $1/\mu$ and increases with c_R/c_D, and N^*/μ are almost the same for small $1/\mu$. Compared Table 3 with Table 1, if $c_T = c_R$, when $c_R/c_D = 0.5$, $1/\mu^* < N^*/\mu$ and $C_R(N^*) > C_R(\mu^*)$ as $1/\mu$ is big enough, other wise, $C_R(N^*) < C_R(\mu^*)$; If $c_R \approx c_T/2$, when $c_R/c_D = 0.5$ and $c_T/c_D = 1.0$, $1/\mu^* > N^*/\mu$ and $C_R(N^*) < C_R(\mu^*)$.

Table 3   Optimal N^* and $C_R(N^*)/c_D$ when $q = 0.9$ and $\lambda = 1$.

$1/\mu$	$c_R/c_D = 0.5$		$c_R/c_D = 1.0$	
	N^*	$C_R(N^*)/c_D$	N^*	$C_R(N^*)/c_D$
0.01	80	1.5129	108	2.3894
0.02	40	1.5206	54	2.3981
0.05	16	1.5435	22	2.4241
0.10	8	1.5812	11	2.4666
0.20	4	1.6553	6	2.5522
0.50	2	1.8667	2	2.8222
1.00	1	2.1667	1	3.2222

3   Model II

3.1   *Periodic and Random Inspections*

We apply the above inspection methods into recovery and backup techniques in a computer system. Suppose that a computer system is checked at periodic times kT ($k = 1, 2, \cdots$) for a specified $T > 0$ and also at successive random times S_j ($j = 1, 2, \cdots$) such as working and processing times. When a fault occurs, it is detected immediately, and rollback operation is executed until the latest checking time, and then backup is made for data in this computer. For simplicity, we call such a rollback and backup processes is backup operation in the following sections.

It is assumed that c_T and c_R be the respective costs for periodic and random checks. In addition, when a fault occurs at time t between kT and $(k + 1)T$ or S_{j+1}, we carry out backup operation from fault point to the latest checking time kT. This incurs a loss cost $c_D(t - kT)$ which includes all costs for processing time and rollback operation from t to kT. While, when a fault occurs at time t between S_j and $(k + 1)T$ or S_{j+1}, this incurs a loss cost $c_D(t - S_j)$.

The probability that the process performs rollback to periodic check is

$$\sum_{k=0}^{\infty} \int_{kT}^{(k+1)T} \left[\sum_{j=0}^{\infty} \int_{0}^{kT} \overline{G}(t - x) \mathrm{d}G^{(j)}(x) \right] \mathrm{d}F(t), \qquad (23)$$

and the probability that the process performs rollback to random check is

$$\sum_{k=0}^{\infty} \int_{kT}^{(k+1)T} \left[\sum_{j=0}^{\infty} \int_{kT}^{t} \overline{G}(t - x) \mathrm{d}G^{(j)}(x) \right] \mathrm{d}F(t), \qquad (24)$$

where (23)+(24)=1.

Therefore, the total expected cost until backup operation is

$$C(T) = \sum_{k=0}^{\infty} \int_{kT}^{(k+1)T} \left\{ \sum_{j=0}^{\infty} \int_{0}^{kT} [c_T k + c_R j + c_D(t - kT)] \right.$$

$$\left. \times \overline{G}(t - x) \mathrm{d}G^{(j)}(x) \right\} \mathrm{d}F(t) + \sum_{k=0}^{\infty} \int_{kT}^{(k+1)T} \left\{ \sum_{j=0}^{\infty} \int_{kT}^{t} [c_T k \right.$$

$$\left. + c_R j + c_D(t - x)] \overline{G}(t - x) \mathrm{d}G^{(j)}(x) \right\} \mathrm{d}F(t)$$

$$= c_T \sum_{k=1}^{\infty} \overline{F}(kT) + c_R \int_{0}^{\infty} M(t) \mathrm{d}F(t) + c_D \mu$$

$$- c_D \left\{ \sum_{k=0}^{\infty} (kT) \int_{kT}^{(k+1)T} \left[\sum_{j=0}^{\infty} \int_{0}^{kT} \overline{G}(t - x) \mathrm{d}G^{(j)}(x) \right] \mathrm{d}F(t) \right.$$

$$\left. + \sum_{k=0}^{\infty} \int_{kT}^{(k+1)T} \left[\sum_{j=0}^{\infty} \int_{kT}^{t} x \overline{G}(t - x) \mathrm{d}G^{(j)}(x) \right] \mathrm{d}F(t) \right\}. \qquad (25)$$

Clearly, when $G(t) \equiv 0$, i.e., the system is checked only at periodic times kT $(k = 1, 2, \cdots)$,

$$C_P(T) = (c_T - c_D T) \sum_{k=1}^{\infty} \overline{F}(kT) + c_D \mu, \qquad (26)$$

which agrees with (5.55) of (Nakagawa, 2008). When $T = \infty$, i.e., the system is checked only at random times S_j $(j = 1, 2, \cdots)$,

$$C_R(G) = c_D \left\{ \mu - \int_{0}^{\infty} \left[\sum_{j=0}^{\infty} \int_{0}^{t} x \overline{G}(t - x) \mathrm{d}G^{(j)}(x) \right] \mathrm{d}F(t) \right\}$$

$$+ c_R \int_{0}^{\infty} M(t) \mathrm{d}F(t), \qquad (27)$$

which agrees with (Nakagawa, 2008).

When $G(t) = 1 - \mathrm{e}^{-\mu t}$, the total expected cost in (25) is

$$C(T) = c_T \sum_{k=1}^{\infty} \overline{F}(kT) + c_R \frac{\mu}{\lambda} + \frac{c_D}{\mu} \sum_{k=0}^{\infty} \int_{kT}^{(k+1)T} [1 - \mathrm{e}^{-\mu(t - kT)}] \mathrm{d}F(t).$$

$$(28)$$

In particular, when $F(t) = 1 - \mathrm{e}^{-\lambda t}$,

$$C(T) = \frac{c_T}{\mathrm{e}^{\lambda T} - 1} + \frac{c_R \mu}{\lambda} + \frac{c_D}{\mu} \left[1 - \frac{\lambda}{\mu + \lambda} \frac{1 - \mathrm{e}^{-(\mu + \lambda)T}}{1 - \mathrm{e}^{-\lambda T}} \right]. \qquad (29)$$

Clearly,

$$C(0) \equiv \lim_{T \to 0} C(T) = \infty,$$

$$C(\infty) \equiv \lim_{T \to \infty} C(T) = \frac{c_R \mu}{\lambda} + \frac{c_D}{\mu + \lambda}.$$

We find an optimal T^* $(0 < T^* \le \infty)$ which minimizes $C(T)$ in (29). Differentiating $C(T)$ with respect to T and setting it equal to zero,

$$\frac{1 - e^{-\mu T}}{\mu} - \frac{1 - e^{-(\mu + \lambda)T}}{\mu + \lambda} = \frac{c_T}{c_D}, \tag{30}$$

whose left-hand increases strictly with T from 0 to $\lambda/[\mu(\mu + \lambda)]$. Therefore, if $\lambda/(\mu + \lambda) > c_T/(c_D/\mu)$ then there exists a finite and unique T^* $(0 < T^* < \infty)$ which satisfies (30). In addition, the left-hand side of (30) increases strictly with $1/\mu$, i.e., optimal T^* decreases with $1/\mu$.

3.2   *Comparison of Periodic and Random Inspections*

We compare periodic and random inspections when $F(t) = 1 - e^{-\lambda t}$ $(0 < \lambda < \infty)$ and $G(t) = 1 - e^{-\mu t}$ $(0 < \mu < \infty)$. In this case, the total expected cost in (26) is

$$C_P(T) = \frac{c_T - c_D T}{e^{\lambda T} - 1} + \frac{c_D}{\lambda}. \tag{31}$$

Clearly,

$$C_P(0) \equiv \lim_{T \to 0} C_P(T) = \infty,$$

$$C_P(\infty) \equiv \lim_{T \to \infty} C_P(T) = \frac{c_D}{\lambda}.$$

Thus, there exists an optimal T^* $(0 < T^* \le \infty)$ which minimizes (31). Differentiating $C_P(T)$ with respect to T and setting it equals to zero,

$$\lambda T - (1 - e^{-\lambda T}) = \frac{c_T}{c_D/\lambda}, \tag{32}$$

whose left-hand increases strictly from 0 to ∞. Thus, there exists a finite and unique T^* $(0 < T^* < \infty)$ which satisfies (32), and the resulting cost is

$$\frac{C_P(T^*)}{c_D/\lambda} = 1 - e^{-\lambda T^*}. \tag{33}$$

Similarly, the total expected cost in (27) is

$$C_R(\mu) = \frac{c_R \mu}{\lambda} + \frac{c_D}{\mu + \lambda}. \tag{34}$$

An optimum μ^* which minimizes $C_R(\mu)$ is easily given by

$$\frac{\lambda}{\mu + \lambda} = \sqrt{\frac{\lambda c_R}{c_D}}, \tag{35}$$

and the resulting cost is

$$\frac{C_R(\mu^*)}{c_D/\lambda} = \frac{\lambda}{\mu^* + \lambda} \left(\frac{\mu^*}{\mu^* + \lambda} + 1 \right). \tag{36}$$

Clearly, if $\lambda c_R/c_D \geq 1$, then $1/\mu^* = \infty$, and $c_R(0) = c_D/\lambda$.

Table 4 presents T^*, $1/\mu^*$ and their costs $C_P(T^*)/c_D$, $C_R(\mu^*)/c_D$ for c_T/c_D when $c_T = c_R$ and $\lambda = 1$. This indicates that $T^* > 1/\mu^*$ when c_T/c_D is small, and $C_P(T^*) < C_R(\mu^*)$, i.e., the periodic inspection is better than the random policy. When

$$T - \left(\frac{T}{1+T} \right)^2 = 1 - e^{-T},$$

$T^* = 1/\mu^* = 0.694$. In this case,

$$\frac{c_T}{c_D} = \left(\frac{T^*}{1+T^*} \right)^2 = 0.168.$$

That is, when $c_T/c_D = 0.168$, $T^* = 1/\mu^* = 0.694$, and $T^* > 1/\mu^*$ for $c_T/c_D < 0.168$ and $T^* < 1/\mu^*$ for $c_T/c_D > 0.168$.

Table 4   Optimum T^*, $1/\mu^*$ and their cost rates when $c_T = c_R$ and $\lambda = 1$.

c_T/c_D	T^*	$C_P(T^*)/c_D$	$1/\mu^*$	$C_R(\mu^*)/c_D$
0.001	0.045	0.044	0.033	0.063
0.002	0.064	0.062	0.047	0.088
0.005	0.102	0.097	0.076	0.136
0.010	0.145	0.135	0.111	0.190
0.020	0.207	0.187	0.165	0.263
0.050	0.334	0.284	0.288	0.397
0.100	0.483	0.383	0.463	0.532
0.200	0.707	0.507	0.809	0.694
0.500	1.198	0.698	2.414	0.914
1.000	1.841	0.841	∞	1.000

It has been assumed that $c_T = c_R$ in Table 4. Next we compute a modified random checking cost $\widehat{c}_R$ when the expected costs of two inspections are the same. From Table 4, we compute $1/\widehat{\mu}$ for c_T/c_D when

$$\frac{C_P(T^*)}{c_D} = 1 - e^{-T^*} = \frac{2\widehat{\mu} + 1}{(\widehat{\mu} + 1)^2},$$

which decreases strictly with $\widehat{\mu}$ from 1 to 0, and compute

$$\frac{\widehat{c}_R}{c_D} = \left(\frac{1}{\widehat{\mu}+1}\right)^2.$$

Table 5 presents $1/\widehat{\mu}$, $\widehat{c}_R/c_D$ and $\widehat{c}_R/c_T$, and indicates that $\widehat{c}_R$ is a little less than the half of c_T. In other words, when the random checking cost is the half of the periodic one, two expected costs are almost the same. For example, $C_P(T^*)/c_T$ when $c_T/c_D = 0.002, 0.010, 0.020, 0.100, 0.200, 1.000$, are almost equal to $C_R(\mu^*)/c_T$ when $c_T/c_D = 0.001, 0.005, 0.010, 0.050, 0.100, 0.500$, respectively, in Table 4.

Table 5   Values of $1/\widehat{\mu}$, $\widehat{c}_R/c_D$ and $\widehat{c}_R/c_T$ when $\lambda = 1$.

c_T/c_D	$1/\widehat{\mu}$	$\widehat{c}_R/c_D$	$\widehat{c}_R/c_T$
0.001	0.023	0.0005	0.5000
0.002	0.032	0.0010	0.5000
0.005	0.052	0.0025	0.5000
0.010	0.075	0.0049	0.4900
0.020	0.109	0.0097	0.4850
0.050	0.182	0.0237	0.4740
0.100	0.273	0.0460	0.4600
0.200	0.424	0.0887	0.4435
0.500	0.820	0.2030	0.4060
1.000	1.511	0.3623	0.3623

3.3   *N th Random Inspection I*

Suppose that the computer system is checked at every Nth $(N = 1, 2, \cdots)$ random process times, i.e., $S_N, S_{2N}, \cdots$. By replacing $G(t)$ with $G^{(N)}(t)$ in (27) formally, the total expected cost until backup operation is

$$C_{R1}(N) = c_R \int_0^\infty M^{(N)}(t)\mathrm{d}F(t) + c_D\mu$$

$$- c_D \int_0^\infty \left[\sum_{j=0}^\infty \int_0^t x[1 - G^{(N)}(t-x)]\mathrm{d}G^{(jN)}(x)\right] \mathrm{d}F(t), \quad (37)$$

where $M^{(N)}(t) \equiv \sum_{j=1}^\infty G^{(jN)}(t)$.

In particular, when $F(t) = 1 - \mathrm{e}^{-\lambda t}$ and $G(t) = 1 - \mathrm{e}^{-\mu t}$,

$$C_{R1}(N) = c_R \frac{A^N}{1 - A^N} + \frac{c_D}{\lambda} \frac{(1-A)^2}{A(1-A^N)} \sum_{j=1}^N jA^j \quad (N = 1, 2, \cdots), \quad (38)$$

where $A \equiv \mu/(\mu + \lambda)$. From the inequality $C_{R1}(N+1) - C_{R1}(N) \geq 0$,

$$(1 - A) \sum_{j=1}^{N} (1 - A^j) \geq \frac{c_R}{c_D/\lambda} \quad (N = 1, 2, \cdots). \tag{39}$$

Therefore, there exists a finite and unique minimum N_1^* $(1 \leq N_1^* < \infty)$ which satisfies (39). If $[\lambda/(\mu + \lambda)]^2 \geq c_R/(c_D/\lambda)$ then $N_1^* = 1$. Note that N_1^* decreases with A, i.e., N_1^* decreases with $1/\mu$ from ∞ to a minimum N integer such that $N \geq c_R/(c_D/\lambda)$.

Next, we obtain the expected cost until N processes have been completed. First, suppose that $N = 1$. When system fault occurs between S_j and S_{j+1}, we carry out backup operation to the latest checking time S_j. Then, the expected cost until completion of one process is given by the renewal equation

$$\widetilde{C}_R(1) = c_R \int_0^\infty \overline{F}(t) dG(t) + \int_0^\infty [c_D t + \widetilde{C}_R(1)] \overline{G}(t) dF(t). \tag{40}$$

The expected cost until completion of N processes is, by replacing $G(t)$ with $G^{(N)}(t)$ formally,

$$\widetilde{C}_R(N) = c_R + \frac{c_D \int_0^\infty t[1 - G^{(N)}(t)] dF(t)}{\int_0^\infty G^{(N)}(t) dF(t)} \quad (N = 1, 2, \cdots). \tag{41}$$

As an approximate objective function, we adopt the expected cost per one process given by

$$C_{R2}(N) \equiv \frac{\widetilde{C}_R(N)}{N} = \frac{c_R}{N} + \frac{c_D}{N} \frac{\int_0^\infty t[1 - G^{(N)}(t)] dF(t)}{\int_0^\infty G^{(N)}(t) dF(t)} \quad (N = 1, 2, \cdots). \tag{42}$$

In particular, when $F(t) = 1 - e^{-\lambda t}$ and $G(t) = 1 - e^{-\mu t}$,

$$C_{R2}(N) = \frac{c_R}{N} + \frac{c_D}{\lambda} \frac{(1-A)^2}{N A^{N+1}} \sum_{j=1}^{N} j A^j \quad (N = 1, 2, \cdots). \tag{43}$$

From the inequality $C_{R2}(N+1) - C_{R2}(N) \geq 0$,

$$\frac{1-A}{A^{N+1}} \sum_{j=1}^{N} (1 - A^j) \geq \frac{c_R}{c_D/\lambda} \quad (N = 1, 2, \cdots), \tag{44}$$

whose left-hand side increases strictly from $(\lambda/\mu)^2$ to ∞. Therefore, there exists a finite and unique N_2^* $(1 \leq N_2^* < \infty)$ which satisfies (44). If $(\lambda/\mu)^2 \geq c_R/(c_D/\lambda)$, then $N_2^* = 1$, i.e., we should place checking times at every completion of process. Clearly, N_2^* decreases with $1/\mu$ from ∞ to 1. Compared to (39) and (44), $N_1^* \geq N_2^*$.

3.4 *Nth Random Inspection II*

It has been assumed until now that fault is detected immediately. Suppose that the fault is detected only at random checking times and backup operation is executed until the latest checking time. Then, the total expected cost until backup operation is

$$C_M(1) = c_D \sum_{j=0}^{\infty} \int_0^{\infty} \left\{ \int_0^{\infty} x[F(t+x) - F(t)]dG(x) \right\} dG^{(j)}(t)$$

$$+ c_R \sum_{j=0}^{\infty} \int_0^{\infty} G^{(j)}(t)dF(t). \tag{45}$$

Suppose that the system is checked at every Nth ($N = 1, 2, \cdots$) random times, i.e., $S_N, S_{N+1}, \cdots$. By replacing $G(t)$ with $G^{(N)}(t)$ in (45), the total expected cost until backup operation is

$$C_{M1}(N) = c_D \sum_{j=0}^{\infty} \int_0^{\infty} \left\{ \int_0^{\infty} x[F(t+x) - F(t)]dG^{(N)}(x) \right\} dG^{(Nj)}(t)$$

$$+ c_R \sum_{j=0}^{\infty} \int_0^{\infty} G^{(Nj)}(t)dF(t) \quad (N = 1, 2, \cdots). \tag{46}$$

In particular, when $F(t) = 1 - e^{-\lambda t}$ and $G(t) = 1 - e^{-\mu t}$,

$$C_{M1}(N) = c_R \sum_{j=0}^{\infty} (A^N)^j + c_D \left(\frac{N}{\mu} - \frac{N}{\mu+\lambda} A^N \right) \sum_{j=0}^{\infty} (A^N)^j$$

$$= \frac{c_R}{1 - A^N} + \frac{c_D}{\lambda} \frac{N(1-A)(1-A^{N+1})}{A(1-A^N)} \quad (N = 1, 2, \cdots).$$

From the inequality $C_{M1}(N+1) - C_{M1}(N) \geq 0$,

$$\frac{(1-A)^2}{A} \left[\frac{(1-A^N)(1-A^{N+2})}{(1-A)^2 A^N} - N \right] \geq \frac{c_R}{c_D/\lambda} \quad (N = 1, 2, \cdots), \tag{47}$$

whose bracket on the left-hand increases strictly with N to ∞. Clearly, the left-hand side of (47) decreases with A, i.e., N_3^* decreases with $1/\mu$ from ∞ to 1. Therefore, there exists a finite and unique N_3^* ($1 \leq N_3^* < \infty$) which satisfies (47).

Next, suppose that when system fault occurs between S_j and S_{j+1}, its fault is detected at time S_{j+1} and we carry out the backup operation from S_{j+1} to S_j. Then, the expected cost between one process is given by the renewal equation

$$\tilde{C}_M(1) = c_R \int_0^{\infty} \overline{F}(t)dG(t) + \int_0^{\infty} [c_D t + \tilde{C}_M(1)]F(t)dG(t). \tag{48}$$

The expected cost until the completion of N processes is, by replacing $G(t)$ with $G^{(N)}(t)$,

$$\widetilde{C}_M(N) = c_R + \frac{c_D \int_0^\infty tF(t)\mathrm{d}G^{(N)}(t)}{\int_0^\infty \overline{F}(t)\mathrm{d}G^{(N)}(t)} \quad (N = 1, 2, \cdots). \tag{49}$$

Thus, the expected cost per one process is

$$C_{M2}(N) \equiv \frac{\widetilde{C}_M(N)}{N} = \frac{c_R}{N} + \frac{c_D}{N} \frac{\int_0^\infty tF(t)\mathrm{d}G^{(N)}(t)}{\int_0^\infty \overline{F}(t)\mathrm{d}G^{(N)}(t)}. \tag{50}$$

In particular, when $F(t) = 1 - \mathrm{e}^{-\lambda t}$ and $G(t) = 1 - \mathrm{e}^{-\mu t}$,

$$C_{M2}(N) = \frac{c_R}{N} + \frac{c_D}{\lambda} \frac{(1 - A)(1 - A^{N+1})}{A^{N+1}} \quad (N = 1, 2, \cdots). \tag{51}$$

From the inequality $C_{M2}(N + 1) - C_{M2}(N) \geq 0$,

$$\left(\frac{1 - A}{A}\right)^2 \frac{N(N + 1)}{A^N} \geq \frac{c_R}{c_D/\lambda} \quad (N = 1, 2, \cdots), \tag{52}$$

whose left-hand increases strictly to ∞. Therefore, there exists a finite and unique N_4^* $(1 \leq N_4^* < \infty)$ which satisfies (52). If $\lambda(\mu+\lambda)/\mu^3 \geq c_R/(2c_D/\lambda)$ then $N_4^* = 1$. Clearly, the left-hand side of (52) decreases with A, i.e., N_4^* decreases with $1/\mu$ from ∞ to 1, and $N_3^* \geq N_4^*$.

Table 6 presents optimal N_1^*, N_2^*, N_3^*, and N_4^* when $c_R/(c_D/\lambda) = 0.1$ for λ/μ. This indicates that all N_i^* decreases with λ/μ and $N_1^* \geq N_2^*$ and $N_3^* \geq N_4^*$, and are almost the same for large λ/μ.

Table 6   Optimal N_1^*, N_2^*, N_3^*, and N_4^* when $c_R/(c_D/\lambda) = 0.1$.

λ/μ	N_1^*	N_2^*	N_3^*	N_4^*
0.01	49	39	32	28
0.02	25	20	16	14
0.05	10	8	6	6
0.10	5	4	3	3
0.20	3	2	2	2
0.50	1	1	1	1

4   Conclusions

We have firstly used a standard inspection policy with imperfect check to detect fault in a computer system. For the first model, the system is checked at periodic times and random times to detect faults according to computer processes, which are called periodic inspection and random inspection in this chapter. System fault is detected at the next checking time with a certain probability and undetected fault is detected at the second checking time with the same probability. For the second model, we have applied periodic and random inspections into fault detection, rollback operation, and backup process for a computer system.

For these proposed two models, expected costs of periodic and random inspections have been obtained and the optimal inspection times which minimize them have been derived analytically, when the failure and random times are exponential. It has been shown numerically that when costs for periodic and random checks are the same, periodic inspections are better than random policies. However, it is of interest that if a modified cost for random check is lower than that for periodic policy, random inspection would be better and more practical. Furthermore, we have considered the inspection policy in which the system is checked at the Nth interval and derived analytically optimal N when the fault time is exponential. Two cases when the fault is detected immediately and at the following checking time have been discussed for the model that is an application of inspections.

References

Barlow, R. E. and Proschan, F. (1965). *Mathematical Theory of Reliability* (Wiley, New York).

Castillo, X., McConner, S. R. and Siewiorek, D. P. (1982). Derivation and calibration of a transient error reliability model, *IEEE Transactions on Computers* **C-31**, pp. 658–671.

Fukumoto, S., Kaio, N. and Osaki, S. (1992). A study of checkpoint generations for a database recovery mechanism, *Computer & Mathematics with Applications* **24**, pp. 63–70.

Liu, Y. M. and Traore, I. (2007). Properties for security measures of software products, *Applied Mathematics & Information Sciences* **1**, pp. 129–156.

Malaiya, Y. K. and Su, S. Y. H. (1981). Reliability measure of hardware redundancy fault-tolerant digital systems with intermittent faults, *IEEE Transactions on Computers* **C-30**, pp. 600–604.

Malaiya, Y. K. (1982). Linearly corrected intermittent failures, *IEEE Transactions on Reliability* **R-31**, pp. 211–215.

Maeji, S., Naruse, K. and Nakagawa, T. (2010). Optimal checking models with random working times. In: Chukova, S., Haywood, J. and Dohi, T. (Eds) *Advanced Reliability Modeling IV* (McGraw-Hill, Taiwan), pp. 488–495.

Nakagawa, T., Yasui, K. and Sandoh, H. (1993). An optimal policy for a data transmission system with intermittent faults, *IEICE Transactions on Fundamentals of Electronics, Communications and Computer Sciences* **J76-A**, pp. 1201–1206.

Nakagawa, T. (2005). *Maintenance Theory of Reliability* (Springer, London).

Nakagawa, T. (2008). *Advanced Reliability Models and Maintenance Policies* (Springer, London).

Nakagawa, T. (2011). *Stochastic Process with Applications to Reliability Theory* (Springer, London).

Nakamura, S. and Nakagawa, T. (2010). *Stochastic Reliability Modeling, Optimization and Applications* (World Scientific, Singapore).

Naruse, K., Nakagawa, T. and Maeji, S. (2009). Random checkpoint models with *N* tanem tasks, *IEICE Transactions on Fundamentals of Electronics Communications and Computer Sciences* **E92-A**, pp. 1572–1577.

Qian, W., Yin, X. and Xie, L. (2012). System reliability allocation based on Bayesian network, *Applied Mathematics & Information Sciences* **6**, pp. 681–687.

Reuter, A. (1984). Performance analysis of recovery techniques, *ACM Transaction on Database Systems* **9**, pp. 526–559.

Su, S. Y. H. and Koren, T. and Malaiya, Y. K. (1978). A Continuous-paremeter markov model and detection procedures for intermittent faults, *IEEE Transactions on Computers* **C-27**, pp. 567–570.

Savsar, M. and Aldaihani, M. (2012). A stochastic model for analysis of manufacturing modules, *Applied Mathematics & Information Sciences* **6**, pp. 587–600.

Xiao, Y., Zhang, H., Xu, G. and Wang, J. (2012). A prediction recovery method for supporting real-time data services, *Applied Mathematics & Information Sciences* **6-2S**, 363S–369S.

Yasui, K., Nakagawa, T. and Sandoh, H. (2002). Reliability models in reliability and maintenance. In: S. Osaki (Ed.) *Stochastic Models in Reliability and Maintenance* (Springer, Berlin), pp. 281–301.

Zhao, X. and Nakagawa, T. (2012). Optimization problems of replacement first or last in reliability theory, *European Journal of Operational Research* **223**, pp. 141–149.

Zhang, Y. and Xiao, J. (2012). A software reliability modeling method based on gene expression programming, *Applied Mathematics & Information Sciences* **6**, pp. 125–132.

PART 4
Reliability Applications

Chapter 15

Dynamic Fault Tree Analysis

Tetsushi Yuge and Shigeru Yanagi

Department of Electrical and Electronic Engineering,
National Defense Academy,
Hashirimizu 1-10-20, Yokosuka 239-8686, Japan

1 Introduction

Fault tree analysis (FTA) is one of the oldest, most important logic and probabilistic techniques in industrial applications [Lee *et al.* (1985); Vesely (1981); Stamatelatos (2002)]. It can be simply described as an analytical technique, whereby an undesired state of a system, called a top event (TE), is specified (a state that is critical from a safety or reliability standpoint), and the system is then analyzed in the context of its environment and operation to find all realistic ways in which the TE can occur. The faults can be events that are associated with component hardware failures, human errors, software errors, or any other pertinent events which can lead to the undesired event. A fault tree (FT) thus depicts the logical interrelationships of basic events that lead to the TE.

One of the main aims in FTA is to obtain the exact TE probability, but the exact analysis of a reasonably large-scale FT with a complex structure, such as that for a chemical plant, a nuclear reactor or an airplane, can be expensive, in terms of both the time required to develop the FT model and the time required to solve the model. Several types of dynamic behavior in such a system as well as the scale of the system make the analysis difficult. Examples of dynamic behavior include transient recovery, intermittent errors, and sequence dependence [Dugan *et al.* (1993)]. Dynamic fault tree (DFT) analysis, that is an extension of traditional static FT (SFT) analysis, allows the modeling of dynamic behavior. DFTs take into account not only the combination of failure events but also the order in which they occur. As this last aspect is not taken into account in the Boolean model of failures (which only expresses whether a basic event has occurred or not), a classical Boolean function cannot represent the dynamic relations between the TE and the basic events that exist in a DFT.

The DFT has been continuously studied in the last two decades. We review the recent extensions of the analysis in this chapter. In Sec. 2, a brief introduction on dynamic gates used in a DFT is presented. The representative techniques to solve DFT in terms of quantitative analysis are introduced in Sec. 3. The recent extension of the algebraic approach in terms of both qualitative and quantitative evaluations is presented in Sec. 4.

2 Dynamic Gates

DFT defines special gates that capture a variety of failure sequences and functional dependences. Six dynamic gates are proposed in [Dugan *et al.* (1993)]: priority AND (PAND) gate, functional dependency (FDEP) gate, hot spare (HSP) gate, warm spare (WSP) gate, cold spare (CSP) gate, and sequence enforcing (SEQ) gate.

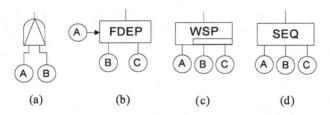

(a)   (b)   (c)   (d)

Fig. 1   Dynamic gates: (a) PAND gate, (b) FDEP gate, (c) WSP gate and (d) SEQ gate.

The first dynamic gate, PAND gate, was introduced in 1976 to model sequences of failures [Fussell *et al.* (1976)]. It is logically equivalent to an AND gate, of which input events must occur in a specific order for the output event to occur. The output of a PAND gate with two inputs depicted in Fig. 1 (a) becomes true if and only if both basic events A and B have occurred and basic event A has occurred before basic event B.

As a second dynamic gate, FDEP gate was introduced in 1990 to model common cause failures [Dugan *et al.* (1990)]. The FDEP gate in Fig. 1 (b) has a single trigger input, i.e., event A, a non-dependent output (reflecting the status of the trigger event) and one or more dependent basic events. The dependent events are functionally dependent on the trigger event. When the trigger event occurs, dependent events B and C are forced to occur. An FDEP gate can be used to reflect such dependence explicitly. The occurrence of any of the dependent events has no effect on the trigger event.

As a derivative gate, the probabilistic dependent (PDEP) gate is proposed [Montani *et al.* (2005)]. The trigger event causes other dependent events with probability $p_{dep} \leq 1$ in the PDEP gate. When $p_{dep} = 1$, the gate consists with the FDEP gate.

The CSP and HSP gates were also introduced in [Dugan *et al.* (1990)]. The WSP gate generalizes the CSP and HSP gates, and introduced in 1998 [Ren and Dugan (1998)]. Figure 1 (c) shows a WSP gate with two warm spares. Basic event A corresponds to a primary unit which is originally powered on, and the other inputs specify the components used as replacements for the primary unit. The failure rates of spare units are reduced by a factor α, called the dormancy factor in standby mode. If $\alpha = 0$, the gate is a CSP gate. If $\alpha = 1$, it is regarded as an HSP gate. The WSP gate has one output that becomes true after all the input events occur.

SEQ gate was introduced in 1993 [Dugan *et al.* (1993)]. The SEQ gate in Fig. 1 (d) forces events to occur in a particular order. The input events are constrained to occur in the left-to-right order, i.e., the leftmost event must occur before the event on its immediate right which must occur before the event on its immediate right is allowed to occur, etc.

Many authors have described the actual examples of DFT. Table 1 gives the information of the examples and helps us to understand DFT and sequence dependency.

3 Analysis of Dynamic Fault Tree

Several approaches have been used to analyze DFTs. These approaches can be either modular or global [Merle *et al.* (2011b)]. Global approaches consist in solving the whole DFT directly, whereas modular approaches consist in:

- Dividing the DFT into independent static and dynamic subtrees (or modules) prior to analysis. The linear time algorithm [Dutuit and Rauzy (1996)] is a useful method to search modules of FT, where dynamic gates

Table 1   Examples of DFT.

Fault Tolerant Parallel Processor	[Dugan *et al.* (1990)], [Dugan *et al.* (1992)], [Dugan *et al.* (1993)]
Mission Avionic System	[Tang and Dugan (1994)], [Dugan (2000)]
Active Heat Rejection System	[Boudali and Dugan (2005a)], [Montani *et al.* (2008)]
Multiprocessor Computing System	[Stamatelatos (2002)], [Montani *et al.* (2006)], [Zhang *et al.* (2009)], [Merle *et al.* (2011a)]
Redundant system with switch	[Fussell *et al.* (1974)], [Stamatelatos (2002)], [Yuge and Yanagi (2008)], [Merle *et al.* (2010)]
Hypothetical Cardiac Assist System	[Ren and Dugan (1998)], [Boudali and Dugan (2005b)], [Montani *et al.* (2005)], [Merle *et al.* (2011b)]
Vehicle Management System	[Stamatelatos (2002)]

can be treated in the same way as static gates. If a subtree contains static gates only, it is considered as static, whereas, a subtree contains at least one dynamic gate, it is considered as dynamic.

- Solving the modules separately.
- Combining the results of the various modules to obtain the overall result for the entire tree.

Generally, modular approaches have better calculation efficiency than global approaches. In modular approach, solving static modules can be done by using Binary Decision Diagrams (BDDs) or other combinatorial techniques. On the other hand, the methods of solving dynamic modules are restrictive because a classical Boolean function cannot represent the dynamic behaviors. Followings are the proposed methods to analyze dynamic modules. Table 2 classifies the references.

3.1 *Markov Analysis*

Markov chain (MC) is a well-known tool to capture the behaviors of a system which contain various types of dependences between basic events. MCs and their extensions have been proved to be a versatile tool for modeling DFTs [Dugan *et al.* (1992, 1993); Dugan (2000); Ou and Dugan (2004); Gulati and Dugan (1997); Ren and Dugan (1998)]. These can be applied to both entire trees and subtrees. However, the MC-based approaches are beset with two well-known problems:

Table 2   Classification of references.

Static fault tree	
survey	[Lee *et al.* (1985)], [Vesely (1981)], [Stamatelatos (2002)]
minimal cut set	[Fussell *et al.* (1974)]
modularization	[Dutuit and Rauzy (1996)]
Bayesian network	[Pearl (2000)], [Bobbio *et al.* (2001)], [Weber and Jouffe (2003)]
sum of disjoint products	[Abraham (1979)], [Locks (1987)]

Dynamic fault tree	
Markov analysis	[Fussell *et al.* (1976)], [Dugan *et al.* (1992)], [Dugan *et al.* (1993)], [Gulati and Dugan (1997)], [Ren and Dugan (1998)], [Dugan (2000)], [Ou and Dugan (2004)]
Algebraic approach	[Fussell *et al.* (1976)], [Tang and Dugan (1994)], [Yuge and Yanagi (2007)], [Yuge and Yanagi (2008)], [Yoneda *et al.* (2010)], [Merle *et al.* (2010)], [Merle *et al.* (2011a)], [Merle *et al.* (2011b)]
Bayesian network	[Weber and Jouffe (2003)], [Montani *et al.* (2005)], [Boudali and Dugan (2005a)], [Boudali and Dugan (2005b)], [Boudali and Dugan (2006)], [Montani *et al.* (2006)], [Montani *et al.* (2008)], [Yuge and Yanagi (2012)]
Monte Carlo & Petri-net	[Long *et al.* (2002)], [Bobbio and Codetta-Raiteri (2004)], [Zhang *et al.* (2009)]
Repairable events	[Dugan *et al.* (1990)], [Dugan *et al.* (1993)], [Bobbio and Codetta-Raiteri (2004)], [Yuge *et al.* (2012)]

(1) Ineffectiveness in solving large dynamic modules, i.e. the number of states grows exponentially as the number of basic events in the module increases.

(2) Lack of modeling power capabilities, i.e., the failure time distribution of a basic event is limited to the exponential distribution.

Concerning the problem (1), the existence of a repeated event causes a large module. A repeated event is defined as an event connected to several gates. If one of the input events of a dynamic gate is shared by another gate, we have to tackle a large module. In a dynamic module, the dimensionality of MC reaches to the total number of basic events in the module for the sake of capturing event sequences precisely. Therefore, the construction and analysis of an MC are tedious and require huge computational resources as the number of basic events in a module increases (see Table 3). Even if the modularization is adopted, in many cases, the size of a single module remains significant, potentially leading to an unreasonably long computa-

Table 3 Number of states in dynamic module.

number of events in a dynamic module	number of states to be considered in MC
2	5
4	65
6	19,557
8	109,601
10	986,410
15	3.55×10^{12}

tion time. Hence, the applicability of the modularization technique is not always evident. The problem (2) is intrinsic in MC, we have to try another method to avoid the problem.

3.2 *Algebraic Approach*

As a method of extending the modeling capability, i.e., as an alternative method to deal with problem (2) in MC, the algebraic approach has been proposed. Merle *et al.* proposed a complete algebraic expression of DFTs with priority dynamic gates [Merle *et al.* (2010)]. Priority dynamic gates indicate the PAND and FDEP gates, which express semantics of priority. They introduced new temporal operators to define the sequence dependence of such gates. By using the operators and conventional Boolean operators, the occurrence time of a TE is represented as a sum of the product canonical form. The terms of this form correspond to the minimal cut sequence set of the DFT. The TE probability is calculated from the cut sequence set by an inclusion-exclusion technique. This method does not require a restriction on the failure time distribution. However, to obtain the probability of a cut sequence, one needs to solve multiple integrations[1]. This may restrict the sequence size to handle in the analysis. Furthermore, as the number of minimal cut sequence sets in a DFT tends to be large compared with that in a static FT, even if the scales (number of basic events or gates) of the FTs are almost the same, the application of their inclusion-exclusion-based method to the calculation of a TE probability is limited to very small FTs. They continuously studied algebraic expressions for DFTs with spare gates [Merle *et al.* (2011a,b)]. The details of this approach and the related studies are discussed in Sec. 4.

---

[1]The complexity of the calculation may restrict the usage of the failure distribution except in the case of performing the numerical calculation.

3.3 *Bayesian Network*

The use of a Bayesian network (BN) [Pearl (2000)] in DFT analysis is expected to be an effective method to overcome both problems in MC. The BN directly takes into account the repeated events without a specific computation. This makes the BN efficient for obtaining exact results. An SFT can be transformed to an equivalent BN without losing the one-to-one match to the actual structure of the system [Bobbio *et al.* (2001)].

In [Boudali and Dugan (2005b)], Boudali and Dugan proposed an interval-based discrete-time BN (DTBN) for the reliability analysis of DFTs. They introduced mission time and divided it into n intervals. The variable assigned to a node in the DTBN has $n + 1$ states. The r.v. X is in state x if and only if the node has occurred in the x-th time interval. The last state represents the survival (no occurrence) for the duration of mission time. Arcs that connect pairs of nodes in the DTBN represent the causal probabilistic relationship between nodes. These causal probabilities are specified by defining $n + 1$ dimensional conditional probability tables (CPTs). The DTBN is generated using a standard BN inference algorithm. However, this is an approximate solution and requires huge memory resources to obtain the joint probability distribution (JPD) accurately.

Boudali and Dugan sequentially proposed a continuous-time BN (CTBN) for the reliability analysis of DFTs [Boudali and Dugan (2006)]. If the number of states, n, tends to infinity, the DTBN becomes a CTBN. The CTBN gives an exact closed-form solution and does not require the CPT of a node. The marginal probability distribution (MPD) of an r.v. representing a gate of a DFT is expressed by the multiple integral form using the failure distributions of input events. In a CTBN, the CPTs cannot be used for evaluating the conditional probabilities of nodes. This restricts the hierarchical features of BN and makes the analysis complex, because one has to derive all the conditional probabilities analytically.

Montani *et al.* [Montani *et al.* (2008)] proposed a translation of the DFT into a dynamic Bayesian network (DBN). Their method was on the basis of the 2-time-slice DBN [Montani *et al.* (2006); Weber and Jouffe (2003)]. The 2-time-slice DBN is equivalent to an MC, i.e., they both possess the Markov property. Thus the DBN model is applicable exclusively to Markov processes. Furthermore, it should be mentioned that DBN analysis is classified as a simulation method, thus, the result of the calculation gives the approximated probability.

3.4   *Monte Carlo and Petri-net*

Monte Carlo approaches [Long *et al.* (2002)] are also available for solving DFTs that do not require the assumption of an exponential distribution for the failure times. Stochastic Petri net (PN) is also applied to DFT [Bobbio and Codetta-Raiteri (2004); Zhang *et al.* (2009)]. By transforming a DFT model into a PN model, we can avoid the both of the serious problems in Markov analysis. However, modeling with Monte Carlo/PN is very complex and in order to achieve high precision, it is necessary to increase the number of cycles, which result in longer computation time. Nevertheless, these methods are useful to derive the probabilistic results in a complicated DFT analysis.

3.5   *Repairable Dynamic Fault Tree*

The conventional FT usually considers only failure occurrence as an input event. However, considering repair improves the analysis capability of FT. This FT is referred to as a recoverable/repairable FT (RFT). MC is suitable to deal with RFT even if dynamic gates are included in. Actually, some studies by using MC dealt with RFT [Dugan *et al.* (1990, 1993)]. The simulation based methods are also suitable to deal with repair event [Bobbio and Codetta-Raiteri (2004)]. Reference [Yuge *et al.* (2012)] tried to handle the state transitions in MC as equivalent event occurrences by using the concept of renewal process. That is, the occurrence and restoration of a gate output were regarded as following an alternative renewal process. Then, the steady state probability of an FT was derived from the limit theorem of a renewal process.

4   Algebraic Expression of DFT

4.1   *Temporal Functions and Operators*

The Boolean model, which realizes the complete algebraic expression of SFT, cannot render the order of occurrence of events defined by dynamic gates. In order to take into account the priority relations, Merle *et al.* consider the temporal functions and temporal operators [Merle *et al.* (2010)]. All the basic events and the intermediate events are defined as temporal functions which are piecewise right-continuous on $\mathbb{R}^+ \cup \{+\infty\}$, and whose range are $\mathbb{B} = \{0, 1\}$. The unique date of occurrence of event a is defined as $d(a)$. The identity elements equivalent to 0 and 1 are denoted by $\bot$ and $\top$ to which these dates can be assigned as;

$$d(\bot) = +\infty, \qquad d(\top) = 0.$$

They also introduced three operators, BEFORE ($\lhd$), SIMULTANEOUS ($\triangle$) and INCLUSIVE BEFORE ($\unlhd$), based on occurrence of a and b, as follows;

$$a \lhd b = \begin{cases} a & \text{if } d(a) < d(b) \\ \bot & \text{if } d(a) > d(b) \\ \top & \text{if } d(a) = d(b) \end{cases}, \quad a \triangle b = \begin{cases} \bot & \text{if } d(a) < d(b) \\ \bot & \text{if } d(a) > d(b) \\ a & \text{if } d(a) = d(b) \end{cases},$$

$$a \unlhd b = a \lhd b + a \triangle b.$$

Hereafter, $+$ and $\cdot$ mean logical disjunction and conjunction, respectively. They introduced the following theorems for any non-recoverable events a, b, and c (see [Merle *et al.* (2010)] for a complete set of theorems).

$$a \lhd a = \bot \tag{1}$$
$$a \lhd (b + c) = (a \lhd b) \cdot (a \lhd c) \tag{2}$$
$$a \lhd (b \cdot c) = (a \lhd b) + (a \lhd c) \tag{3}$$
$$a \lhd (b \lhd c) = (a \lhd b) + a \cdot b \cdot ((c \lhd b) + (c \triangle b)) \tag{4}$$
$$(a + b) \lhd c = (a \lhd c) + (b \lhd c) \tag{5}$$
$$(a \cdot b) \lhd c = (a \lhd c) \cdot (b \lhd c) \tag{6}$$
$$(a \lhd b) \lhd c = (a \lhd b) \cdot (a \lhd c) \tag{7}$$
$$(a \lhd b) \cdot (b \lhd c) \cdot (a \lhd c) = (a \lhd b) \cdot (b \lhd c) \tag{8}$$
$$a + (a \lhd b) = a \tag{9}$$
$$(a \lhd b) + b = a + b \tag{10}$$
$$a \cdot (a \lhd b) = a \lhd b \tag{11}$$
$$(a \lhd b) \cdot (b \lhd a) = \bot \tag{12}$$

Note that Eq. (2)–Eq. (11) hold for INCLUSIVE BEFORE($\unlhd$) operator too. Equation (1)–Eq. (12), which are used to simplify the top event representation, provide the relations between events. Suppose that the basic events are s-independent and cannot occur simultaneously. Hence, for any two basic events e_i and e_j, the following relation holds.

$$e_i \triangle e_j = 0 \tag{13}$$

4.2 *Minimal Cut Sequences Set*

4.2.1 *DFT with PAND gates*

Let us consider the algebraic expressions of dynamic gates by using the temporal functions and operators. The output of the PAND gate in Fig. 1 (a), say Q, is given in [Merle *et al.* (2010)] as

$$Q = B(A \lhd B).$$

Note that the omission of a logical conjunction symbol ($\cdot$) is admitted hereafter. For a PAND gate having inputs $e_1, e_2, \ldots, e_n$ and a condition that the order of the input events is specified in this order, the output Q is expressed as follows,

$$Q = e_n(e_1 \lhd e_2)(e_2 \lhd e_3) \cdots (e_{n-1} \lhd e_n). \tag{14}$$

Let us consider the FT in Fig. 2. The output of the top event is expressed, using the relations in Eq. (1)–Eq. (14)), as follows,

$$
\begin{aligned}
TE &= G_2(G_1 \lhd G_2) = (e_3 + e_4)(G_1 \lhd G_2) = e_3(G_1 \lhd G_2) + e_4(G_1 \lhd G_2) \\
&= e_3((G_3(e_1 \lhd G_3)) \lhd G_2) + e_4((G_3(e_1 \lhd G_3)) \lhd G_2) \\
&= e_3(G_3 \lhd G_2)((e_1 \lhd G_3) \lhd G_2) + e_4(G_3 \lhd G_2)((e_1 \lhd G_3) \lhd G_2) \\
&= e_3(G_3 \lhd G_2)(e_1 \lhd G_3)(e_1 \lhd G_2) + e_4(G_3 \lhd G_2)(e_1 \lhd G_3)(e_1 \lhd G_2) \\
&= e_3((e_2 + e_3) \lhd G_2)(e_1 \lhd G_3)(e_1 \lhd G_2) \\
&\quad + e_4((e_2 + e_3) \lhd G_2)(e_1 \lhd G_3)(e_1 \lhd G_2) \\
&= e_3(e_2 \lhd G_2)(e_1 \lhd G_3)(e_1 \lhd G_2) + e_3(e_3 \lhd G_2)(e_1 \lhd G_3)(e_1 \lhd G_2) \\
&\quad + e_4(e_2 \lhd G_2)(e_1 \lhd G_3)(e_1 \lhd G_2) + e_4(e_3 \lhd G_2)(e_1 \lhd G_3)(e_1 \lhd G_2) \\
&= e_3(e_2 \lhd e_3)(e_2 \lhd e_4)(e_1 \lhd e_2)(e_1 \lhd e_3)(e_1 \lhd e_3)(e_1 \lhd e_4) \\
&\quad + e_3(e_3 \lhd e_3)(e_3 \lhd e_4)(e_1 \lhd e_2)(e_1 \lhd e_3)(e_1 \lhd e_3)(e_1 \lhd e_4) \\
&\quad + e_4(e_2 \lhd e_3)(e_2 \lhd e_4)(e_1 \lhd e_2)(e_1 \lhd e_3)(e_1 \lhd e_3)(e_1 \lhd e_4) \\
&\quad + e_4(e_3 \lhd e_3)(e_3 \lhd e_4)(e_1 \lhd e_2)(e_1 \lhd e_3)(e_1 \lhd e_3)(e_1 \lhd e_4) \\
&= e_3(e_2 \lhd e_3)(e_2 \lhd e_4)(e_1 \lhd e_2) + e_3(e_3 \lhd e_4)(e_1 \lhd e_2)(e_1 \lhd e_3) \\
&\quad + e_4(e_2 \lhd e_3)(e_2 \lhd e_4)(e_1 \lhd e_2) + e_4 e_3(e_3 \lhd e_4)(e_1 \lhd e_2)(e_1 \lhd e_3) \\
&= e_3(e_2 \lhd e_3)(e_2 \lhd e_4)(e_1 \lhd e_2) + (e_3 \lhd e_4)(e_1 \lhd e_2)(e_1 \lhd e_3) \\
&\quad + e_4(e_2 \lhd e_3)(e_2 \lhd e_4)(e_1 \lhd e_2) + e_4(e_3 \lhd e_4)(e_1 \lhd e_2)(e_1 \lhd e_3) \tag{15}
\end{aligned}
$$

Finally, four product terms are derived from the DFT. Equation (15) is the sum of canonical form of the DFT. In the case of static FTs, such form provides the cut sets of an FT. In the case of DFTs, the concept of cut set corresponds to cut sequence representing the ordered failure sequence of events that causes the TE to occur. Therefore, Eq. (15) shows the cut

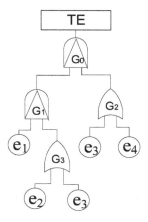

Fig. 2   DFT with two PAND gates.

sequence set (CSS) of the DFT. Since the fourth CSS in Eq. (15) is included in the second CSS, it can be removed. In general, CSS_i is included in other CSS and removed if the following equation holds,

$$CSS_i \cdot \sum_{j \neq i} CSS_j = CSS_i. \tag{16}$$

As a result of this minimization, the minimal canonical form of the TE is expressed as follows,

$$
\begin{aligned}
TE = {}& e_3(e_2 \lhd e_3)(e_2 \lhd e_4)(e_1 \lhd e_2) + (e_3 \lhd e_4)(e_1 \lhd e_2)(e_1 \lhd e_3) \\
& + e_4(e_2 \lhd e_3)(e_2 \lhd e_4)(e_1 \lhd e_2).
\end{aligned} \tag{17}
$$

This form also offers the minimal cut sequence set (MCSS) of the DFT as follows,

$$
\begin{aligned}
MCSS = {}& \{MCSS_1, MCSS_2, MCSS_3\} \\
= {}& \{e_3(e_2 \lhd e_3)(e_2 \lhd e_4)(e_1 \lhd e_2), \ (e_3 \lhd e_4)(e_1 \lhd e_2)(e_1 \lhd e_3), \\
& e_4(e_2 \lhd e_3)(e_2 \lhd e_4)(e_1 \lhd e_2)\}.
\end{aligned} \tag{18}
$$

Note that each MCSS in Eq. (18) is not a single cut sequence, but an algebraic expression that may contain more than one cut sequence providing a sufficient condition on the order of basic event failures that lead to the top event. For instance, $MCSS_1$ in Eq. (18) contains three cut sequences, $\bar{e}_4[e_1, e_2, e_3]$, $[e_1, e_2, e_3, e_4]$ and $[e_1, e_2, e_4, e_3]$, here, $[\dots]$ means the ordered cut sequence.

4.2.2 *DFT with WSP gates*

The output of a WSP gate in Fig. 1 (c), say Q, will occur if A, B, and C fail according to sequences $[A, B, C]$, $[A, C, B]$, $[B, A, C]$, $[B, C, A]$, $[C, A, B]$, or $[C, B, A]$ [Merle *et al.* (2011a)]. It is important to note that B and C will not have the same distribution function in the six sequences. For instance, in sequence $[A, B, C]$, both B and C fail during their active mode (denoted by B_a and C_a), whereas in sequence $[B, C, A]$, both B and C fail during their dormant mode (denoted by B_d and C_d). The algebraic model of the WSP gate can hence be expressed as follows [Merle *et al.* (2011a)],

$$
\begin{aligned}
Q = {} & C_a(A \lhd B_a)(B_a \lhd C_a) + B_a(A \lhd C_d)(C_d \lhd B_a) \\
& + C_a(B_d \lhd A)(A \lhd C_a) + A(B_d \lhd C_d)(C_d \lhd A) \\
& + B_a(C_d \lhd A)(A \lhd B_a) + A(C_d \lhd B_d)(B_d \lhd A).
\end{aligned} \tag{19}
$$

As B and C cannot be both in an active state and in a dormant state, we have

$$
B_d B_a = C_d C_a = \bot. \tag{20}
$$

Equation (19) implies the minimal cut sequences of the WSP gate. The number of minimal cut sequences is a factorial of the number of inputs events.

Next, let us consider a DFT in Fig. 3. The intermediate events and TE are expressed using the theorems presented in the previous section as

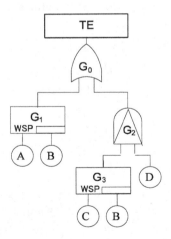

Fig. 3   Example of DFT.

follows

$$G_1 = B_a(A \vartriangleleft B_a) + A(B_d \vartriangleleft A) + A(C \vartriangleleft A)$$
$$G_3 = B_a(C \vartriangleleft B_a) + C(B_d \vartriangleleft C) + C(A \vartriangleleft C)$$
$$G_2 = D(G_3 \vartriangleleft D)$$
$$= (C \vartriangleleft B_a)(B_a \vartriangleleft D) + (B_d \vartriangleleft C)(C \vartriangleleft D) + (A \vartriangleleft C)(C \vartriangleleft D)$$
$$TE = G_0 = G_1 + G_2$$
$$= B_a(A \vartriangleleft B_a) + A(B_d \vartriangleleft A) + A(C \vartriangleleft A) + D(C \vartriangleleft B_a)(B_a \vartriangleleft D)$$
$$+ D(B_d \vartriangleleft C)(C \vartriangleleft D) + D(A \vartriangleleft C)(C \vartriangleleft D) \tag{21}$$

Here, the last term of G_1 is added to represent the condition of shared event. The spare event B cannot be in use when event A fails, if event C has already failed. The last term of G_3 is added for the same reason.

Finally, TE is represented as the sum of six product terms. Each of which represents the CSS of the DFT. After removing redundant cut sequences set by Eq. (16), the MCSS are generated. All six terms in Eq. (21) are MCSSs.

4.2.3 *FEDP and SEQ gates*

As already noticed in [Merle *et al.* (2010)], the algebraic formalization proves that the FDEP gate can be represented by Boolean OR gates. That is we can represent basic event B and C in Fig. 1 (b) as variables $B' = A+B$ and $C' = A + C$.

The SEQ gate can be expressed as a specific case of WSP gate, i.e., $\alpha = 0$. Any expression containing B_d and C_d in Eq. (19) can be removed.

4.3 *Sequence Probability*

As a first accomplishment concerned with the probabilistic analysis of event sequences, Fussell *et al.* [Fussell *et al.* (1976)] formulated the occurrence probability of PAND gate in 1976. The sequence probability of n inputs PAND gate, $[e_1, e_2, \ldots, e_n]$, is obtained by

$$\Pr([e_1, e_2, \ldots, e_n])(t)$$
$$= \int_0^t f_n(x_n) \int_0^{x_n} f_{n-1}(x_{n-1}) \int_0^{x_{n-1}} \cdots \int_0^{x_2} f_1(x_1) dx_1 dx_2 \ldots dx_n, \tag{22}$$

where, x_i indicates the occurrence time of e_i and $f_i(t)$ is a failure time distribution of event e_i. They assumed an exponential distribution with parameter λ_i for $f_i(t)$ and derived the following equation [Fussell *et al.* (1976)].

$$Pr([e_1, e_2, \ldots, e_n])(t) = \prod_{i=1}^{n} \lambda_i \sum_{k=0}^{n} \frac{e^{-a_k t}}{\prod_{j=0, j \neq k}^{n} (a_j - a_k)} \tag{23}$$

where,

$$a_0 = 0, \quad a_i = \sum_{k=i}^{n} \lambda_k \quad \text{for} \quad i > 0. \tag{24}$$

Yuge and Yanagi [Yuge and Yanagi (2012)] discussed a probability of event sequence including several spare events. Let us consider an event sequence composed with n events, $e_1, e_2, \ldots, e_n$. An event in the sequence is denoted by e_j^i, which means that the event that failed in the j-th order of the sequence is designated a spare of an event that failed in the i-th order. e_j^0 denotes an event that was originally in active mode. $e_j^i, (i \neq 0)$ has a dormancy factor $0 \leq \alpha_j \leq 1$. On the basis of the notation, the sequence is represented as $[e_1^{i1}, e_2^{i2}, \ldots, e_n^{in}]$. The events in the sequence are classified into the following sets [Yuge and Yanagi (2012)]:

- S_a : Events that were originally in active mode, i.e., e_j^0. The events in S_a are divided into the following two sets:

  - S_{ar} : Events in which the function is taken over by another one after the failure.
  - S_{an} : Events that have no spare or that cannot use their spare because of the failure of the designated spare.

- S_s : Events that were originally in standby mode. The events in S_s are divided into the following three sets:

  - S_{ss} : Events that fail in standby mode before the replacement. For $e_j^i \in S_{ss}, i > j$.
  - S_{sar} : Owing to the failure of the designated event, the operation mode changes to active mode then fails in active mode. After the failure, the function is continuously taken over by another spare.
  - S_{san} : Events that have no inheritable spare after their failure in active mode.
    Let $S_{sa} = \{S_{sar}, S_{san}\}$. For $e_j^i \in S_{sar}, i < j$ and $e_k^j \in S_{sa}$. For $e_j^i \in S_{san}, i < j$ and $e_k^j, k = 1, \ldots, n$, does not exist in the sequence.

The sequence probability of $[e_1^{i1}, e_2^{i2}, \ldots, e_n^{in}]$ can be derived using the n-tuple integration as

$$Pr([e_1^{i1}, e_2^{i2}, \ldots, e_n^{in}])(t) = \int_0^t \int_0^{x_n} \cdots \int_0^{x_2} \prod_{e_j^i \in S_a} f_j(x_j) \prod_{e_j^i \in S_{ss}} f_{j\alpha}(x_j)$$

$$\times \prod_{e_j^i \in S_{sa}} \bar{F}_{j\alpha}(x_i) f_j(x_j - x_i) dx_1 dx_2 \ldots dx_n, \quad (25)$$

where, $f_j(x)$ is the failure density function of e_j^i and $\bar{F}_{j\alpha}(x)$ is the survival function of e_j^i in standby mode.

Equation (25) is a closed form solution applicable to general failure distributions. However, to obtain the solution, one needs to go through a series of symbolic integrations. This process is time-consuming and feasible only for relatively small sequences. If the failure distributions are assumed to be exponential, the following theorem can be applied [Yuge and Yanagi (2012)].

Theorem 15.1. Let $[e_1^{i1}, e_2^{i2}, \ldots, e_n^{in}]$ be an event sequence containing spare events $e_j^i \in \{S_{sa}, S_{ss}\}$ with the dormancy factor, α_j, $0 \leq \alpha_j \leq 1$. When the failure time of e_j^i in active mode follows an exponential distribution with a mean $1/\lambda_j$, the sequence probability is

$$Pr([e_1^{i1}, e_2^{i2}, \ldots, e_n^{in}])(t) = \prod_{e_j^i \in S_{ss}} \alpha_j \cdot L^{-1} \left\{ \frac{1}{s} \prod_{i=1}^n \left(\frac{\lambda_i}{s + a_i} \right) \right\} \quad (26)$$

where,

$$a_i = \sum_{k=i}^n \lambda_k - \sum_{\substack{k=i \\ e_k^i \in S_{ss}}}^n (1 - \alpha_k) \lambda_k - \sum_{\substack{k=i \\ e_j^k \in S_{sa}}}^n (1 - \alpha_j) \lambda_j \quad \text{for} \quad i > 0, \quad (27)$$

and L^{-1} is the inverse Laplace transform operator.

If an event sequence is specified, we can transform the inverse Laplace expression in Eq. (26). Without the specification, it is not easy to obtain the sequence probability with a time domain.

As a special case, if every a_i in Eq. (27) is distinct from the others, the sequence probability is

$$Pr([e_1^{i1}, e_2^{i2}, \ldots, e_n^{in}])(t) = \prod_{e_j^i \in S_{ss}} \alpha_j \prod_{i=1}^n \lambda_i \sum_{k=0}^n \frac{e^{-a_k t}}{\prod_{j=0, j \neq k}^n (a_j - a_k)}, \quad (28)$$

where $a_j, j > 0$, are given by Eq. (27) and $a_0 = 0$.

If e_i, $2 \leq i \leq n$, works as a cold spare of e_{i-1} in the sequence, i.e., $[e_1^0, e_2^1, \ldots, e_n^{n-1}]$, Eq. (26) is in agreement with the occurrence probability of an n inputs CSP gate. Furthermore, in i.i.d. case ($\lambda_1 = \ldots = \lambda_n = \lambda$), the occurrence probability is

$$Pr([e_1^0, e_2^1, \ldots, e_n^{n-1}])(t) = 1 - e^{-\lambda t} \left\{ 1 + \frac{\lambda t}{1!} + \ldots + \frac{(\lambda t)^{n-1}}{(n-1)!} \right\}. \quad (29)$$

This is an Erlang distribution with phase n and parameter λ.

When all the spare events in a sequence are hot spare events, the sequence probability is in agreement with the occurrence probability of an n inputs PAND gate, whose occurrence condition is $[e_1, e_2, \ldots, e_n]$, presented in Eq. (23).

4.4 *Top Event Probability of DFT*

In the case of static FTs, a widespread way to calculate top event probabilities is to follow an inclusion-exclusion (IE) rule if all the minimal cut sets are given. This technique can be applied to DFTs [Merle *et al.* (2010)]. However, the number of MCSSs in DFT tends to be enormous compared with that of a static FT even if the scale of the FTs almost the same. Furthermore, the calculation of product terms generated from the technique is more complex than static FT cases. Therefore, applying an IE based technique is limited to small size DFTs.

Another popular method for obtaining the reliability of a coherent system is making use of the sum of disjoint products(SDP) algorithm [Abraham (1979)]. The algorithm was proposed in order to analyze network reliability. It starts with the Boolean sum of products corresponding to each of the simple paths between the pair of nodes, then transforms this into sum of disjoint products, from which the reliability expression can be directly obtained. This method can be applied to obtaining the top event probability of static FTs when the minimal cut sets are given. In this case, the algorithm bases on the following theorem [Abraham (1979)].

Theorem 15.2. (Abraham's theorem) Let S_j be a Boolean product corresponding to a minimal cut set and P_i be any product which implies some minimal cut set.

a) If there is at least one variable which exists in S_j and complemented in P_i, then S_j and P_i are disjoint.
b) If S_j and P_i are not disjoint, let $X \equiv \{x_a, x_b, \ldots, x_c\}$ be the set of variables which exist in S_j and which do not exist in P_i. Then,

i) If $X = \phi$, then $S_j \cup P_i = S_j$.

ii) If $X \neq \phi$ then $S_j \cup P_i = S_j \cup \bar{x}_a P_i \cup x_a \bar{x}_b P_i \cup \ldots \cup x_a x_b \ldots \bar{x}_c P_i$, and all the products are mutually disjoint.

Several rapid algorithms for calculating the disjoint sum have been proposed [Locks (1987)]. We expand the concept of the SDP algorithm to DFTs. In a DFT case, each product term of minimal canonical form contains basic events and ordered pairs of basic events linked by BEFORE operator. The ordered pair of basic events (for example, $e_i \lhd e_j$) can be treated as if it is a single event meaning event e_i occurs before basic event e_j. As the result, we can apply the SDP algorithm to the DFT expression. Again, let's consider the DFT in Fig. 3. At first, in order to make event occurrences clear, Eq. (21) is rewritten,

$$
\begin{aligned}
TE &= e_3(e_2 \lhd e_3)(e_2 \lhd e_4)(e_1 \lhd e_2) + (e_3 \lhd e_4)(e_1 \lhd e_2)(e_1 \lhd e_3) \\
&\quad + e_4(e_2 \lhd e_3)(e_2 \lhd e_4)(e_1 \lhd e_2) \\
&= e_1 e_2 e_3(e_2 \lhd e_3)(e_2 \lhd e_4)(e_1 \lhd e_2) + e_1 e_3(e_3 \lhd e_4)(e_1 \lhd e_2)(e_1 \lhd e_3) \\
&\quad + e_1 e_2 e_4(e_2 \lhd e_3)(e_2 \lhd e_4)(e_1 \lhd e_2)
\end{aligned}
$$

Then, the canonical form is transformed to the following equation by the Abraham's theorem.

$$
\begin{aligned}
TE &= e_1 e_2 e_3(e_2 \lhd e_3)(e_2 \lhd e_4)(e_1 \lhd e_2) + e_1 \bar{e}_2 e_3(e_3 \lhd e_4)(e_1 \lhd e_3) \\
&\quad + e_1 e_2 e_3(e_3 \lhd e_4)(e_1 \lhd e_2)(e_1 \lhd e_3)\overline{(e_2 \lhd e_3)} \\
&\quad + e_1 e_2 e_3(e_3 \lhd e_4)(e_1 \lhd e_2)(e_1 \lhd e_3)(e_2 \lhd e_3)\overline{(e_2 \lhd e_4)} \\
&\quad + e_1 e_2 \bar{e}_3 e_4(e_2 \lhd e_4)(e_1 \lhd e_2)
\end{aligned}
$$

These terms are already disjoint each other. Next, the negative ordered event can be decomposed by using the following relation,

$$
\overline{(e_i \lhd e_j)} = (e_j \lhd e_i) + \bar{e}_i e_j + \bar{e}_i \bar{e}_j.
$$

At last,

$$
\begin{aligned}
TE &= e_1 e_2 e_3(e_2 \lhd e_3)(e_2 \lhd e_4)(e_1 \lhd e_2) + e_1 \bar{e}_2 e_3(e_3 \lhd e_4)(e_1 \lhd e_3) \\
&\quad + e_1 e_2 e_3(e_3 \lhd e_4)(e_1 \lhd e_2)(e_1 \lhd e_3)(e_2 \lhd e_3) \\
&\quad + e_1 e_2 \bar{e}_3 e_4(e_2 \lhd e_4)(e_1 \lhd e_2).
\end{aligned}
$$

This is a disjoint expression of the TE. The first one contains three cut sequences, $\bar{e}_4[e_1, e_2, e_3]$, $[e_1, e_2, e_3, e_4]$ and $[e_1, e_2, e_4, e_3]$. The second has $\bar{e}_2[e_1, e_3, e_4]$ and $\bar{e}_2 \bar{e}_4[e_1, e_3]$. The third has three terms, $[e_1, e_3, e_4, e_2]$,

$[e_1, e_3, e_2, e_4]$ and $\bar{e}_4[e_1, e_3, e_2]$. The last shows $\bar{e}_4[e_1, e_3, e_2]$. The probability of each cut sequence is given by the method in Sec. 4.3. Therefore, the top event probability is given by adding the probabilities of these 9 cut sequences.

5   Conclusion

The DFT analysis has been continuously studied in the last two decades. We reviewed the recent extesions of the analysis in this chapter. Algebraic approach of DFT analysis is a method of promise to extend the capability of modeling and analyzing in terms of both qualitative and quantitative standpoints. The TE of a DFT was represented algebraically as a sum of product canonical form. The probability of a cut sequence containing spare events was formulated under the assumption that all the event occurrences in the sequence follow exponential failure distributions. The probability of TE was derived by an SDP based methods. Without the assumption, we have to manage a series of symbolic integrations to obtain the sequence probability and to execute our SDP procedure. Obtaining the sequence probability without the assumption is our future work.

References

Abraham, J. A. (1979). *An improved algorithm for network reliability, IEEE Trans. on Reliab.* **28**, 1, pp. 58–61.

Bobbio, A., Portinale, L., Minichino, M. and Ciancamerla, E. (2001). *Inproving the analysis of dependable systems by mapping fault trees into Bayesian networks, Reliab. Eng. Sys. Safety* **71**, 3, pp. 249–260.

Bobbio, A. and Codetta-Raiteri, D. (2004). *Parametric fault trees with dynamic gates and repair boxes, in Reliab. and Mainta. Sympo.*, pp. 459–465.

Boudali, H. and Dugan, J. B. (2005). *A new Bayesian network approach to solve dynamic fault trees, in Reliab. and Mainta. Sympo.*, pp. 451–456.

Boudali, H. and Dugan, J. B. (2005). *A discrete-time Bayesian network reliability modeling and analysis framework, Reliab. Eng. Sys. Safety,* **87**, 3, pp. 337–349.

Boudali, H. and Dugan, J. B. (2006). *A continuous-time Bayesian network reliability modeling, and analysis framework, IEEE Trans. on Reliab.,* **55**, 1, pp. 86–97.

Dugan, J., Bavuso, S. and Boyd, M. (1990). *Fault Trees and Sequence Dependencies, in Reliab. and Mainta. Sympo. (RAMS 1990)*, pp. 286–293.

Dugan, J. B., Bavuso, S. J. and Boyd, M. A. (1992). *Dynamic fault-tree models for fault-tolerant computer systems, IEEE Trans. on Reliab.*, **41**, 3, pp. 363–377.

Dugan, J. B., Bavuso, S. J. and Boyd, M. A. (1993). *Fault trees and Markov models for reliability analysis of fault-tolerant digital systems, Reliab. Eng. Sys. Safety*, **39**, 3, pp. 291–307.

Dugan, J. B. (2000). *Galileo: a tool for dynamic fault tree analysis*, (Springer Berlin/Heidelberg).

Dutuit, Y. and Rauzy, A. (1996). *A linear time algorithm to find modules of fault trees, IEEE Trans. on Reliab.*, **45**, 3, pp. 422–425.

Fussell, J. B., Henry, E. B. and Marshall, N. H. (1974). *MOCUS – a computer program to obtain minimal cut sets from fault trees, ANCR-1156.*

Fussell, J. B., Aber, E, F and Rahl, R. G. (1976). *On the quantitative analysis of priority-AND failure logic, IEEE Trans. on Reliab.* **25**, 5, pp. 324–326.

Gulati, R. and Dugan, J. B. (1997). *A modular approach for analyzing static and dynamic fault trees, in Reliab. and Mainta. Sympo.*, pp. 57–63.

Lee, W. S., Grosh, D. L., Tillman, F. A. and Lie, C. H. (1985). *Fault tree analysis, methods, and applications—A review, IEEE Trans. on Reliab.* **34**, 3, pp. 194–203.

Locks, M. O. (1987). *A minimizing algorithm for sum of disjoint products, IEEE Trans. on Reliab.*, **36**, 4, pp. 445–453.

Long, W., Zhang, T.L., Lu, Y.F. and Oshima, M. (2002). *On the quantitative analysis of sequential failure logic using Monte Carlo method for different distributions, Probabilistic Safety Assessment and Management (PSAM6)*, Elsevier, New York, pp. 391–396.

Merle, G., Roussel, J.M. , Lesage, J.J. and Bobbio, A. (2010). *Probabilistic algebraic analysis of fault trees with priority dynamic gates and repeated events, IEEE Trans. on Reliab.*, **59**, 1, pp. 250–261.

Merle, G.,Roussel, J.M. and Lesage, J. J. (2011). *Dynamic fault tree analysis based on the structure function, in Reliab. and Mainta. Sympo.*, pp. 1–6.

Merle, G.,Roussel, J.M. and Lesage, J. J. (2011). *Algebraic determination of the structure function of dynamic fault trees, Reliab. Eng. Sys. Safety*, **96**, 2, pp. 267–277.

Montani, S., Portinale, L. and Bobbio, A. (2005). *Dynamic Bayesian networks for modeling advanced fault tree features in dependability analysis, in ESREL 2005*, pp. 1414–1422.

Montani, S., Portinale, L., Bobbio, A., Varesio, M. and Codetta-Raiteri, D. (2006). *A tool for automatically translating dynamic fault trees into dynamic Bayesian networks, in Reliab. and Mainta. Sympo.*, pp. 434–441.

Montani, S., Portinale, L., Bobbio, A. and Codetta-Raiteri, D. (2008). *RADYBAN: a tool for reliability analysis of dynamic fault trees through conversion into dynamic Bayesian networks, Reliab. Eng. Sys. Safety*, **93**, 7, pp. 922–932.

Ou, Y. and Dugan, J. B. (2004). *Modular solution of dynamic multi-phase systems, IEEE Trans. on Reliab.*, **53**, 4, pp. 499–508.

Pearl, J. (2000). *Causality: Models, Reasoning, and Inference* (Cambridge University Press).

Ren, Y. and Dugan, J. B. (1998). *Design of reliable systems using static and dynamic fault trees, IEEE Trans. on Reliab.*, **47**, 3, pp. 234–244.

Stamatelatos, M. (2002). *Fault tree handbook with aerospace applications, NASA.*

Tang, Z. and Dugan, J. B. (1994). *Minimal cut set/sequence generation for dynamic fault trees, in Reliab. and Mainta. Sympo.*, pp. 207–213.

Vesely, W. E. (1981). *Fault tree handbook, NRC, NUREG-0492.*

Weber, P. and Jouffe, L. (2003). *Reliability modelling with dynamic Bayesian networks, in 5th IFAC Sympo. on Fault Detection, Supervision and Safety of Technical Processes (SAFEPROCESS'03).*

Yoneda, T., Yuge, T., Tamura, N. and Yanagi, S. (2010). *Minimal cut sequences and top event probability of dynamic fault trees, in: Proc. 4th Asia-Pacific Int Sympo on Advanced Reliab. and Mainte. Modeling (APARM 2010)*, Wellington, pp. 788–795.

Yuge, T. and Yanagi, S. (2007). *Minimal cut set/sequences of a fault tree with priority and gates, in Proc. of the 13th Int. Conf. on Reliab. and Quality in Design (ISSAT)*, pp. 352–356.

Yuge, T. and Yanagi, S. (2008). *Quantitative analysis of a fault tree with priority and gates, Reliab. Eng. Sys. Safety*, **93**, 11, pp. 1577–1583.

Yuge, T. and Yanagi, S. (2012). *Bayesian network modeling for dynamic fault tree, in Proc. 18th Int. Conf. on Reliab. and Quality in Design (ISSAT)*, Boston, pp. 111–115.

Yuge, T. Tamura, N. and Yanagi, S. (2012). *Repairable fault tree analysis using renewal intensities, Quality Technology & Quantitative Management*, **9**, 3, pp. 231–241.

Zhang, X., Miao, Q., Fan, X. and Wang, D. (2009). *Dynamic fault tree analysis based on Petri nets, in Reliab. and Mainta. Sympo.*, pp. 138–142.

Chapter 16

Reliability Analysis and Modeling Technique for an Open Source Solution

Yoshinobu Tamura[1] and Shigeru Yamada[2]

[1] *Graduate School of Science and Engineering, Yamaguchi University, Tokiwadai 2-16-1, Ube-shi 755-8611, Japan*
[2] *Department of Social Management Engineering, Tottori University, Minami 4-101, Koyama, Tottori 680-8552, Japan*

1 Introduction

At present, there is growing interest in the next-generation software development paradigm by using network computing technologies such as a cloud computing. Considering the software development environment, one has been changing into new development paradigms such as concurrent distributed development environment and the so-called open source project by using network computing technologies [Umar (1993)].

The successful experience of adopting the distributed development model in such open source projects includes GNU/Linux operating system, Apache HTTP server, and so on [E-Soft Inc. (2012)]. However, the poor handling of the quality and customer support prohibits the progress of OSS. We focus on the problems of software quality, which prohibit the progress of OSS. Especially, a large-scale open source solution is now attracting attention as the next-generation software development paradigm. Also, the large-scale open source solution is composed of several OSS's.

Many software reliability growth models (SRGM's) [Yamada (1994)] have been applied to assess the reliability for quality management and testing-progress control of software development. On the other hand, the effective method of dynamic testing management for new distributed development paradigm as typified by the open source project has only a few presented [MacCormack *et al.* (2006); Kuk (2006); Zhoum and Davis (2005); Shaw *et al.* (2004)]. In case of considering the effect of the debugging process on entire system in the development of a method of reliability assessment for open source solution, it is necessary to grasp the situation of registration for bug tracking system, the combination status of OSS's, the degree of maturation of OSS, and so on.

In this chapter, we focus on an open source solution developed under several OSS's. We discuss a useful method of software reliability assessment in open source solution as a typical case of next-generation distributed development paradigm. Then, we introduce a method of software reliability assessment based on a jump diffusion model by using stochastic differential equations in order to consider the active state of the open source project and the component collision of OSS. Then, we assume that the software failure intensity depends on the time, and the software fault-report phenomena on the bug tracking system keep an irregular state. Also, we analyze actual software fault-count data to show numerical examples of software reliability assessment for the open source solution. Especially, we derive several reliability assessment measures from our model. Then, we show that the introduced model can assist improvement of quality for open source solution developed under several OSS.

2 Modeling Technique

2.1 *Stochastic Differential Equation*

Let $S(t)$ be the number of detected faults in the open source solution by testing time t ($t \geq 0$). Suppose that $S(t)$ takes on continuous real values. Since latent faults in the open source solution are detected and eliminated during the testing-phase, $S(t)$ gradually increases as the testing procedures go on. Thus, under common assumptions for software reliability growth modeling, we consider the following linear differential equation:

$$\frac{dS(t)}{dt} = \lambda(t)S(t), \tag{1}$$

where $\lambda(t)$ is the intensity of inherent software failures at testing time t and is a non-negative function.

Generally, it is difficult for users to use all functions in open source solution, because the connection state among open source components is unstable in the testing-phase of open source solution. Considering the

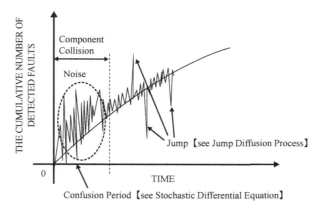

Fig. 1   Fault reporting phenomena in open source solution.

characteristic of open source solution, the software fault-report phenomena keep an irregular state in the early stage of testing-phase. Moreover, the addition and deletion of software components are repeated under the development of an OSS system, i.e., we consider that the software failure intensity depends on the time.

Therefore, we suppose that $\lambda(t)$ in Eq. (1) has the irregular fluctuation. That is, we extend Eq. (1) to the following stochastic differential equation [Arnold (1974); Wong (1971)]:

$$\frac{dS(t)}{dt} = (\lambda(t) + \sigma\mu(t)\gamma(t))\, S(t), \tag{2}$$

where σ is a positive constant representing a magnitude of the irregular fluctuation, $\gamma(t)$ a standardized Gaussian white noise, and $\mu(t)$ the collision level of open source component as shown in Fig. 1.

We extend Eq. (2) to the following stochastic differential equation of an Itô type:

$$dS(t) = \left\{\lambda(t) + \frac{1}{2}\sigma^2\mu(t)^2\right\} S(t)dt + \sigma\mu(t)S(t)dW(t), \tag{3}$$

where $W(t)$ is a one-dimensional Wiener process which is formally defined as an integration of the white noise $\gamma(t)$ with respect to time t. The Wiener process is a Gaussian process and it has the following properties:

$$\Pr[W(0) = 0] = 1, \tag{4}$$

$$E[W(t)] = 0, \tag{5}$$

$$E[W(t)W(t')] = \mathrm{Min}[t, t']. \tag{6}$$

By using Itô's formula [Arnold (1974); Wong (1971)], we can obtain the solution of Eq. (3) under the initial condition $S(0) = v$ as follows [Yamada *et al.* (1994)]:

$$S(t) = v \cdot \exp\left[\int_0^t \lambda(s)ds + \sigma\mu(t)W(t)\right], \tag{7}$$

where v is the total number of detected faults for the applied OSS's. Using solution process $S(t)$ in Eq. (7), we can derive several software reliability measures.

Moreover, we define the intensity of inherent software failures in case of $\lambda(t) = \lambda_1(t)$ and $\lambda(t) = \lambda_2(t)$, and the collision level function $\mu(t)$ as follows:

$$\int_0^t \lambda_1(s)ds = (1 - \exp[-\alpha t]), \tag{8}$$

$$\int_0^t \lambda_2(s)ds = (1 - (1 + \alpha t)\exp[-\alpha t]), \tag{9}$$

$$\mu(t) = \exp[-\beta t], \tag{10}$$

where α is an acceleration parameter of the intensity of inherent software failures, and β the growth parameter of the stability of open source solution. Eqs. (8) and (9) mean the exponential and S-shaped curves. The proposed model can widely describe the growth curves by using Eqs. (8) and (9). Also, we define that $\mu(t)$ decreases as the testing procedures go on, i.e., the open source solution becomes stable as the testing procedures go on.

2.2 *Jump Diffusion Process*

The jump term can be added to the introduced stochastic differential equation models in order to incorporate the irregular state around the time t by a change in the level of component collision. Then, the jump-diffusion process [Merton (1976)] is given as follows.

$$dS_j(t) = \left\{\lambda(t) + \frac{1}{2}\sigma^2\mu(t)^2\right\} S(t)dt + \sigma\mu(t)S(t)dW(t)$$

$$+ d\left\{\sum_{i=1}^{M_t(\lambda)} (V_i - 1)\right\}, \tag{11}$$

where $M_t(\lambda)$ is a Poisson point process with parameter λ at operation time t. Also, $M_t(\lambda)$ the number of occurred jumps, λ the jump rate. $M_t(\lambda)$,

$W(t)$, and V_i are assumed to be mutually independent. Moreover, V_i is i-th jump range. $dS_j(t)$ means that the jump term is added to $dS(t)$, i.e., $dS_j(t)$ has the mixed distribution composed of both the Wiener process and the jump diffusion process.

By using Itô's formula [Arnold (1974); Wong (1971)], the solution of the former equation can be obtained as follows:

$$S_j(t) = v \cdot \exp\left[\int_0^t \lambda(s)ds + \sigma\mu(t)W(t) - \sum_{i=1}^{M_t(\lambda)} \log V_i\right]. \tag{12}$$

3 Parameter Estimation

3.1 Method of Maximum-likelihood

In this section, the estimation method of unknown parameters α, β and σ in Eq. (7) is presented. Let us denote the joint probability distribution function of the process $S(t)$ as

$$P(t_1, \ y_1; \ t_2, \ y_2; \ \cdots; \ t_K, \ y_K)$$
$$\equiv \Pr[N(t_1) \leq y_1, \ \cdots, \ N(t_K) \leq y_K$$
$$\mid S(t_0) = v], \tag{13}$$

where $S(t)$ is the cumulative number of faults detected up to the testing time t $(t \geq 0)$, and denote its density as

$$p(t_1, \ y_1; \ t_2, \ y_2; \ \cdots; \ t_K, \ y_K)$$
$$\equiv \frac{\partial^K P(t_1, \ y_1; \ t_2, \ y_2; \ \cdots; \ t_K, \ y_K)}{\partial y_1 \partial y_2 \cdots \partial y_K}. \tag{14}$$

Since $S(t)$ takes on continuous values, we construct the likelihood function l for the observed data $(t_k, y_k)(k = 1, 2, \cdots, K)$ as follows:

$$l = p(t_1, \ y_1; \ t_2, \ y_2; \ \cdots; \ t_K, \ y_K). \tag{15}$$

For convenience in mathematical manipulations, we use the following logarithmic likelihood function:

$$L = \log l. \tag{16}$$

The maximum-likelihood estimates α^*, β^* and σ^* are the values making L in Eq. (16) maximize. These can be obtained as the solutions of the following simultaneous likelihood equations [Yamada *et al.* (1994)]:

$$\frac{\partial L}{\partial \alpha} = \frac{\partial L}{\partial \beta} = \frac{\partial L}{\partial \sigma} = 0. \tag{17}$$

3.2 *Estimation of Jump-diffusion Parameters*

Generally, it is difficult to estimate the jump-diffusion parameters of stochastic differential equation model because of the complicated likelihood function, mixed distribution, etc. The estimation methods of jump-diffusion parameters are introduced by several researchers. However, the effective method of estimation has only a few presented. We focus on the estimation methods performed in two stages [Honoré (1998)]. A genetic algorithm (GA) in order to estimate the jump-diffusion parameters of the introduced model is used in this section. The procedure of GA algorithm is given in the following [Holland (1975)].

It is assumed that the introduced jump-diffusion model includes the parameters λ, μ, and τ. The parameters μ and τ mean the parameters included in i-th jump range V_i.

Step 1 The initial individuals are randomly generated. Also, the set of initial individual is converted to the binary digit.

Step 2 Two parental individuals are selected, and new individuals are produced by the crossover recombination.

Step 3 The value of fitness is calculated from the evaluated value of each individual. The following value of fitness as the error between the estimated and the actual values is defined in this paper:

$$\min_{\boldsymbol{\theta}} \; F_i(\boldsymbol{\theta}),$$

$$F_i = \sum_{i=0}^{K} \{S_j(i) - y_i\}^2, \tag{18}$$

where $S_j(i)$ is the number of detected faults at operation time i in the introduced jump-diffusion model, y_i the number of actual detected faults. Also, $\boldsymbol{\theta}$ means the set of parameters λ, μ, and τ.

Step 4 Step 2 and Step 3 are continued until reaching the specific size.

The jump-diffusion parameters λ, μ, and τ are estimated by using above mentioned steps.

4 Software Reliability Assessment Measures

4.1 *Stochastic Differential Equation Model*

We consider the expected number of faults detected up to testing time t. The density function of $W(t)$ is given by:

$$f(W(t)) = \frac{1}{\sqrt{2\pi t}} \exp\left[-\frac{W(t)^2}{2t}\right]. \tag{19}$$

Information on the cumulative number of detected faults in the system is important to estimate the situation of the progress on the software testing procedures. Since $S(t)$ is a random variable in our model, its expected value can be a useful measure. We can calculate them from Eq. (7) as follows [Yamada *et al.* (1994)]:

$$E[S(t)] = v \cdot \exp\left[\int_0^t \lambda(s)ds + \frac{\sigma^2 \mu(t)^2}{2}t\right], \tag{20}$$

where $E[S(t)]$ is the expected number of faults detected up to time t. Also, the expected number of remaining faults at time t can obtain as follows:

$$E[N(t)] = v \cdot e - E[S(t)], \tag{21}$$

where v is the total number of detected faults for the applied OSS's, and e means $\exp(1)$.

4.2 *Jump-diffusion Model*

Similarly, the cumulative number of detected faults in the system is important to estimate the situation of the progress on the software debugging procedures. Since $N_j(t)$ is a random variable in the introduced model, it is calculated as Eq. (12) [Yamada *et al.* (1994)].

Also, it is important for software managers to assess the number of latent faults according to the change of specification of OSS. The number of remaining faults based on the jump-diffusion model considering the change of requirements specification can be obtained as follows:

$$N_j(t) \simeq v \cdot e - S_j(t). \tag{22}$$

Also, the mean time between software failures is useful to measure the property of the frequency of software failure-occurrences. Then, the cumulative MTBF(denoted by MTBF_C) is approximately given by:

$$MTBF_{Cj}(t) \simeq \frac{t}{S_j(t)}. \tag{23}$$

5   Numerical Illustrations

5.1   *Data for Numerical Illustrations*

We focus on a large-scale open source solution based on the Apache HTTP Server [The Apache HTTP Server Project (2012)], Apache Tomcat [Apache Tomcat (2012)], MySQL [MySQL (2012)] and JSP (JavaServer Pages). The fault-count data used in this paper are collected in the bug tracking system on the website of each open source project.

5.2   *Reliability Assessment Results*

The estimated expected cumulative numbers of detected faults in Eq. (20), $\widehat{E}[S(t)]$'s, in case of $\lambda(t) = \lambda_1(t)$ and $\lambda(t) = \lambda_2(t)$ are shown in Figs. 2 and 3, respectively. Also, the sample paths of the estimated numbers of detected faults in Eq. (7), $\widehat{S}(t)$'s, in case of $\lambda(t) = \lambda_1(t)$ and $\lambda(t) = \lambda_2(t)$ are shown in Figs. 4 and 5, approximately. Moreover, the sample paths of the estimated numbers of detected faults in Eq. (7), $\widehat{S_j}(t)$'s, in case of $\lambda(t) = \lambda_1(t)$ and $\lambda(t) = \lambda_2(t)$ are shown in Figs. 4 and 5, approximately. Furthermore, the estimated expected cumulative numbers of remaining faults in Eq. (21), $\widehat{E}[N(t)]$'s, in case of $\lambda(t) = \lambda_1(t)$ and $\lambda(t) = \lambda_2(t)$ are shown in Figs. 8 and 9, respectively.

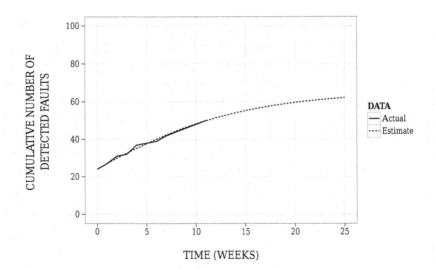

Fig. 2   Estimated cumulative number of detected faults, $\widehat{E}[S(t)]$, in case of $\lambda(t) = \lambda_1(t)$.

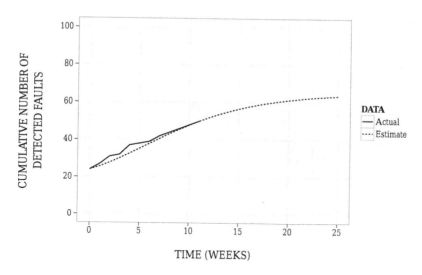

Fig. 3   Estimated cumulative number of detected faults, $\hat{E}[S(t)]$, in case of $\lambda(t) = \lambda_2(t)$.

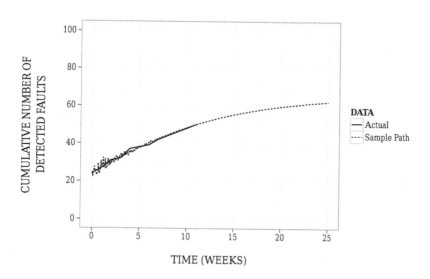

Fig. 4   Sample path of the estimated number of detected faults, $\widehat{S}(t)$, in case of $\lambda(t) = \lambda_1(t)$.

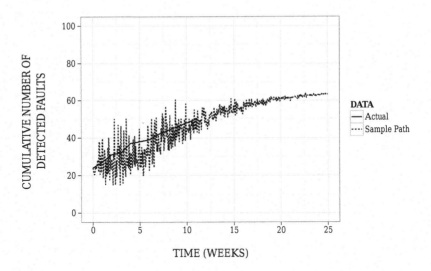

Fig. 5 Sample path of the estimated number of detected faults, $\widehat{S}(t)$, in case of $\lambda(t) = \lambda_2(t)$.

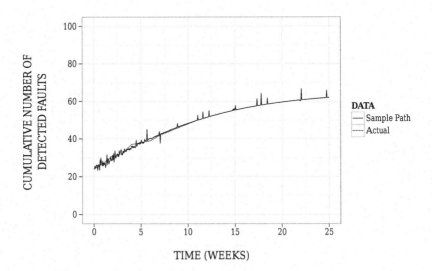

Fig. 6 Sample path of the estimated number of detected faults, $\widehat{S_j}(t)$, in case of $\lambda(t) = \lambda_1(t)$.

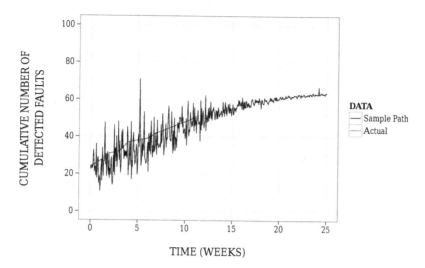

Fig. 7   Sample path of the estimated number of detected faults, $\widehat{S_j}(t)$, in case of $\lambda(t) = \lambda_2(t)$.

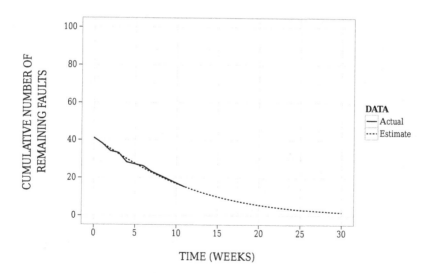

Fig. 8   Estimated cumulative number of remaining faults, $\hat{E}[S(t)]$, in case of $\lambda(t) = \lambda_1(t)$.

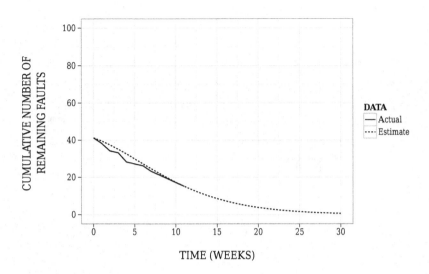

Fig. 9 Estimated cumulative number of remaining faults, $\hat{E}[S(t)]$, in case of $\lambda(t) = \lambda_2(t)$.

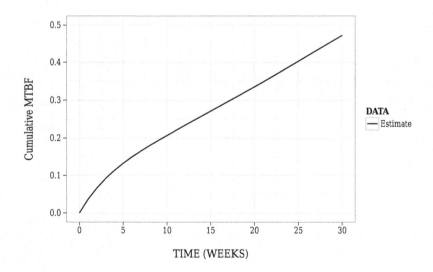

Fig. 10   Estimated MTBF$_C$, $\widehat{MTBF}_C(t)$, in case of $\lambda(t) = \lambda_1(t)$.

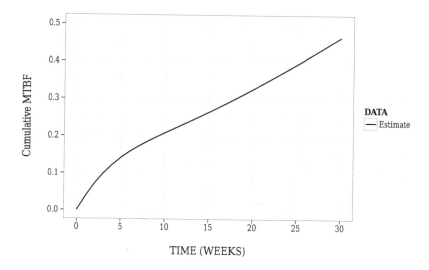

Fig. 11   Estimated MTBF$_C$, $\widehat{MTBF}_C(t)$, in case of $\lambda(t) = \lambda_2(t)$.

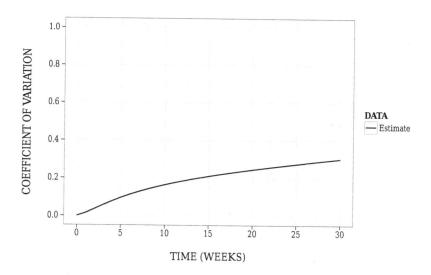

Fig. 12   Estimated coefficient of variation, $CV(t)$, in case of $\lambda(t) = \lambda_1(t)$.

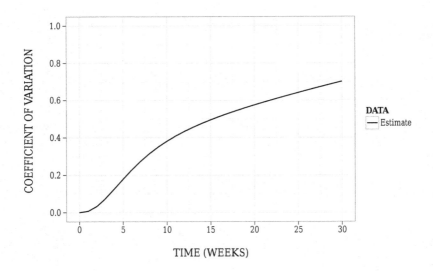

Fig. 13   Estimated coefficient of variation, $CV(t)$, in case of $\lambda(t) = \lambda_2(t)$.

The estimated $\mathrm{MTBF_C}$'s in case of $\lambda(t) = \lambda_1(t)$ and $\lambda(t) = \lambda_2(t)$ are plotted in Figs. 10 and 11, respectively. These figures show that the $\mathrm{MTBF_C}$ increase as the testing procedures go on. Also, $CV(t)$'s in case of $\lambda(t) = \lambda_1(t)$ and $\lambda(t) = \lambda_2(t)$ are shown in Figs. 12 and 13, respectively. From Figs. 12 and 13, we can confirm that the estimated coefficient of variation approaches the constant value. We consider that the coefficient of variation is useful measure to compare several fault data sets in the past system development projects.

From above mentioned results, we can confirm that the our model can cover the noise of collision level of open source solution by using the Wiener process and the jump diffusion process.

6   Concluding Remarks

In this chapter, we have focused on the open source solution which is known as the large-scale software system, and discussed the method of reliability assessment for the open source solution developed under several OSS's.

Moreover, we have introduced a software reliability growth model based on a jump diffusion process by using stochastic differential equations in order to consider the active state of the open source project and the collision among the open source components. Especially, we have assumed that the software failure intensity depends on the time, and the software fault-report phenomena on the bug tracking system keep an irregular state. Also, we have analyzed actual software fault-count data to show numerical examples of software reliability assessment for the large-scale open source solution. Moreover, we have derived several reliability assessment measures from our model.

At present, a new paradigm of distributed development typified by such open source project will evolve at a rapid pace in the future. Especially, it is difficult for the software testing managers to assess the reliability for the large-scale open source solution as a typical case of next-generation distributed development paradigm. Our method may be useful as the method of reliability assessment for the large-scale open source solution.

Acknowledgments

This work was supported in part by the Grant-in-Aid for Scientific Research (C), Grant No. 24500066 and No. 25350445 from the Ministry of Education, Culture, Sports, Science, and Technology of Japan.

References

Apache Tomcat, The Apache Software Foundation. [Online].
    Available: http://tomcat.apache.org/
Arnold, L. (1974). *Stochastic Differential Equations-Theory and Applications*, John Wiley & Sons, New York.
E-Soft Inc., Internet Research Reports. [Online]. Available:
    http://www.securityspace.com/s_survey/data/
Holland, J. H. (1975). *Adaptation in Natural and Artificial Systems*, University of Michigan Press.
Honoré, P. (1998). *Pitfalls in estimating jump-diffusion models*, Working Paper Series 18, University of Aarhus, School of Business.
Kuk, G. (2006). Strategic interaction and knowledge sharing in the KDE developer mailing list, *Informs Journal of Management Science*, vol. 52, no. 7, pp. 1031–1042.
Li P., Shaw, M., Herbsleb, J., Ray, B., and Santhanam P. (2004). Empirical evaluation of defect projection models for widely-deployed production software systems, *Proceeding of the 12th International Symposium on the Foundations of Software Engineering (FSE-12)*, pp. 263–272.

MacCormack, A., Rusnak, J., and Baldwin, C. Y. (2006). Exploring the structure of complex software designs: an empirical study of open source and proprietary code, *Informs Journal of Management Science*, vol. 52, no. 7, pp. 1015–1030.

Merton, R. C. (1976). Option pricing when underlying stock returns are discontinous, *Journal of Financial Economics*, vol. 3, pp. 125–144.

MySQL, Oracle Corporation and/or its affiliates. [Online].
  Available: http://www.mysql.com/

The Apache HTTP Server Project, The Apache Software Foundation. [Online].
  Available: http://httpd.apache.org/

Umar, A. (1993). *Distributed Computing and Client-Server Systems*, Prentice Hall, Englewood Cliffs, New Jersey.

Wong, E. (1971). *Stochastic Processes in Information and Systems*, McGraw-Hill, New York, 1971.

Yamada, S. (1994). *Software Reliability Models: Fundamentals and Applications* (in Japanese), JUSE Press, Tokyo.

Yamada, S., Kimura, M., Tanaka, H., and Osaki, S. (1994). Software reliability measurement and assessment with stochastic differential equations, *IEICE Transactions on Fundamentals*, vol. E77–A, no. 1, pp. 109–116.

Zhoum, Y. and Davis, J. (2005). Open source software reliability model: an empirical approach, *Proceedings of the Workshop on Open Source Software Engineering (WOSSE)*, vol. 30, no. 4, pp. 67–72.

Chapter 17

Maintenance Models of Miscellaneous Systems

Kodo Ito

Institute of Consumer Sciences and Human Life,
Kinjo Gakuin University,
1723 Omori 2-chome, Moriyama-ku, Nagoya 463-8521, Japan

1 Introduction

Miscellaneous systems such as social infrastructure and mass transportation, sustain our comfortable daily lives and economic activities. For the stable operation of these systems without any severe faults, suitable maintenances have to undergo. However, maintenance costs become extraordinarily pricy in most advanced countries because of expensive personnel expenditure. Today, the conflict between the austerity budget and the inevitable sustenance demand becomes a serious social issue and the establishing cost-effective maintenance has become a top priority social matter.

The maintenance is classified into preventive maintenance (PM) and corrective maintenance (CM): PM is a maintenance policy in which we undergoes some maintenance on a specific schedule before failure, and CM is a maintenance policy after failure [Barlow and Proschan (1965); Nakagawa (2005)]. Many researchers have studied optimal PM policies because the CM cost is usually much higher than the PM one and the optimal PM policy differs in each individual system. Therefore, target system characteristics must be investigated minutely for the consideration of cost-effective PM policies.

In this chapter, we survey optimal maintenance models for two different systems such as the aged fossil-fired power plant and the civil aircraft based on our original works.

In Section 2, we consider the aged fossil-fired power plant maintenance: Aged fossil-fired power systems, which need the maintenance for
their steady operations, are on the great increase in Japan. The preventive
maintenance and/or repair of such systems are indispensable to prevent
the serious trouble such as the emergency stop of operation. Because the
cumulative damage of system parts remains, the condition of system after
repair cannot return to brand-new. Such repair degradation of system have
to be considered when the maintenance plan is established. In Section 2,
a system is repaired at prespecified schedule when the cumulative damage
level is below a managerial level. When the cumulative damage level exceeds a certain critical level, the system fails and such critical level lowers
at every repair. The expected cost per unit of time between maintenances
is secured, and the optimal maintenance policy is derived.

Section 3 is devoted to the comparison of three cumulative damage
models: (1) A unit is subjected shocks and suffers some damage due to
shocks (Model 1). (2) The amount of damage due to shocks is measured
only at periodic times (Model 2). (3) The amount of damage increases
linearly with time (Model 3). The total damage is additive and the unit
fails when the total damage has exceeded a failure level for three models.
Models 2 and 3 would be actually used as the approximated ones of Model
1. As the preventive replacement policy, the unit is replaced before failure
at a planned time. The expected cost rates of each model are obtained,
and optimal policies that minimize them are derived.

Finally, Section 4 takes up the civil aircraft maintenance: As the accumulation of tiny stress causes the failure of airframe, its maintenance is
indispensable to operate aircraft without any serious troubles. Four echelons preventive maintenance (PM) is undergone for commercial aviation
and is the imperfect maintenance. The imperfect maintenance assumes the
reduction rate of cumulative hazard functions after PMs and how to settle
this rate is a problem. In Section 4, a standard imperfect preventive maintenance policy (Model 1) is discussed. And it is modified to Model 2 which
is useful in actual design. The airframe failure during operation is not so
serious and the failure rate remains undisturbed by repair. The expected
cost rate of airframe maintenance is secured and the optimal PM number
which minimizes it is discussed.

We obtain the expected costs or the availability of each model as an
objective function and derive optimal maintenance policies which minimize
them, using reliability techniques. Furthermore, we give shortly some comments regarding the limitation and possible extensions of the above models.

2 Optimal Maintenance Policy for a Damage System with Repair

A number of aged fossil-fired power plants are increasing in Japan. For example, 33% of these plants are currently operated from 150,000 to 199,999 hours (from 17 to 23 years), and 26% of them are above 200,000 hours (23 years) [Hisano (2000)]. Although Japanese government eliminates regulations of electric power industry, most industries restrain from the investment for new plants and are prefer to operate current plants efficiently because of the long-term recession and the reducing of operating atomic reactor number in Japan.

The deliberative maintenance plans are indispensable to operate these aged plants without serious trouble such as the emergency stop of operation. The importance of maintenance for aged plants is much higher than that for new ones because occurrence probabilities of severe troubles increase and new failure phenomena might appear according to the degradation of plants. Furthermore, actual life spans of plant components are mostly different from predicted ones because they are affected by various kinds of factors such as material qualities and operational circumstances [Hisano (2001)]. So, maintenance plans should be established considering occurrence probabilities of miscellaneous components.

The occurrence of failure is discussed by utilizing the cumulative damage model. It has be well-known that the following stochastic reliability model is called shock or cumulative damage model : A unit is subjected to shocks and suffers some amount of damage such as wear, fatigue, crack growth, creep, and dielectric at each shock. The total damage due to shocks is additive. This can be described theoretically by cumulative process [Cox (1962)] and compound Poisson processes [Çinlar (1975); Ross (1983)]. Some reliability quantities were obtained in [Esary, Marshall and Proschan (1973); Nakagawa and Osaki (1974)]. Markov models of such models were studied and their life data were collected in [Bogdanoff and Kozin (1985)]. Several kinds of damage models were analyzed, using cumulative processes, and their maintenance policies were summarized and optimal policies were discussed analytically in [Nakagawa (2007)].

A plant consists of a wide variety of mechanical parts such as power boiler, compressor, combustor, steam and gas turbines. Some parts suffers high temperature at operation and such thermal damages accumulated in these parts. PM is performed periodically before these damages cause serious failures. The condition of system after PM cannot return to the brand-new condition because the cumulative fatigue damage of system parts remains after PM [Kosugiyama, Takizuka, Kunitomi, Yan, Katanishi and Takada (2003)].

In past PM studies and cumulative damage models, the condition of system after PM is supposed to be brand-new. In the actual plant maintenance, the remaining damage after PM should be considered. We considered a cumulative damage PM model in which a system is repaired at prespecified schedule when the cumulative damage level is below a managerial level and the damage remains after the repair [Ito, Nakagawa and Teramoto (2006)]. We extended it the optimal operation cencoring policy which mamimizes the expected profit of system [Ito and Nakagawa (2006, 2007)].

In this section, the cumulative damage PM model is considered [Ito, Nakagawa and Teramoto (2006)]. When the cumulative damage level exceeds a certain critical level, the system fails and the critical level degrades at every repair. The expected cost per unit of time between maintenance is considered and the optimal maintenance policy is derived.

2.1 *Model 1*

We consider the following maintenance policy:

1) The system is operating continuously and shocks during operation occur at a Poisson process. The probability that the j-th shock occurs during $(0, t]$ is $H_j(t) = [(\lambda t)^j/j!]e^{-\lambda t}$ $(j = 0, 1, 2, \cdots)$ [Osaki (1992)]. Thus, the probability that shocks occur more than j-times during $(0, t]$ is $F_j(t) = \sum_{i=j}^{\infty} H_i(t)$.

2) The damage caused by each shock has an identical probability distribution $G(x) \equiv Pr\{Y_j \leq x\}$ $(j = 1, 2, \cdots)$ with finite mean, and each damage is additive. Then, the total damage $Z_j \equiv \sum_{i=1}^{j} Y_i$ to the j-th shock where $Z_0 \equiv 0$ has a distribution $Pr\{Z_j \leq x\} = G^{(j)}(x)$ $(j = 1, 2, \cdots)$, where $\Phi^{(j)}(x)$ $(j = 1, 2, \cdots)$ denotes the j-fold Stieltjes convolution of $\Phi(x)$ with itself and $\Phi^{(0)}(x) \equiv 1$ for $x \geq 0$. $M_\Phi(x) \equiv \sum_{j=1}^{\infty} \Phi^{(j)}(x)$ is a renewal function of any distribution $\Phi(x)$.

3) The total damage is below a managerial level k during $(T_i, T_{i+1}]$ $(i = 0, 1, 2, \cdots)$ where $T_0 \equiv 0$, the system is repaired at time T_{i+1} and its cost is c_0. The system degrades at every repair.

4) When the total damage exceeds a failure level K_i $(i = 0, 1, \cdots)$, the system fails and its maintenance cost is c_{K_i}, where K_i declines, *i.e.*, $K_0 >$

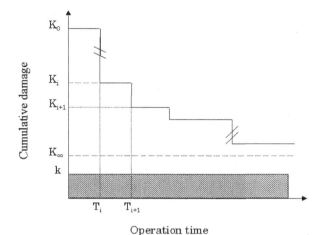

Fig. 1 Schematic diagram of Model 1.

$K_1 > \cdots > K_{i-1} > K_i > \cdots$ because the damage of system parts remains at every repair (see Fig. 1).

5) When the total damage is between k and K_i, the system is overhauled and its cost is c_1 $(c_{K_i} > c_1 > c_0)$.

In this model, the system operates until its damage exceeds k which is less than $K_i, i.e., \lim_{i \to \infty} K_i > k$. The probability P_{k_i} that the system undergoes overhaul when the total damage exceeds k is

$$P_{k_i} = \sum_{j=1}^{\infty}[G^{(j-1)}(k) - G^{(j)}(k)] \int_{T_i}^{T_{i+1}} \mathrm{d}F_j(t)\,, \tag{1}$$

and it is obvious that $\sum_{i=0}^{\infty} P_{k_i} = 1$. The probability P_{K_i} that the system fails during $(T_i, T_{i+1}]$ when the total damage exceeds K_i, is

$$P_{K_i} = \sum_{j=1}^{\infty}\int_0^k[1 - G(K_i - x)]\mathrm{d}G^{(j-1)}(x)\int_{T_i}^{T_{i+1}} \mathrm{d}F_j(t)\,. \tag{2}$$

Let $E\{U\}$ denote the mean time to some maintenance. From (1) and (2),

$$E\{U\} = \sum_{i=0}^{\infty}\sum_{j=1}^{\infty}[G^{(j-1)}(k) - G^{(j)}(k)] \int_{T_i}^{T_{i+1}} t\mathrm{d}F_j(t)$$

$$= \frac{1}{\lambda}[1 + M_G(k)]\,. \tag{3}$$

Further, the total expected cost $E\{C\}$ to some maintenance is

$$E\{C\} = \sum_{i=0}^{\infty} [(c_1 + ic_0)(P_{k_i} - P_{K_i}) + (c_{K_i} + ic_0)P_{K_i}] \,. \tag{4}$$

Therefore, from (3) and (4), the expected cost rate is

$$\frac{C_1(k)}{\lambda} = \frac{\sum_{i=0}^{\infty} \sum_{j=1}^{\infty} \left\{ (c_1 + ic_0)[G^{(j-1)}(k) - G^{(j)}(k)] \int_{T_i}^{T_{i+1}} \mathrm{d}F_j(t) \atop + (c_{K_i} - c_1) \int_0^k [1 - G(K_i - x)]\mathrm{d}G^{(j-1)}(x) \int_{T_i}^{T_{i+1}} \mathrm{d}F_j(t) \right\}}{1 + M_G(k)} \,. \tag{5}$$

2.1.1 *Case 1*

Suppose that $G(x) = 1 - \exp(-\mu x)$, i.e., $G^{(j)}(x) = \sum_{i=j}^{\infty}[(\mu x)^i/i!]\mathrm{e}^{-\mu x}$ and $M_G(x) = \mu x$. Then, (4) is

$$E\{C\} = c_1 - c_0 + \sum_{i=0}^{\infty} \sum_{j=0}^{\infty} [(A_i - A_{i-1})\mathrm{e}^{\mu k} + c_0]G^{(j)}(k)H_j(T_i) \,, \tag{6}$$

where $A_i \equiv (c_{K_i} - c_1)\mathrm{e}^{-\mu K_i}$ $(i = 0, 1, 2, \cdots)$ and $A_{-1} \equiv 0$. Therefore, from (3) and (6), the expected cost rate is

$$\frac{C_1(k)}{\lambda} = \frac{c_1 - c_0 + \sum_{i=0}^{\infty}[(A_i - A_{i-1})\mathrm{e}^{\mu k} + c_0] \sum_{j=0}^{\infty} G^{(j)}(k)H_j(T_i)}{1 + \mu k} \,. \tag{7}$$

Differentiating $C_1(k)$ with respect to k and putting it to zero,

$$\sum_{i=0}^{\infty} \sum_{j=0}^{\infty} \left\{ \left[(A_i - A_{i-1})\mathrm{e}^{\mu k} + c_0 \right] (1 + \mu k)G^{(j-1)}(k) \right.$$

$$\left. - \left[(A_i - A_{i-1})\mathrm{e}^{\mu k} + c_0(2 + \mu k) \right] G^{(j)}(k) \right\} H_j(T_i) = c_1 - c_0 \,, \tag{8}$$

where $G^{(j)}(k) \equiv 0 \, (j < 0)$. Letting denote the left-hand side of (8) by $L_1(k)$,

$$L_1(0) = \sum_{i=0}^{\infty} \{ [(A_i - A_{i-1}) + c_0](\lambda T_i - 1) - c_0 \} \mathrm{e}^{-\lambda T_i} \,, \tag{9}$$

$$L_1(K_\infty) = \sum_{i=0}^{\infty} \sum_{j=0}^{\infty} \left\{ \left[(A_i - A_{i-1})\mathrm{e}^{\mu K_\infty} + c_0 \right] (1 + \mu K_\infty)G^{(j-1)}(K_\infty) \right.$$

$$\left. - \left[(A_i - A_{i-1})\mathrm{e}^{\mu K_\infty} + c_0(2 + \mu K_\infty) \right] G^{(j)}(K_\infty) \right\} H_j(T_i) \,. \tag{10}$$

Thus, if $L_1(0) < c_1 - c_0 < L_1(K_\infty)$, then there exists a finite k^* $(0 < k^* < K_\infty)$ which minimizes $C_1(k)$.

2.1.2 *Case 2*

When $i = 0$, *i.e.*, the failure level is K_0 and it does not change, (7) and (8) are rewritten as, respectively,

$$\frac{C_1(k)}{\lambda} = \frac{c_1 + (c_{K_0} - c_1)e^{-\mu(K_0 - k)}}{1 + \mu k}, \tag{11}$$

$$ke^{\mu k} = \frac{c_1}{\mu(c_{K_0} - c_1)e^{-\mu K_0}}. \tag{12}$$

It is evidently seen that the left-hand side of (12) is strictly increasing from 0 to $K_0 e^{\mu K_0}$. Therefore, if $\mu K_0 > c_1/(c_{K_0} - c_1)$, then there exist a unique k^* which minimizes $C_1(k)$.

2.2 *Model 2*

We consider the following maintenance policy which has the same assumptions except 5) (see Fig. 2):

5)' When the total damage is between k and K_i, or the operation time exceeds time T_n, whichever occurs first, the system is overhauled and its cost is c_1 $(c_{K_i} > c_1 > c_0, i = 0, 1, \cdots)$.

The probability that the total damage is below a managerial level k until time T_n is

$$P_{T_n} = \sum_{j=0}^{\infty} G^{(j)}(k) H_j(T_n). \tag{13}$$

Then, the mean time to some maintenance or time T_n is

$$E\{U\} = \sum_{i=0}^{n-1} \sum_{j=1}^{\infty} [G^{(j-1)}(k) - G^{(j)}(k)] \int_{T_i}^{T_{i+1}} t\,\mathrm{d}F_j(t) + T_n P_{T_n}$$

$$= \sum_{j=0}^{\infty} G^{(j)}(k) \int_0^{T_n} H_j(t)\mathrm{d}t. \tag{14}$$

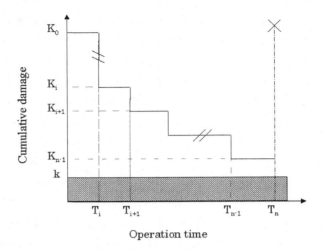

Fig. 2   Schematic diagram of Model 2.

Next, the total expected cost to some maintenance or time T_n is

$$E\{C\} = [c_1 + (n-1)c_0]P_{T_n} + \sum_{i=0}^{n-1}(c_1 + ic_0)(P_{k_i} - P_{K_i})$$

$$+ \sum_{i=0}^{n-1}(c_{K_i} + ic_0)P_{K_i}$$

$$= c_1 - c_0 + \sum_{i=0}^{n}[(A_i - A_{i-1})e^{\mu k} + c_0]\sum_{j=0}^{\infty}G^{(j)}(k)H_j(T_i)$$

$$- A_{n-1}e^{\mu k}\sum_{j=0}^{\infty}G^{(j)}(k)H_j(T_n)\,. \tag{15}$$

Therefore, from (14) and (15), the expected cost rate is

$$C_2(n,k) = \frac{\begin{aligned}&c_1 - c_0 + \sum_{i=0}^{n}[(A_i - A_{i-1})e^{\mu k} + c_0]\\&\times \sum_{j=0}^{\infty}G^{(j)}(k)H_j(T_i) - A_{n-1}e^{\mu k}\sum_{j=0}^{\infty}G^{(j)}(k)H_j(T_n)\end{aligned}}{\sum_{j=0}^{\infty}G^{(j)}(k)\int_0^{T_n}H_j(t)\mathrm{d}t}\,, \tag{16}$$

which agrees with (7) as $n \to \infty$.

3 Comparison of Three Cumulative Damage Models

In Section 2, the cumulative damage model is utilized for establishing the optimal PM policies of aged fossil-fired power plant. In this section, we expand the cumulative damage model [Ito and Nakagawa (2008)].

First, we consider a standard cumulative damage model where the unit suffers some damage due to shocks and the total damage is additive. The unit fails when the total damage has exceeded a failure level K. The probability that the unit fails in time t and its properties are obtained by cumulative processes [Nakagawa (2007)]. However, it might be impossible to estimate and know occurrences of shocks and the total damage every at each shock. Secondly, the amount of damage due to shocks is measured only at periodic times by inspections, irrespective occurences of shocks. Thirdly, the total damage increases linearly with time t. The distributions of total damage and its means for three models are obtained.

Next, we consider an age replacement policy where the unit is replaced when the total damage has exceeded a failure level K or at time T for Models 1 & 3 and NT_0 for Model 2, whichever occurs first. The expected cost rates of three models are obtained and optimal policies that minimize them are derived.

3.1 *Three Models*

We consider the following three cumulative damage models for an operating unit [Cox (1962); Nakagawa (2007)]:

3.1.1 *Model 1: Standard model*

The unit is subjected to shocks and suffers some damage at each shock. Let random variables X_j $(j = 1, 2, \cdots)$ denote a sequence of interarrival times between successive shocks, and random variables W_j $(j = 1, 2, \cdots)$ denote the damage produced by the jth shock, where $W_0 \equiv 0$.

Further, let $N(t)$ denote the random variable that is the total number of shocks up to time t. Then, define a random variable

$$Z(t) \equiv \sum_{j=0}^{N(t)} W_j \quad (N(t) = 0, 1, 2, \cdots), \tag{17}$$

that represents the total damage at time t.

It is assumed that $F(t) \equiv \mathrm{P_r}\{X_j \le t\}$ with finite mean $1/\lambda$ and variance σ_F^2 and $G(x) \equiv \mathrm{P_r}\{W_j \le x\}$ with finite mean μ and variance σ_G^2 for $t, x \ge 0$.

Then, the probability that shocks occur exactly j times in $[0, t]$ is

$$\Pr\{N(t) = j\} = F^{(j)}(t) - F^{(j+1)}(t) \quad (j = 0, 1, 2, \cdots), \tag{18}$$

$$\Pr\{Z(t) \le x\} = \sum_{j=0}^{\infty} G^{(j)}(x)[F^{(j)}(t) - F^{(j+1)}(t)]. \tag{19}$$

Thus, the mean total damage at time t is

$$E\{Z(t)\} = \mu \sum_{j=1}^{\infty} F^{(j)}(t) = \mu M_F(t), \tag{20}$$

and its approximate variance of $Z(t)$ is [Nakagawa (2007)]

$$V\{Z(t)\} \approx \lambda^3 t \left(\mu^2 \sigma_F^2 + \frac{\sigma_G^2}{\lambda^2} \right). \tag{21}$$

3.1.2 *Model 2: Periodic model*

Each amount W_n $(n = 1, 2, \cdots)$ of damage is measured only at periodic times nT $(n = 1, 2, \cdots)$ for a given T_0 $(0 < T_0 < \infty)$ and has an identical distribution $G_T(x) \equiv Pr\{W_n \le x\}$ with mean μ_T and variance σ_T^2. Then,

$$\Pr\{Z(nT_0) \le x\} = G_{T_0}^{(n)}(x), \tag{22}$$

and

$$E\{Z(nT_0)\} = n\mu_T, \qquad V\{Z(nT_0)\} = n\sigma_T^2. \tag{23}$$

3.1.3 *Model 3: Continuous model*

The total damage $Z(t)$ usually increases with time t. Suppose that $Z(t) = A_t t + B_t$ for $A_t \ge 0$. Consider two cases : When $A_t \equiv a$ (constant) and B_t has a distribution $L_B(x)$ with mean 0 and variance σ_B^2.

$$\Pr\{Z(t) \le x\} = \Pr\{B_t \le x - at\} = L_B(x - at), \tag{24}$$

and

$$E\{Z(t)\} = at, \qquad V\{Z(t)\} = \sigma_B^2. \tag{25}$$

Next, when A_t has a distribution $L_A(x)$ with mean a and variance σ_A^2/t and $B_t \equiv 0$,

$$\Pr\{Z(t) \le x\} = \Pr\{A_t \le x/t\} = L_A(x/t), \tag{26}$$

and

$$E\{Z(t)\} = at, \qquad V\{Z(t)\} = \sigma_A^2 t. \tag{27}$$

3.2 *Optimal Replacement Policies*

Suppose that the unit fails when the total damage has excceeded a failure level K. As the preventive replacement policy, the unit is also replaced before failure at time T. For the above replacement model, costs c_1 and c_2 are the replacement costs at failure and at time T, where $c_1 > c_2$.

The expected cost rate for Model 1 is [Nakagawa (2007)]

$$C_1(T) = \frac{c_1 - (c_1 - c_2)\sum_{j=0}^{\infty}[F^{(j)}(T) - F^{(j+1)}(T)]G^{(j)}(K)}{\sum_{j=0}^{\infty}G^{(j)}(K)\int_0^T[F^{(j)}(t) - F^{(j+1)}(t)]dt}. \quad (28)$$

The optimal time T_1^* that minimizes $C_1(T)$ is as follows : Let $f(t)$ be a density function of $F(t)$ and

$$Q_1(T) \equiv \frac{\sum_{j=0}^{\infty}f^{(j+1)}(T)[G^{(j)}(K) - G^{(j+1)}(K)]}{\sum_{j=0}^{\infty}[F^{(j)}(T) - F^{(j+1)}(T)]G^{(j)}(K)}.$$

Then, if $Q_1(T)$ increases strictly with T and $Q_1(\infty)[1+M_G(K)] > \lambda c_1/(c_1 - c_2)$, then there exists a finite and unique T_1^* ($0 < T_1^* < \infty$) that satisfies

$$Q_1(T)\sum_{j=0}^{\infty}G^{(j)}(K)\int_0^T[F^{(j)}(t) - F^{(j+1)}(t)]dt$$

$$-\sum_{j=0}^{\infty}F^{(j+1)}(T)[G^{(j)}(K) - G^{(j+1)}(K)] = \frac{c_2}{c_1 - c_2}. \quad (29)$$

Next, for Model 2, the unit is replaced at time NT_0 or at failure, whichever occurs first. Then, the expected cost rate is

$$C_2(N) = \frac{c_1 - (c_1 - c_2)G^{(N)}(K)}{T_0\sum_{j=0}^{N-1}G^{(j)}(K)} \quad (N = 1, 2, \cdots). \quad (30)$$

The optimal number N^* that minimizes $C_2(N)$ is as follows : Let

$$Q_2(N) \equiv \frac{G^{(N-1)}(K) - G^{(N)}(K)}{G^{(N-1)}(K)} \quad (N = 1, 2, \cdots).$$

Then, if $Q_2(N)$ increases strictly with N and $Q_2(\infty)[1+M_G(K)] > c_1/(c_1 - c_2)$, then there exists a finite and unique N^* ($1 \leq N^* < \infty$) that satisfies

$$Q_2(N+1)\sum_{j=0}^{N-1}G^{(j)}(K) - [1 - G^{(N)}(K)] \geq \frac{c_2}{c_1 - c_2} \quad (N = 1, 2, \cdots). \quad (31)$$

Finally, for Model 3, the unit is replaced at time T or at failure, whichever occurs first. When $Z(t) = A_t t$, the expected cost rate is

$$C_3(T) = \frac{c_1 - (c_1 - c_2)L_A(K/T)}{\int_0^T L_A(K/t)\mathrm{d}t}. \tag{32}$$

The optimal time T_3^* that minimizes $C_3(T)$ is : Let $l_A(t)$ be a density function of $L_A(t)$ and $r(t)$ be the failure rate of $L_A(t)$, *i.e.*, $r(t) \equiv l_A(t)/L_A(t)$. Then, if $r(t)$ increases with t, there exists a finite and unique T_3^* that satisties

$$r(K/T)\int_0^T L_A(K/t)\mathrm{d}t + L_A(K/T) = \frac{c_1}{c_1 - c_2}. \tag{33}$$

4 Optimal Imperfect Maintenance of Aircraft

The airframe of commercial aviation has to be lightweight because its weight affects the range and fuel-efficiency of aircraft. Coincidentally, the airframe has to be damage tolerant during operation because airframe suffers serious statistical and dynamic mechanical stress which is caused by the variation of environmental air pressure and temperature, vibration of engine and aerodynamic turbulence and shocks when the aircraft takes off and lands. Although the ferrous alloy is common metallic material for mechanical structure, aluminum alloys 2024 and 7075 are utilized for the major material of airframe structure because the tensile strength of aluminum alloy is almost as same as that of ferrous alloy and the specific weight of aluminum alloy is from one third to one fourth of ferrous alloy. Furthermore, structural consideration for lightweight is performed when airframe is designed. The stressed skin monocoque structure is adopted and its stress is analyzed strictly using finite element method (FEM) for reducing weight [Paul, Kelly and Venkayya (2002)].

Under cyclic stress condition, the consumed life of mechanical structures is assessed by S-N curve, and the ferrous alloy structure has infinite lifetime when it is designed to hold stress below threshold of endurance limit. Whereas, non-ferrous alloy such as aluminum has no such distinct limit and fails finally under slight stress condition. Thus, the lifetime of aluminum alloy airframe is finite and the accumulation of tiny stresses will cause serious damage in a long period. Therefore, the maintenance of airframe is indispensable to operate aircraft without any serious troubles.

In Section 1, we denoted that detailed investigations of target system characteristics must be performed for the consideration of cost-effective

PM policies. To undergoing the cost-effective PM, various kinds of PMs of which time periods and maintenance contents are different, are assembled appropriately. We call such maintenance as the multi-echelon PM and studied optimal maintenance policies for the fossil-fired power plant with multi-echelon risks [Nakagawa and Ito (2008)]. Following four echelons PMs are undergone for commercial aviation [FAA Advisory Circular (1998)]:

1) *A check*: Inspection of landing gear, control surfaces, fluid levels, oxygen systems, lighting and auxiliary power system is performed. It occurs every three to five days at airport gate.
2) *B check*: A check topics plus inspection of internal control systems, hydraulic systems and energy equipment is performed. It occurs every eight months at airport gate.
3) *C check*: Aircraft is opened up extensively and inspection of wear, corrosion and cracks is performed. It occurs every 12 to 17 months at the maintenance hangar of airline's hub airport.
4) *D check*: Aircraft is disassembled and overhauled perfectly. It occurs every 22,500 flight hours at specialized facility.

A and B checks are categorized as the minor level maintenance. Whereas, C and D checks are heavy maintenance. Especially, D check affects aircraft availability and lifecycle cost.

The airframe has to damage-tolerant during designed operation period. To assure the airframe damage-tolerance, the fatigue test using full-scale airplane structure is required by Federal Aviation Administration (FAA) regulation [Dixon (2006); Faderal Aviation Administration (1998)]. The regulation directs that the sufficient full-scale test evidence has to be more than two times of the prespecified operation interval to guarantee the operation safety. Because catastrophic troubles of airframe during operation can be avoidable when appropriate D checks are undergone, periods of A, B, and C checks can be determined considering their costs.

A, B, C, and D checks are the imperfect maintenance and such maintenance assumes that cumulative hazard functions after PMs reduce at a certain rate. The basic imperfect PM policy was introduced [Nakagawa (2005)], and it extended to the three echelons one [Ito and Nakagawa (2009)], the N echelons one [Ito and Nakagawa (2010)], the practical useful one [Ito and Nakagawa (2011)]. Utilizing the practical useful imperfect PM policy, we considered the optimal operation censoring policy of aircraft which maximizes the expected profit [Ito and Nakagawa (2012)].

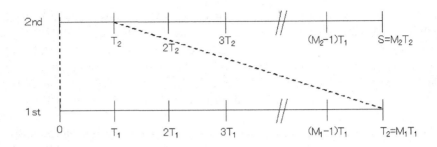

Fig. 3 Preventive maintenance schedule of airframe.

From the practical view point, the reduction rate of cumulative hazard function should be defined precisely because the small change of reduction rate causes the great change of calculation results. In this section, the traditional imperfect maintenance policy [Ito and Nakagawa (2009)] and its practically reviced one [Ito and Nakagawa (2011)] of civil aircraft are discussed.

4.1 Model 1

4.1.1 Model and assumptions

Following imperfect preventive maintenance is assumed:

1) The airframe is maintained preventively at times iT_1 ($i = 1, 2, 3, \cdots, M_1 - 1$), $jT_2 \equiv jM_1T_1$ ($j = 1, 2, 3, \cdots, M_2 - 1$) and M_2T_2. M_1T_1 and M_2T_2 are defined as T_2 and S, respectively (see Fig. 3).

2) Cumulative hazard functions $H(t)$ after PMs reduce to $a^i H(iT_1)$ ($0 < a \leq 1$) at times iT_1 and $a^{jM_1+i}b^j H(jT_2 + iT_1)$ ($0 < b \leq 1$) at times $jT_2 + iT_1 = (jM_1 + i)T_1$ (see Fig. 4). The airframe after PM at time $S = NT$ becomes as good as new. The airframe failure during operation is not so serious and the hazard rate $h(t) \equiv dH(t)/dt$ remains undisturbed by repair.

3) PM costs are c_1 at times iT_1, c_2 at times jT_2 and c_s ($c_s > c_2 > c_1$) at time S. The average repair cost at airframe failure is c_r.

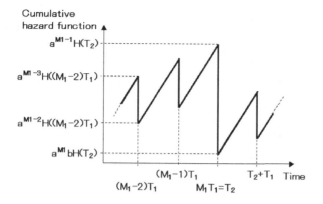

Fig. 4   Variance of cumulative hazard function with time of Model 1.

When $M_2 = 1$, the total expected cost of airframe maintenance is [2]

$$C_1(M_1, 1) = c_r \sum_{i=0}^{M_1-1} a^i \int_{iT_1}^{(i+1)T_1} h(t)\mathrm{d}t + (M_1 - 1)c_1 + c_s$$

$$= c_r \left[(1-a) \sum_{i=1}^{M_1-1} a^{i-1} H\left(\frac{iT_2}{M_1}\right) + a^{M_1-1} H(T_2) \right]$$

$$+ (M_1 - 1)c_1 + c_s .\tag{34}$$

When $M_2 = 2$, the total expected cost is

$$C_1(M_1, 2) = \left[c_r \sum_{i=0}^{M_1-1} a^i \int_{iT_1}^{(i+1)T_1} h(t)\mathrm{d}t + \sum_{i=0}^{M_1-1} a^{M_1+i} b \int_{iT_1}^{(i+1)T_1} h(T_2+t)\mathrm{d}t \right]$$

$$+ 2(M_1 - 1)c_1 + c_2 + c_s$$

$$= c_r \left\{ (1-a) \sum_{i=1}^{M_1-1} a^{i-1} \left[H\left(\frac{iT_2}{M_1}\right) + a^{M_1} b H\left(\left(1 + \frac{i}{M_1}\right) T_2\right) \right] \right.$$

$$\left. + a^{2M_1-1} b H(2T_2) + a^{M_1-1}(1-ab)H(T_2) \right\}$$

$$+ 2(M_1 - 1)c_1 + c_2 + c_s .\tag{35}$$

When $M_2 = 3$, the total expected cost is

$C_1(M_1, 3)$

$$
= \left[c_r \sum_{i=0}^{M_1-1} a^i \int_{iT_1}^{(i+1)T_1} h(t)\mathrm{d}t + \sum_{i=0}^{M_1-1} a^{M_1+i} b \int_{iT_1}^{(i+1)T_1} h(T_2 + t)\mathrm{d}t \right.
$$

$$
\left. + \sum_{i=0}^{M_1-1} a^{2M_1+i} b^2 \int_{iT_1}^{(i+1)T_1} h(2T_2 + t)\mathrm{d}t \right] + 3(M_1 - 1)c_1 + 2c_2 + c_s
$$

$$
= c_r \left\{ (1 - a) \sum_{i=1}^{M_1-1} a^{i-1} \left[H\left(\frac{iT_2}{M_1}\right) + a^{M_1} b H\left(\left(1 + \frac{i}{M_1}\right) T_2 \right) \right. \right.
$$

$$
\left. + a^{2M_1} b^2 H\left(\left(2 + \frac{i}{M_1}\right) T_2 \right) \right] + a^{3M_1-1} b^2 H(3T_2)
$$

$$
\left. + a^{2M_1-1}(1 - ab) H(2T_2) + a^{M_1-1}(1 - ab) H(T_2) \right\}
$$

$$
+ 3(M_1 - 1)c_1 + 2c_2 + c_s . \tag{36}
$$

From (34), (35) and (36), $C_1(M_1, M_2)$ is

$$
C_1(M_1, M_2) = c_r \sum_{j=0}^{M_2-1} \sum_{i=0}^{M_1-1} a^{jM_1+i} b^j \int_{iT_1}^{(i+1)T_1} h(jT_2 + t)\mathrm{d}t
$$

$$
+ M_2(M_1 - 1)c_1 + (M_2 - 1)c_2 + c_s
$$

$$
= c_r \sum_{j=0}^{M_2-1} a^{jM_1} b^j \left[(1 - a) \sum_{i=1}^{M_1} a^{i-1} H\left(\left(j + \frac{i}{M_1}\right) T \right) \right.
$$

$$
\left. + a^{M_1} H((j + 1)T_2) - H(jT_2) \right]
$$

$$
+ M_2(M_1 - 1)c_1 + (M_2 - 1)c_2 + c_s . \tag{37}
$$

4.1.2 *Optimal maintenance policies*

The time S is fixed because the huge cost is necessary for its undergoing. Optimal M_1^*s which minimize total expected costs $C_1(M_1, M_2)$ are searched when $M_2 = 1, 2, 3, \cdots$. Forming the inequality $C_1(M_1 + 1, M_2) - C_1(M_1, M_2) \geq 0$,

$$
\frac{M_2 c_1}{c_r(1-a)} \geq \sum_{j=0}^{M_2-1} a^{jM_1} b^j \left\{ \sum_{i=1}^{M_1} a^{i-1} \left[H \left(\left(j + \frac{i}{M_1} \right) T_2 \right) \right.\right.
$$
$$
\left.\left. - a^j H \left(\left(j + \frac{i}{M_1 + 1} \right) T_2 \right) \right] \right.
$$
$$
\left. - \frac{1-a^j}{1-a} \left[H(jT_2) - a^{M_1} H((j+1)T_2) \right] \right\}. \tag{38}
$$

Suppose that $H(t) = \lambda t$, (37) and (38) are rewritten as, respectively,

$$
C_1(M_1, M_2) = \frac{c_r \lambda T_2}{M_1} \frac{1 - a^{M_1}}{1 - a} \frac{1 - (a^{M_1} b)^{M_2}}{1 - a^{M_1} b}
$$
$$
+ M_2(M_1 - 1)c_1 + (M_2 - 1)c_2 + c_s, \tag{39}
$$

$$
\frac{M_2 c_1}{c_r \lambda T_2} \geq \frac{1}{M_1} \frac{1 - a^{M_1}}{1 - a} \frac{1 - (a^{M_1} b)^{M_2}}{1 - a^{M_1} b}
$$
$$
- \frac{1}{M_1 + 1} \frac{1 - a^{M_1+1}}{1 - a} \frac{1 - (a^{M_1+1} b)^{M_2}}{1 - a^{M_1+1} b}. \tag{40}
$$

When $M_2 = 1$, (39) and (40) are rewritten as, respectively,

$$
C_1(M_1, 1) = \frac{c_r \lambda T_2}{M_1} \frac{1 - a^{M_1}}{1 - a} + (M_1 - 1)c_1 + c_s, \tag{41}
$$

$$
\frac{c_1}{(1-a)\lambda T_2 c_0} \geq \frac{1}{M_1(M_1 + 1)} \sum_{i=1}^{M_1} i a^{i-1}. \tag{42}
$$

By letting $L(M_1)$ be the right-hand side of (42),

$$
L_1(1) = \frac{1}{2}, \qquad \lim_{M_1 \to \infty} L_1(M_1) = 0, \tag{43}
$$

$$
L_1(M_1) - L_1(M_1 + 1)
$$
$$
= \frac{2}{M_1(M_1 + 1)(M_1 + 2)} \sum_{i=1}^{M_1} i a^{i-1}(1 - a^{M_1+i-1}) > 0. \tag{44}
$$

Therefore, $L_1(M_1)$ decreases with M_1 and we have the following optimal policy when $M_2 = 1$:

1) If $c_1/((1 - a)\lambda T_2 c_r) \geq 1/2$, then $M_1^* = 1$.
2) If $c_1/((1-a)\lambda T_2 c_r) < 1/2$, then there exists a finite and unique M_1^* ($1 < M_1^* < \infty$).

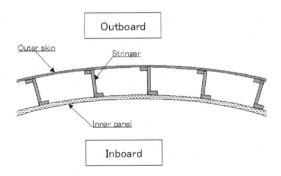

Fig. 5 An example of fuselage skin cross section.

4.2 *Model 2*

For applying Model 1, parameters a and b have to be determined and it may create confusion in practice because the slight change of a and b cause the great change of PM policy and how to determe a and b with high accuracy is a problematic issue. From the practical view point, settling such sensible parameters should be eliminated.

Figure 5 denotes an example of fuselage skin cross section and we can easily find that stringers cannot be inspected except that the inner panel or the outer skin is stripped off at the maintenance hangar and specialized facility. When the total cumulative hazard function is constituted with discrete cumulative hazard functions of inner panel, outer skin and stringers, cumulative hazard functions of inner panel and outer skin can be renewed at A and B *checks* and the stringer hazard function can be renewed at C and D *checks*. Using above sheme, Model 1 can be expanded as follows.

4.2.1 *Model and assumptions*

We consider the following maintenance policy which has the same assumptions except 2):

2') If the PM is not made, the cumulative hazard function of airframe is constituted with four parts, *i.e.*,

$$H(t) = H_0(t) + H_1(t) + H_2(t) + H_s(t), \qquad (45)$$

where $H_1(t)$ becomes new at times iT_1 because all failures of this part are discovered and are repaired, $H_2(t)$ becomes new at times jT_2

because all failures of this part are discovered and are repaired, $H_s(t)$ becomes new at time S because the system is opened up, inspected and repaired extensively, and $H_0(t)$ is unchanged at any maintenance and overhaul.

Cumulative hazard functions $H(t)$ before $(-)$ and after $(+)$ PM at time iT_1 are denoted as follows, respectively.

$$
\begin{aligned}
H_-(iT_1) &= H_1(T_1) + H_2(iT_1) + H_s(iT_1) + H_0(iT_1)\,, \\
H_+(iT_1) &= H_2(iT_1) + H_s(iT_1) + H_0(iT_1)\,,
\end{aligned}
\tag{46}
$$

and $H(t)$ before $(-)$ and after $(+)$ PM at time jT_2 are denoted as follows, respectively.

$$
\begin{aligned}
H_-(jT_2) &= H_1(T_1) + H_2(T_2) + H_s(jT_2) + H_0(jT_2)\,, \\
H_+(jT_2) &= H_s(jT_2) + H_0(jT_2)\,.
\end{aligned}
\tag{47}
$$

Similarly, $H(S)$ before $(-)$ and after $(+)$ overhaul are denoted as follows, respectively,

$$
\begin{aligned}
H_-(S) &= H_1(T_1) + H_2(T_2) + H_s(S) + H_0(S)\,, \\
H_+(S) &= H_0(S)\,.
\end{aligned}
\tag{48}
$$

The expected number of failures in $[0, S]$ is

$$
\begin{aligned}
\sum_{j=0}^{M_2-1} &\left\{ \sum_{i=0}^{M_1-1} \left[\int_{iT_1+jT_2}^{(i+1)T_1+jT_2} h(t)\mathrm{d}t - H_1(T_1) \right] - H_2(T_2) \right\} \\
&= H(S) - M_2 M_1 H_1(T_1) - M_2 H_2(M_1 T_1) \\
&\equiv H_{012}(S) - M_2 M_1 H_1(T_1) - M_2 H_2(M_1 T_1)\,,
\end{aligned}
\tag{49}
$$

where $H_{012}(t) \equiv H_0(t) + H_1(t) + H_2(t) + H_s(t)$. The expected cost rate of airframe maintenance in $[0, S]$ is

$$
\begin{aligned}
C_2(M_1, M_2) = \frac{1}{M_2 M_1 T_1} \Big\{ &c_r \Big[H_{012}(M_2 M_1 T_1) - M_2 M_1 H_1(T_1) \\
&- M_2 H_2(M_1 T_1) \Big] + c_1 M_2 M_1 + c_2 M_2 + c_s \Big\}\,.
\end{aligned}
\tag{50}
$$

4.2.2 *Optimal maintenance policies*

We find optimal M_1^* and M_2^* which minimize $C_2(M_1, M_2)$ in (50). Forming the inequality $C_2(M_1, M_2 + 1) - C_2(M_1, M_2) \geq 0$,

$$M_2(M_2 + 1)M_1T_1 \left[\frac{H_{012}((M_2 + 1)M_1T_1)}{(M_2 + 1)M_1T_1} - \frac{H_{012}(M_2M_1T_1)}{M_2M_1T_1} \right] \geq \frac{c_s}{c_r}.$$

(51)

Suppose that $h_{012}(t)$ and $h_2(t)$ are strictly increasing functions with t and $h_i(\infty) = \infty$. Denoting the left-hand side of (51) by $L_{21}(M_2)$, $L_{21}(M_2)$ is always positive, $L_{21}(\infty) = \infty$ and $L_{21}(M_2)$ is strictly increasing function with M_2. Therefore, we have the following optimal policy:

1) If $L_{21}(1) \geq c_s/c_r$, then $M_2^* = 1$.
2) If $L_{21}(1) < c_s/c_r$, then there exists a finite and unique minimum M_2^* $(1 < M_2^* < \infty)$ which satisfies (51).

Forming the inequality $C_2(M_1 + 1, M_2) - C_2(M_1, M_2) \geq 0$,

$$M_2M_1(M_1 + 1)T_1$$

$$\times \left[\frac{H_{012}(M_2(M_1 + 1)T_1)}{M_2(M_1 + 1)T_1} - \frac{H_{012}(M_2M_1T_1)}{M_2M_1T_1} + \frac{H_2(M_2(M_1 + 1)T_1)}{M_2(M_1 + 1)T_1} \right.$$

$$\left. - \frac{H_2(M_2M_1T_1)}{M_2M_1T_1} - \frac{H_2((M_1 + 1)T_1)}{(M_1 + 1)T_1} + \frac{H_2(M_1T_1)}{M_1T_1} \right]$$

$$\geq \frac{c_2M_2 + c_s}{c_r}.$$

(52)

Because the same argument can be applied for the term including H_{012} of the left-hand side of (52), we notice the term including $H_2(t)$ of the left-hand side of (52) and have

$$L_{23}(M_1) \equiv M_2M_1(M_1 + 1)T_1 \left[\frac{H_2(M_2(M_1 + 1)T_1)}{M_2(M_1 + 1)T_1} - \frac{H_2(M_2M_1T_1)}{M_2M_1T_1} \right.$$

$$\left. - \frac{H_2((M_1 + 1)T_1)}{(M_1 + 1)T_1} + \frac{H_2(M_1T_1)}{M_1T_1} \right]$$

$$= M_2M_1(M_1 + 1)T_1 \left[L_3(M_2, (M_1 + 1)T_1) - L_3(M_2, M_1T_1) \right],$$

(53)

where $L_3(n, t) \equiv [H(nt) - nH(t)]/nt$. When $h(t)$ is a strictly increasing function,

$$L_3(n, t) = \frac{H(nt) - nH(t)}{nt}$$

$$= \frac{1}{nt} \left[\int_0^{nt} h(u)\mathrm{d}u - n \int_0^t h(u)\mathrm{d}u \right]$$

$$= \frac{1}{nt} \sum_{i=1}^n \left[\int_{(i-1)t}^{it} h(u)\mathrm{d}u - \int_0^t h(u)\mathrm{d}u \right] > 0, \qquad (54)$$

$$L_3(n, \infty) = \lim_{t \to \infty} \frac{H(nt) - nH(t)}{nt}$$

$$= \lim_{t \to \infty} \left[\frac{H(nt)}{nt} - \frac{H(t)}{t} \right]$$

$$= \lim_{t \to \infty} [h(nt) - h(t)] = \infty, \qquad (55)$$

$$\frac{\mathrm{d}L_3(n, t)}{\mathrm{d}t} = \frac{[h(nt)n - nh(t)]t - [H(nt) - nH(t)]}{nt^2}$$

$$= \frac{1}{nt^2} [nth(nt) - H(nt) - nth(t) + nH(t)]$$

$$= \frac{1}{nt^2} \sum_{j=1}^n \left\{ \int_{(j-1)t}^{jt} [h(nt) - h(u)]\mathrm{d}u - \int_0^t [h(t) - h(u)]\mathrm{d}u \right\}$$

$$= \frac{1}{nt^2} \sum_{j=1}^n \int_0^t \{h(nt) - h[u + (j-1)t] - h(t) + h(u)\}\,\mathrm{d}u > 0,$$

$$(56)$$

for $n > 1$. Thus, $L_{23}(M_1)$ is always positive.

Furthermore,

$$L_{23}(M_1 + 1) - L_{23}(M_1)$$

$$= (M_1 + 1) \left\{ \int_{M_2(M_1+1)T_1}^{M_2(M_1+2)T_1} h_2(u)\mathrm{d}u - \int_{M_2 M_1 T_1}^{M_2(M_1+1)T_1} h_2(u)\mathrm{d}u \right.$$

$$\left. - M_2 \left[\int_{(M_1+1)T_1}^{(M_1+2)T_1} h_2(u)\mathrm{d}u - \int_{M_1 T_1}^{(M_1+1)T_1} h_2(u)\mathrm{d}u \right] \right\}. \qquad (57)$$

Thus, $L_{23}(M_1)$ is strictly increasing function with M_1 when (57) is positive. Therefore, we have the following optimal policy when $L_{22}(M_1)$ is denoted by the left-hand side of (52) :

1) If $L_{22}(1) \geq (c_2 M_2 + c_s)/c_r$, then $M_1^* = 1$.
2) If $L_{22}(1) < (c_2 M_2 + c_s)/c_r$, then there exists a finite and unique minimum M_1^* $(1 < M_1^* < \infty)$ which satisfies (52).

5   Conclusions

In this chapter, optimal maintenance policies for the aged fossil-fired power plant and the civil aircraft are discussed.

In Section 2, the optimal PM policies for a damage system with repair have considered. The system fails when the cumulative damage exceeds a certain critical level and the critical level degrades at every repair. Two models are considered and expected costs of these models are derived. In Model 1, the system has no operation time-limit and it undergoes overhaul when the cumulative damage is between a managerial level and the critical level. In Model 2, the system has operation time-limit and it undergoes overhaul when the cumulative damage is between a managerial level and the critical level, or when the operation time exceeds the prespecified time, whichever occurs first. Optimal pocilies which minimize costs are discussed.

In Section 3, three cumulative damage models have considered. In case of Model 1 (Standard Model), the operating unit is subjected shocks and suffers some damage at each shock. In case of Model 2 (Periodic Model), each amount of damage is measured only at periodic times and has an identical distribution. And in case of Model 3 (Continuous Model), the total damage usually increases with time. The expected cost rates of each model are derived and optimal policies which minimize them are considered.

Finally, in Section 4, two imperfect PM models of airframe has considered. Model 1 is the conventional imperfect PM model and the reduction rate of cumulative hazard functions after PM is assumed. Model 2 is the improved imperfect PM model and a part of the cumulative hazard function after PMs disappeared. As the cumulative hazard function of airframe is concerned with mechanical parts such as rivets of airframe skin lap-joints, the changes of it can be defined easily. Three echelons PMs are secured and optimal policies which minimize expected cost rates of airframe are considered.

References

Barlow, R. E. and Proschan, F. (1965). *Mathematical Theory of Reliability* (John Wiley & Sons).

Bogdanoff, J. L. and Kozin, F. (1985). *Probabilistic Models of Cumulative Damage* (John Wiley & Sons).

Çinlar, E. (1975). *Introduction to Stochastic Processes* (Prentice-Hall).

Cox, D. R. (1962). *Renewal Theory* (Methuen).

Dixon, M. (2006). *The maintenance Costs of Aging Aircract: Insights from Commercial Aviation* (RAND Corporation).

Esary, J. D., Marshall, A. W. and Proschan, F. (1973). *Shock models and wear processes*, Annals of Probability, Vol. 1, pp. 627–649.

Faderal Aviation Administration (1998). *Fatigue Evaluation of Structure; Final Rule.* March 31, 63, 61: 15708-15715. (14 CFR Part 25, Docket No. 27358, Amendment No. 25-96).

FAA Advisory Circular (1998). *Damage Tolerance and Fatigue Evaluation of Structure.* AC 25, 571-1C.

Hisano, K. (2000). *Preventive Maintenance and Residual Life Evaluation Technique for Power Plant (I.Preventive Maintenance)* (in Japanese), *The Thermal and Nuclear Power*, Vol. 51, No. 4, pp. 491–517.

Hisano, K. (2001). *Preventive Maintenance and Residual Life Evaluation Technique for Power Plant (V.Review of Future Advances in Preventive Maintenance Technology)* (in Japanese), *The Thermal and Nuclear Power* Vol. 52, No. 3, pp. 363–370.

Ito, K., Nakagawa, T. and Teramoto, K. (2006). *Optimal Maintenance Policy for a System with Damage Repair, Proceedings of the 2nd Asian International Workshop (AIWARM 2006) Advanced Reliability Modeling II, Reliability Testing and Improvement*, Busan, Korea, pp. 235–242.

Ito, K. and Nakagawa, T. (2006). *Optimal Operation Censoring Policy for a Damaged System, Proceedings of International Workshop on Recent Advances in Stochastic Operations Research II (RASOR Nanzan)*, Nagoya, Japan, pp. 106–111.

Ito, K. and Nakagawa, T. (2007). *Optimal Operation Censoring Policy for a System with Damage Repair, Proceedings of the 13th ISSAT International Conference on Reliability and Quality in Design*, Seattle, Washington, U.S.A, pp. 216–220.

Ito, K. and Nakagawa, T. (2008). *Comparison of Three Cumulative Damage Models, Proceedings of the 3rd Asian International Workshop (AIWARM 2008)*, Taichung, Taiwan, pp. 332–338.

Ito, K. and Nakagawa, T. (2009). *Optimal Imperfect Maintenance of Aircraft, Proceedings of the 15th ISSAT International Conference Reliability and Quality in Design*, San Francisco, California, U.S.A., pp. 215–218.

Ito, K. and Nakagawa, T. (2010). *Optimal Multi-echelon Maintenance of Aircraft, Proceedings of the 4th Asia-Pacific International Symposium (APARM*

2010), Advanced Reliability Modeling IV, Wellington, New Zealand, pp. 257–264.

Ito, K. and Nakagawa, T. (2011). *Optimal Maintenance of Aircraft, Proceedings of The 7th International Conference on "Mathematical Methods in Reliability": Theory. Methods. Applications.(MMR 2011)*, Beijing, China, pp. 859–864.

Ito, K. and Nakagawa, T. (2012). *Optimal Operation Censoring Policy of Aircraft, Proceedings of the Asia-Pacific International Symposium (APARM2012)*, Nanjing, China, pp. 184–191.

Kosugiyama, S., Takizuka, T., Kunitomi, K., Yan, X., Katanishi, S. and Takada, S. (2003). *Basic Policy of Maintenance for the Power Conversion System of the Gas Turbine High Temperature Reactor 300 (GTHTR300)* (in Japanese), *Journal of Nuclear Science and Technology*, Vol. 2, No. 3, pp. 105–117.

Nakagawa, T. and Osaki, S. (1974). *Some aspects of damage model, Microelectronics and Reliability*, Vol. 13, pp. 253–257.

Nakagawa, T. (2005). *Maintenance Theory of Reliability* (Springer-Verlag).

Nakagawa, T. (2007). *Shock and Damage Models in Reliability Theory* (Springer Verlag).

Nakagawa, T. and Ito, K. (2008). *Optimal Maintenance Policies for a System With Multiechelon Risks, IEEE Transactions on Systems, Man, and Cybernetics, Part A: Systems and Humans* Vol. 38, pp. 461–469.

Osaki, S. (1992). *Applied Stochastic Systems Modeling* (Springer Verlag).

Paul, D., Kelly, L. and Venkayya, V. (2002). *Evolution of U.S. Military Aircraft Structures Technology, Journal of Aircraft*, Vol. 39, No. 1, pp. 18–29.

Ross, S. M. (1983). *Stochastic Processes* (John Wiley & Sons).

Chapter 18

Reliability Analysis of a System Connected with the Radio Link

Mitsuhiro Imaizumi

College of Contemporary Management, Aichi Gakusen University,
1 Shiotori, Ohike-cho, Toyota 471-8532, Japan

1   Introduction

As mobile devices have been widely used and radio communication technology has remarkably developed, the demand for improvement of the reliability of radio communication has greatly increased. Generally compared with the wired link, packet loss occurs with high probability on the radio link due to radio interference such as phasing [Miyoshi, Sugano and Murata (2002); Yuki, Yamamoto, Sugano, Murata, Miyahara and Hatauchi (2002); Takahashi, Saito, Aida, Tobe and Tokuda (2005); Itaya, Hasegawa, Hasegawa, Davis, Kadowaki and Obana (2005)]. Therefore, how we improve the throughput efficiency on the radio link has become a problem. In this chapter, we pay attention to a system connected with the radio link, and treat three stochastic models. Using the theory of Markov renewal processes, we derive the measures such as the mean time until the transmission of packets succeeds.

The theory of Markov renewal processes [Osaki (1992)] is used to analyze the system. Barlow and Proschan [Barlow and Proschan (1965)] gave a table of applicable stochastic process associated repairman problems. Nakagawa and Osaki [Nakagawa and Osaki (1979)] analyzed two-unit systems using a unique modification of the regeneration point techniques of Markov renewal processes.

In Section 2 and 3, we formulate a stochastic model of a network system with retransmission control scheme using duplicated ACK (positive acknowledgment) packets. In the TCP protocol, the high reliability of the data transmission is realized by returning an ACK packet from a receiver when a sender transmits data packets, but on the radio link, it is not rare to lose an ACK packet. In order to cope with this problem, several schemes have been considered [Matsuda and Yamamoto (2004); Zhang, Shirazi and Tanaka (2004); Saito and Teraoka (2004); Balakrishnan and Katz (1998); Bakre and Badrinath (1995)]. As one of schemes to cope with this problem, the retransmission scheme using duplicated ACK packets where a sender returns redundant ACK packets has been proposed [Miyoshi, Sugano and Murata (2001)].

In Section 4, we formulate a stochastic model of a network system with selective repeat data transmission error recovery scheme. In Section 2 and 3, we assume that if a sender has failed to transmit data packets, it retransmits all packets. On the other hand, in Section 4, we assume that if a sender has failed to transmit data packets, it retransmits packets that the sender has failed to transmit.

Although the simulation about the policy for the system connected with the radio link has already been introduced, there are few formalized stochastic models.

In each Model, the mean time until the transmission succeeds is analytically derived. Further, an optimal policy which maximizes the amount of data transmissions per unit of time until the transmission succeeds is discussed.

2 Model 1

2.1 *Model and Analysis*

We pay attention to only a data transmission terminal connected to the network with the radio link. The Data is transmitted from the fixed terminal to the mobile terminal via the base station. Fig. 1 draws the outline of the model.

(1) A sender transmits the requirement for connection establishment to a receiver. After receiving the requirement for connection establishment, the receiver returns a response. The connection establishment needs the time according to a general distribution $A(t)$ with finite mean a.

(2) The sender transmits n data packets to the receiver where the connection is established. The editing time for each data packet needs

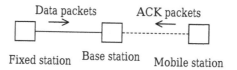

Fig. 1   Outline of the model.

the time according to a general distribution $B_1(t)$ with finite mean b_1. The transmission from the sender to the receiver needs according to a general distribution $D_1(t)$.

(3) After the receiver have received n data packets, it return an ACK (positive acknowledgment) packet if it has received all data packets correctly. The editing time for each ACK packet needs the time according to a general distribution $B_2(t)$ with finite mean b_2. If the sender has not received an ACK packet and it has judged time out, it retransmits n data packets. The time to transmit an ACK packet has a general distribution $D_2(t)$, and denotes that $D(t) \equiv D_1(t) * D_2(t)$ with finite mean d. The asterisk mark denotes the Stieltjes convolution and $\Phi_1(t) * \Phi_2(t) \equiv \int_0^t \Phi_2(t - u)d\Phi_1(u)$. We assume that the time until the sender judges time out has the same distribution $D(t)$, the probability that a data packet loss occurs is p and the probability that an ACK packet loss occurs is q.

(4) When the receiver has received retransmission data packet (the second transmission packet), it returns 2 ACK packets. The transmission succeeds if the sender has received at least an ACK packet. When the sender has judged time out without receiving all ACK packets, it retransmits n packets. Generally, when the receiver has received the $(k-1)$-th time retransmission data packet (the k-th time transmission packet), it returns k ACK packets, and when the sender has judged time out without receiving all ACK packets, it retransmits n data packets.

(5) If the transmission have failed k times, once the sender interrupts the retransmission, and restarts again from the beginning after a constant time g where $G(t) \equiv 0$ for $t < g$ and for $1\ t \geq g$.

Under above assumptions, we define the following states of the system.

State 0: System begins to operate.
State 1: Transmission begins.

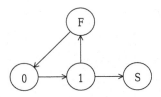

Fig. 2   Transition diagram between system states.

State F: Retransmission fails $k(k = 1, 2, \cdots)$ times and is interrupted.
State S: Transmission of n packets succeeds.

The system states defined above form a Markov renewal process where state S is an absorbing state. Transition diagram between system states is shown in Fig. 2.

Let $Q_{ij}(t)$ $(i = 0, 1, F; j = 0, 1, S, F)$ be one-step transition probabilities of a Markov renewal process. Then, by the similar method of [Yasui, Nakagawa and Sandoh (2002)],

$$Q_{01}(t) = A(t),$$
$$Q_{1S}(t) = \gamma_1 [B_1^{(n)}(t) * B_2(t) * D(t)]$$
$$+ \gamma_2 [B_1^{(n)}(t) * B_2(t) * D(t)] * [B_1^{(n)}(t) * B_2^{(2)}(t) * D(t)]$$
$$+ \cdots$$
$$+ \gamma_k [B_1^{(n)}(t) * B_2(t) * D(t)] * [B_1^{(n)}(t) * B_2^{(2)}(t) * D(t)]$$
$$* \cdots * [B_1^{(n)}(t) * B_2^{(k)}(t) * D(t)],$$
$$Q_{1F}(t) = \left(1 - \sum_{i=1}^{k} \gamma_i\right) [B_1^{(n)}(t) * B_2(t) * D(t)]$$
$$* [B_1^{(n)}(t) * B_2^{(2)}(t) * D(t)] * \cdots * [B_1^{(n)}(t) * B_2^{(k)}(t) * D(t)],$$
$$Q_{F0}(t) = G(t), \tag{1}$$

where $\Phi^{(i)}(t)$ denotes the i-fold Stieltjes convolution of a distribution $\Phi(t)$ with itself, i.e., $\Phi^{(i)}(t) \equiv \Phi^{(i-1)}(t) * \Phi(t), \Phi_1(t) * \Phi_2(t) \equiv \int_0^t \Phi_2(t - u) d\Phi_1(u), \Phi^{(0)}(t) \equiv 1$. The probability that a sender is received ACK packets at k-th transmission without receiving ACK packets at $(k-1)$-th transmission is

$$\gamma_k \equiv \left(1 - \sum_{i=1}^{k-1} \gamma_i\right)(1 - p)^n (1 - q^k) \quad (k = 1, 2, \cdots), \tag{2}$$

where $\sum_{i=1}^{0} \equiv 0$.

Let $\phi(s)$ be the Laplace-Stieltjes (LS) transform of any function $\Phi(t)$, i.e., $\phi(s) \equiv \int_0^\infty e^{-st} d\Phi(t)$ for $Re(s) > 0$. Then, from (1),

$$q_{01}(s) = a(s),$$

$$q_{1S}(s) = \sum_{i=1}^{k} \gamma_i [b_1(s)^n d(s)]^i b_2(s)^{\frac{i(i+1)}{2}},$$

$$q_{1F}(s) = \left(1 - \sum_{i=1}^{k} \gamma_i\right) [b_1(s)^n d(s)]^k b_2(s)^{\frac{k(k+1)}{2}},$$

$$q_{F0}(s) = g(s). \tag{3}$$

We derive the mean time $\ell_1(n)$ until the transmission of n packets succeeds. Let $H_{iS}(t)$ $(i = 0, 1, F)$ be the first-passage time distribution from state i to state S. Then, we have the following renewal equations:

$$
\begin{aligned}
H_{0S}(t) &= Q_{01}(t) * H_{1S}(t), \\
H_{1S}(t) &= Q_{1F}(t) * H_{FS}(t) + Q_{1S}(t), \\
H_{FS}(t) &= Q_{F0}(t) * H_{0S}(t).
\end{aligned} \tag{4}
$$

Taking the LS transform on the both sides of (4) and solving renewal equations,

$$h_{0S}(s) = \frac{q_{01}(s) q_{1S}(s)}{1 - q_{01}(s) q_{1F}(s) q_{F0}(s)}. \tag{5}$$

Hence, the mean time $\ell_1(n)$ until the transmission succeeds is given by

$$\ell_1(n) \equiv \int_0^\infty t \, dH_{0S}(t) = \lim_{s \to 0} \frac{-d[h_{0S}(s)]}{ds}$$

$$= \frac{\sum_{i=1}^{k-1} \gamma_i \left[i(nb_1 + d) + \dfrac{i(i+1)}{2} b_2 \right] + \left(1 - \sum_{i=1}^{k-1} \gamma_i\right) \left[k(nb_1 + d) + \dfrac{k(k+1)}{2} b_2 \right] + a + g}{\sum_{i=1}^{k} \gamma_i} - g. \tag{6}$$

When $k = 1$, $\ell_1(n)$ is given by

$$\ell_1(n) = \frac{nb_1 + b_2 + d + a + g}{(1 - p)^n (1 - q)} - g. \tag{7}$$

2.2　*Optimal Policy*

We discuss an optimal policy which maximizes the transmission of all packets per unit of time until the transmission succeeds. We define the throughput $E_1(n)$, which represents the rate of n packets to their mean transmission times, as the following equation:

$$E_1(n) \equiv \frac{n}{\ell_1(n)}$$

$$= \frac{n\displaystyle\sum_{i=1}^{k}\gamma_i}{\displaystyle\sum_{i=1}^{k-1}\gamma_i\left[i(nb_1+d)+\frac{i(i+1)}{2}b_2\right]+\left(1-\displaystyle\sum_{i=1}^{k-1}\gamma_i\right)\left[k(nb_1+d)+\frac{k(k+1)}{2}b_2\right]+a+g-g\displaystyle\sum_{i=1}^{k}\gamma_i}$$

$$(n=1,2,\cdots). \qquad (8)$$

We seek an optimal n_1^* which maximizes $E_1(n)$. We put formally that $V_1(n) \equiv 1/E_1(n)$ and seek n_1^* which minimizes $V_1(n)$. From the inequality $V_1(n+1)-V_1(n) \geq 0$,

$$nX_1(n+1)-(n+1)X_1(n)+g \geq 0, \qquad (9)$$

where

$$X_1(n) \equiv \frac{\displaystyle\sum_{i=1}^{k-1}\gamma_i\left[i(nb_1+d)+\frac{i(i+1)}{2}b_2\right]+\left(1-\displaystyle\sum_{i=1}^{k-1}\gamma_i\right)\left[k(nb_1+d)+\frac{k(k+1)}{2}b_2\right]+a+g}{\displaystyle\sum_{i=1}^{k}\gamma_i}.$$

Denoting the left-hand side of equation (9) by $L_1(n)$,

$$L_1(n+1)-L_1(n)=(n+1)J_1(n), \qquad (10)$$

where

$$J_1(n) \equiv [X_1(n+2)-X_1(n+1)]-[X_1(n+1)-X_1(n)].$$

Hence, when $X_1(n)$ is a convex function and $J_1(1) > 0$, $L_1(n)$ is strictly increasing in n from $L_1(1)$ to $L_1(\infty)$.

Therefore, when $J_1(1) > 0$, we have the following optimal policy:

(1) If $L_1(1) < 0$ and $L_1(\infty) > 0$ then there exists a finite and unique $n_1^*(> 1)$ which satisfies (9).
(2) If $L_1(\infty) \leq 0$ then $n_1^* = \infty$.
(3) If $L_1(1) \geq 0$ then $n_1^* = 1$.

3 Model 2

3.1 *Model and Analysis*

We modify Model 1. We assume that when the receiver has received the $(k - 1)$-th time retransmission data packet (the k-th time transmission packet), it returns $m(1 \leq m < \infty)$ ACK packets. That is, we assume that the number of ACK packets does not depends on k.

Then, the LS transforms of one-step transition probabilities $Q_{ij}(t)$ are given by the following equations:

$$q_{01}(s) = a(s),$$
$$\begin{aligned}
q_{1S}(s) = {} & (1 - p)^n(1 - q^m)[b_1(s)^n b_2(s)^m d(s)] \\
& + [1 - (1 - p)^n(1 - q^m)] \\
& \times (1 - p)^n(1 - q^m)[b_1(s)^n b_2(s)^m d(s)]^2 \\
& + [1 - (1 - p)^n(1 - q^m)]^2 \\
& \times (1 - p)^n(1 - q^m)[b_1(s)^n b_2(s)^m d(s)]^3 \\
& + \cdots \\
& + [1 - (1 - p)^n(1 - q^m)]^{k-1} \\
& \times (1 - p)^n(1 - q^m)[b_1(s)^n b_2(s)^m d(s)]^k,
\end{aligned}$$
$$q_{1F}(s) = [1 - (1 - p)^n(1 - q^m)]^k[b_1(s)^n b_2(s)^m d(s)]^k,$$
$$q_{F0}(s) = g(s). \tag{11}$$

Then, by the similar analysis of Model 1, we have the mean time $\ell_2(n)$ until the transmission succeeds as follows:

$$\ell_2(n) = \frac{\displaystyle\sum_{i=1}^{k}[1 - (1 - p)^n(1 - q^m)]^{i-1} \times (1 - p)^n(1 - q^m)i(nb_1 + mb_2 + d) + [1 - (1 - p)^n(1 - q^m)]^k[k(nb_1 + mb_2 + d) + g] + a}{1 - [1 - (1 - p)^n(1 - q^m)]^k}. \tag{12}$$

3.2 *Optimal Policy*

In terms of Model 2, we define the throughput $E_2(n)$, which represents the rate of n packets to their mean transmission times, as the following equation:

$$E_2(n) \equiv \frac{n}{\ell_2(n)}$$

$$= \frac{n\{1 - [1 - (1-p)^n(1-q^m)]^k\}}{\displaystyle\sum_{i=1}^{k}[1 - (1-p)^n(1-q^m)]^{i-1}}$$

$$\times (1-p)^n(1-q^m)i(nb_1 + mb_2 + d)$$

$$+[1 - (1-p)^n(1-q^m)]^k[k(nb_1 + mb_2 + d) + g] + a$$

$$(n = 1, 2, \cdots). \qquad (13)$$

We seek an optimal n_2^* which maximizes $E_2(n)$. By similar analysis of Model 1, we can discuss the optimal policy for Model 2.

4 Model 3

4.1 *Model and Analysis*

We modify Model 1. Model 1 and 2 are Go-Back-N type model. On the other hand, Model 3 is Selective-Repeat type model. We have the following assumptions.

After the receiver have received n data packets, it returns an ACK packet or a NAK (negative acknowledgment) packet by depending on whether all data packets have received correctly or not. The editing time for each ACK or NAK packet needs the time according to a general distribution $B_2(t)$ with finite mean b_2.

If the sender has received a NAK packet, it retransmits data packets that the sender has failed to transmit. If the sender has failed to receive an ACK packet or a NAK packet, then it is judged timeout and it retransmits n data packets. The time to transmit an ACK packet or a NAK packet has a general distribution $D_2(t)$, and denotes that $D(t) \equiv D_1(t) * D_2(t)$ with finite mean d. We assume that the time until the sender judges time out has the same distribution $D(t)$ and the probability that a data packet loss occurs is p. Particularly, in terms of Model 3, the probability that an ACK packet loss or a NAK packet loss occurs is $q = 0$.

Then, the LS transforms of one-step transition probabilities $Q_{ij}(t)$ are given by the following equations:

$$q_{01}(s) = a(s),$$

$$q_{1S}(s) = (1-p)^n b_1(s)^n [b_2(s)d(s)]$$

$$+ \sum_{m_1=1}^{n} \binom{n}{m_1} p^{m_1} (1-p)^n b_1(s)^{n+m_1} [b_2(s)d(s)]^2$$

$$+ \sum_{m_1=1}^{n} \binom{n}{m_1} \sum_{m_2=1}^{m_1} \binom{m_1}{m_2} p^{m_1+m_2}(1-p)^n$$

$$\times b_1(s)^{n+m_1+m_2} [b_2(s)d(s)]^3$$

$$+ \cdots$$

$$+ \sum_{m_1=1}^{n} \binom{n}{m_1} \sum_{m_2=1}^{m_1} \binom{m_1}{m_2} \cdots \sum_{m_{k-1}=1}^{m_{k-2}} \binom{m_{k-2}}{m_{k-1}}$$

$$\times p^{m_1+m_2+\cdots+m_{k-1}}(1-p)^n$$

$$\times b_1(s)^{n+m_1+m_2+\cdots+m_{k-1}} [b_2(s)d(s)]^k,$$

$$q_{1F}(s) = \sum_{m_1=1}^{n} \binom{n}{m_1} \sum_{m_2=1}^{m_1} \binom{m_1}{m_2} \cdots \sum_{m_{k-1}=1}^{m_{k-2}} \binom{m_{k-2}}{m_{k-1}}$$

$$\times p^{m_1+m_2+\cdots+m_{k-1}}(1-p)^{n-m_{k-1}}$$

$$\times b_1(s)^{n+m_1+m_2+\cdots+m_{k-2}} [b_2(s)d(s)]^{k-1}$$

$$\times [1-(1-p)^{m_{k-1}}] b_1(s)^{m_{k-1}} [b_2(s)d(s)],$$

$$q_{F0}(s) = g(s). \tag{14}$$

It is difficult to derive explicitly the mean time $\ell_3(n)$ until n data unit transmissions succeed. We consider the particular case $k = 2$ i.e., if the retransmission has failed 2 times, the sender interrupts the retransmission. In this case, from equation (5),

$$\ell_3(n) \equiv \int_0^\infty t \, dH_{0S}(t) = \lim_{s \to 0} \frac{-d[h_{0S}(s)]}{ds}$$

$$= \frac{a + nb_1(1+p) + (b_2+d)[2-(1-p)^n] + \frac{1}{\mu}}{(1-p^2)n} - \frac{1}{\mu}. \tag{15}$$

Similarly, the mean time when $k = 3$, i.e., if the retransmission has failed 3 times, the sender interrupts the retransmission, is given by

$$
\ell_3(n) = \frac{a + nb_1(1 + p + p^2) + (b_2 + d)\left[3 - \sum_{i-1}^{2}(1 - p^i)^n\right] + \frac{1}{\mu}}{(1 - p^3)^n} - \frac{1}{\mu}.
$$

Moreover, from these analyses, the mean time for any $k \geq 1$ is

$$
\ell_3(n) = \frac{a + nb_1[1 + p(1 - p^{k-2}) + (k - 1)p^{k-1}] + (b_2 + d)\left[k - \sum_{i=1}^{k-1}(1 - p^i)^n\right] + \frac{1}{\mu}}{(1 - p^k)^n} - \frac{1}{\mu}. \tag{16}
$$

4.2   *Optimal Policy*

We discuss an optimal policy which maximizes the throughput $E_3(n)$ in terms of Model 3. We define the throughput $E_3(n)$, which presents the ratio of n data units to their mean transmission times, as the following equation:

$$
E_3(n) \equiv \frac{n}{\ell_3(n)}
$$

$$
= \frac{n(1 - p^k)^n}{a + nb_1[1 + p(1 - p^{k-2}) + (k - 1)p^{k-1}] + (b_2 + d)\left[k - \sum_{i=1}^{k-1}(1 - p^i)^n\right] + \frac{1}{\mu} - \frac{1}{\mu}(1 - p^k)^n}
$$

$$
(n = 1, 2, \cdots). \tag{17}
$$

We seek an optimal number n_3^* which maximizes $E_3(n)$. We put formally that $V_3(n) \equiv 1/E_3(n)$ and seek n_3^* which minimizes $V_3(n)$. From the inequality $V_3(n + 1) - V_3(n) \geq 0$,

$$
n\ell_3(n + 1) - (n + 1)\ell_3(n) \geq 0. \tag{18}
$$

Denoting the left side of (18) by $L_3(n)$,

$$L_3(n+1) - L_3(n) = \frac{n+1}{(1-p^2)^{n+2}} J_3(n),$$

$$L_3(1) = \frac{\left[a + \frac{1}{\mu} + k(b_2 + d)\right](-1 + 2p^k) + 2b_1 p^k [1 + p(1 - p^{k-2}) + (k-1)p^{k-1}] - (b_2 + d)\left[\sum_{i=1}^{k-1}(1 - p^i)(-1 + 2p^k - p^i)\right]}{(1 - p^k)^2} + \frac{1}{\mu},$$

$$L_3(\infty) = \infty,$$

where

$$J_3(n) \equiv ap^{2k} + b_1 p^k (2 + np^k)[1 + p(1 - p^{k-2}) + (k-1)p^{k-1}]$$

$$+ (b_2 + d)\left[kp^{2k} - \sum_{i=1}^{k-1}(1 - p^i)^n (p^k - p^i)^2\right] + \frac{1}{\mu}p^{2k},$$

which is strictly increasing in n. That is, when $J_3(1) > 0$, $L_3(n)$ is strictly increasing in n from $L_3(1)$ to ∞.

Therefore, when $J_3(1) > 0$, we have the following optimal policy:

(i) If $L_3(1) < 0$, then there exists a finite and unique $n_3^*(> 1)$ which satisfies (18).

(ii) If $L_3(1) \geq 0$ then $n_3^* = 1$.

5   Numerical Example

We compute numerically the optimal number which maximizes the through-put. Suppose that the mean time a for connection establishment is a unit time in order to the relative tendency of performance measures.

It is assumed that the mean editing time for an ACK packet or a NAK packet is $b_2/a = 1(\times 10^{-1})$, the mean editing time for a data packet is $b_1 = 2b_2$, the mean time to transmit a data packet and an ACK packet or a NAK packet is $d/a = 1 \sim 4(\times 10^{-1})$, the mean time from the sender interrupts retransmission to restart again is $g/a = 2$, the probability that a data packet loss occurs is $p = 0.01 \sim 0.3$ and the probability that an ACK packet loss occurs is $q = 0.01 \sim 0.3$.

Table 1 Optimal number n_1^* to maximize $E(n)$ and $E(n_1^*)$ when $k = 2$.

d/a ($\times 10^{-1}$)	p	q				
		0.01	0.05	0.1	0.2	0.3
1	0.01	21	21	21	21	21
		3.173	3.075	2.953	2.710	2.455
	0.05	8	8	8	8	8
		1.921	1.865	1.794	1.646	1.486
	0.1	5	5	5	5	5
		1.347	1.309	1.259	1.154	1.039
	0.2	3	3	3	3	3
		0.836	0.812	0.782	0.716	0.644
	0.3	2	2	2	2	2
		0.593	0.577	0.555	0.508	0.456
2	0.01	22	22	22	22	22
		3.115	3.019	2.899	2.660	2.410
	0.05	8	8	8	8	8
		1.853	1.800	1.731	1.588	1.434
	0.1	5	5	5	5	5
		1.288	1.251	1.204	1.103	0.994
	0.2	3	3	3	3	3
		0.792	0.770	0.741	0.679	0.611
	0.3	2	2	2	2	2
		0.559	0.544	0.523	0.479	0.431
4	0.01	23	23	23	23	23
		3.010	2.916	2.801	2.570	2.328
	0.05	8	8	8	8	8
		1.732	1.681	1.617	1.484	1.340
	0.1	5	5	5	5	5
		1.184	1.150	1.107	1.015	0.915
	0.2	3	3	3	3	3
		0.718	0.698	0.672	0.616	0.555
	0.3	2	2	2	2	2
		0.502	0.488	0.469	0.430	0.387

Table 1 gives the optimal number n_1^* which maximizes the throughput and $E_1(n_1^*)$ when $k = 2$. For example, when $d/a = 0.1$, $p = 0.05$ and $q = 0.01$, the optimal number is $n_1^* = 8$ and $E_1(8) = 1.921$. This indicates that n_1^* decreases with p, however, increases with d/a. Table 1 also indicates that n_1^* depends little on q.

Next, in terms of the throughput, we compare Model 1 with Model 2. Table 2 gives the throughput $E_1(n_1^*)$ when $k = 3$ and $d/a = 0.1$ in case of Model 1. Table 3 gives the throughput $E_2(n_2^*)$ when $k = 3$, $m = 3$ and $d/a = 0.1$ in case of Model 2.

These indicate that the throughput of Model 1 is larger than that of Model 2. The difference between $E_1(n_1^*)$ and $E_2(n_2^*)$ decreases with p and q. Especially, when p is small, the difference of throughput is large.

Table 2   Throughput $E_1(n_1^*)$ when $k = 3$ in the case of Model 1.

p	q				
	0.01	0.05	0.1	0.2	0.3
0.01	7.322	6.554	5.775	4.631	3.807
0.05	3.034	2.879	2.701	2.382	2.095
0.1	1.953	1.876	1.784	1.610	1.442
0.2	1.144	1.111	1.069	0.984	0.894
0.3	0.807	0.785	0.758	0.702	0.641

Table 3   Throughput $E_2(n_2^*)$ when $k = 3$ and $m = 3$ in the case of Model 2.

p	q				
	0.01	0.05	0.1	0.2	0.3
0.01	3.139	3.139	3.136	3.116	3.062
0.05	1.935	1.935	1.934	1.921	1.888
0.1	1.392	1.392	1.391	1.381	1.356
0.2	0.896	0.896	0.895	0.889	0.872
0.3	0.647	0.647	0.646	0.642	0.629

Table 4   Throughput $E_2(n_2^*)$ when $m = 1$ and $k = 3$ in the case of Model 2.

p	q				
	0.01	0.05	0.1	0.2	0.3
0.01	3.224	3.108	2.958	2.650	2.328
0.05	2.069	1.992	1.894	1.689	1.478
0.1	1.522	1.463	1.388	1.233	1.072
0.2	1.004	0.963	0.911	0.804	0.696
0.3	0.738	0.707	0.668	0.588	0.508

Table 4 gives the throughput $E_2(n_2^*)$ when $m = 1$, $k = 3$ and $d/a = 0.1$ in the case of Model 2. This indicates that the throughput of Table 4 is smaller than that of Table 2. Compared with Table 3, The throughput of Table 4 is larger than that of Table 3 under the same value of p when q is small, conversely, it is smaller than that of Table 3 when q is large.

Table 5   Optimal number n_3^* to maximize $E_3(n)$ when $k = 2$.

p	$d/a(\times 10^{-1})$		
	1	2	4
0.01	15	15	16
0.05	5	5	5
0.1	3	3	3
0.2	2	2	2

Table 5 gives the optimal number n_3^* which maximizes the throughput when $k = 2$ and $d/a = 0.1$ in the case of Model 3. This indicates that n_3^* decreases with p, however, increases with d/a.

6 Conclusion

We have formulated three stochastic models of a system connected with the radio link, and have discussed the optimal policy which maximizes the throughput until the transmission succeeds.

Moreover, we have given numerical examples and have evaluated them for various standard parameters. From numerical examples, we have shown that the optimal number which maximizes the throughput decreases with the time to transmit a packet and the probability that a packet loss occurs and depends little on the probability that an ACK packet loss occurs. If some parameters are estimated from actual data, we could select the best policy.

It would be very important to evaluate and improve the reliability of a network connected with the radio link. The results derived in this paper would be applied in practical fields by making some suitable modification and extensions. Further studies for such subject would be expected.

References

Bakre, A. and Badrinath, B. R. (1995). *I-TCP: Indirect TCP for mobile hosts*, *Proc. 15th ICDCS*, pp. 136–143.

Balakrishnan, H. and Katz, R. H. (1998). *Explicit loss notification and wireless web performance, Proc. IEEE Globecom Internet Mini-Conference*.

Barlow, R. E. and Proschan, F. (1965). *Mathematical Theory of Reliability*, Wiley, New York.

Itaya, S., Hasegawa, J., Hasegawa, A., Davis, P., Kadowaki, N. and Obana, S. (2005). *Stabilization of Large Ad-hoc Wireless Networks in Unstable Radio Environments, IPS Trans.* **46**, pp. 2848–2856.

Matsuda, T. and Yamamoto, M. (2004). *Recent Research Activities in Wireless TCP IEICE J.* **87**, pp. 589–594.

Miyoshi, M., Sugano, M. and Murata, M. (2001). *A Study for Improving TCP Performance on Radio Terminal, IEICE Japan, Mobile Communication Workshop*.

Miyoshi, M., Sugano, M. and Murata, M. (2002). *Performance Evaluation of TCP Throughput with Consideration on Lower Layer Protocols for Wireless Cellular Networks, IEICE Trans.* **J85-B**, pp. 732–743.

Nakagawa, T. and Osaki, S. (1979). *Bibliography for Availability of Stochastic Systems, IEEE Trans. Reliability*, R-25, **4**, pp. 284–287.

Osaki, S. (1992). *Applied stochastic system modeling* (Springer, Berlin).

Saito, S. and Teraoka, F. (2004). *Implementation, Analysis and Evaluation of TCP-J: A New Version of TCP for Wireless Networks, IPS Trans.* **J87-D-I**, pp. 508–515.

Takahashi, H., Saito, M., Aida, H., Tobe, Y. and Tokuda, H. (2005). *Real Environment Evaluations of a Routing Scheme Based on Estimated TCP Throughput for MANET, IPS Trans.* **46**, pp. 2857–2870.

Yasui, K., Nakagawa, T. and Sandoh, H. (2002). *Reliability models in data communication systems, in Stochastic Models in Reliability and Maintenance*, ed. Osaki, S. pp. 281–301 (Springer, Berlin).

Yuki, T., Yamamoto, T., Sugano, M., Murata, M., Miyahara, H. and Hatauchi, T. (2002). *A Study on Performance Improvement of TCP over an Ad Hoc Network, IEICE Trans.* **J85-B**, pp. 2045–2053.

Zhang, B., Shirazi, M. N. and Tanaka, S. (2004). *Improving Wireless TCP Performance with Explicit Wireless Loss Notification Using MAC-layer Information, IPS Trans.* **45**, pp. 1234–1244.

PART 5

Studies on Reliability and Maintenance

Chapter 19

Studies on Reliability and Maintenance

Toshio Nakagawa

Department of Business Administration,
Aichi Institute of Technology,
1247 Yachigusa, Yakusa-cho, Toyota 470-0392 Japan

1 Introduction

I have mainly studied reliability and maintenance theoretically since 1971. Summarizing our research results, several academic monographs and book chapters have been published. Generally, papers are divided into 8 parts such as (1) Maintenance Policies, (2) Reliability Analysis, (3) Damage Models, (4) Reliability Application, (5) Computer Analysis, (6) Management Models, (7) Failure Distributions, and (8) Other Stochastic Models. This chapter makes my publication list of academic monographs, book chapters and main papers in 8 parts in order of date.

Recently, systems have become much more complex, and old plants and structures have increased rapidly in near future. In addition, advanced nations have almost finished public infrastructure. Maintenance policies for industrial systems and public infrastructure should be properly and quickly established according to their occasions. These results of research would be applied practically to actual reliability systems by modifying and extending them.

Hereafter, we have some plans in view such as Proposal of New Maintenance Policies, Analysis of More Random Systems, Formulation of Software Maintenance Theory, Applications to New Management Models, and New Schemes of Reliability Theory. Such research targets would offer new interesting theoretical topics for future studies.

2   Publication List

2.1   *Academic Monographs*

1. Nakagawa T. (2005) Maintenance Theory of Reliability.   Springer, London.
2. Nakagawa T. (2007) Shock and Damage Models in Reliability Theory. Springer, London.
3. Nakagawa T. (2008) Advanced Reliability Models and Maintenance Policies. Springer, London.
4. Nakamura S. and Nakagawa T. (Eds) (2010) Stochastic Reliability Modeling, Optimization and Applications. World Scientific, Singapore.
5. Nakagawa T. (2011) Stochastic Processes with Applications to Reliability Theory. Springer, London.
6. Dohi T. and Nakagawa T. (Eds) (2013) Stochastic Reliability and Maintenance Modeling. Springer, London.
7. Nakagawa T. (2014) Random Maintenance Policies. Springer, London.

2.2   *Book Chapters*

1. Nakagawa T. (2000) Imperfect preventive maintenance models.   In: Ben-Daya M., Duffuaa S.O. and Raouf A. (Eds) Maintenance, Modeling and Optimization. Kluwer Academic Pub., Boston, 201–214.
2. Nakagawa T. (2002) Imperfect preventive maintenance models. In: Osaki S. (Ed) Stochastic Models in Reliability and Maintenance. Springer, Berlin, 125–140.
3. Nakagawa T. (2002) Two-unit redundant models. In: Osaki S. (Ed) Stochastic Models in Reliability and Maintenance. Springer, Berlin, 165–191.
4. Yasui K., Nakagawa T. and Sandoh H. (2002) Reliability models in data communication systems. In: Osaki S. (Ed) Stochastic Models in Reliability and Maintenance. Springer, Berlin, 281–306.
5. Nakagawa T. (2003) Maintenance and optimum policy. In: Pham H. (Ed) Handbook of Reliability Engineering. Springer, London, 367–395.
6. Nakagawa T. (2006) Statistical models on maintenance. In: Pham H. (Ed) Handbook of Engineering Statistics. Springer, London, 835–848.
7. Arafuka M., Nakamura S., Nakagawa T. and Kondo H. (2006) Optimal interval of CRL issue in PKI architecture. In: Pham H. (Ed) Reliability Modeling, Analysis and Optimization. World Scientific, Singapore, 67–79.

8. Sugiura T., Mizutami S. and Nakagawa T. (2006) Optimal random and periodic inspection policies. In: Pham H. (Ed) Reliability Modeling, Analysis and Optimization. World Scientific, Singapore, 393–403.

9. Mizutani S., Nakagawa T. and Ito K. (2006) Optimal inspection policies for a self-diagnosis system with two types of inspections. In: Pham H. (Ed) Reliability Modeling, Analysis and Optimization. World Scientific, Singapore, 417–428.

10. Ito K. and Nakagawa T. (2006) Maintenance of a cumulative damage model and its application to gas turbine engine of co-generation system. In: Pham H. (Ed) Reliability Modeling, Analysis and Optimization. World Scientific, Singapore, 429–438.

11. Nakagawa T. and Ito K. (2007) Optimal availability models of a phased array radar. In: Dohi T., Osaki S. and Sawaki K. (Eds) Recent Advances in Stochastic Operations Research. World Scientific, Singapore, 115–130.

12. Nakamura S., Arafuka M. and Nakagawa T. (2007) Optimal certificate update interval considering communication costs in PKI. In: Dohi T., Osaki S. and Sawaki K. (Eds) Recent Advances in Stochastic Operations Research. World Scientific, Singapore, 235–244.

13. Nakagawa T. (2008) Replacement and preventive maintenance models. In: Misra K.B. (Ed) Handbook of Performability Engineering. Springer, London, 807–823.

14. Nakagawa T. and Mizutani S. (2008) Periodic and sequential imperfect preventive maintenance policies for cumulative damage models. In: Pham H. (Ed) Recent Advances in Reliability and Quality in Design. Springer, London, 85–99.

15. Mizutani S. and Nakagawa T. (2009) Optimal policy for a two-unit system with two types of inspections. In: Dohi T., Osaki S. and Sawaki K. (Eds) Recent Advances in Stochastic Operations Research II. World Scientific, Singapore, 171–181.

16. Ito K. and Nakagawa T. (2009) Optimal censoring policies for the operation of a damage system. In: Dohi T., Osaki S. and Sawaki K. (Eds) Recent Advances in Stochastic Operations Research II. World Scientific, Singapore, 201–210.

17. Naruse K., Nakagawa T. and Maeji S. (2009) Optimal sequential checkpoint intervals for error detection. In: Dohi T., Osaki S. and Sawaki K. (Eds) Recent Advances in Stochastic Operations Research II. World Scientific, Singapore, 213–224.

18. Nakamura S., Nakagawa T. and Kondo H. (2009) Optimal backup in-

terval of a database system using a continuous damage model. In: Dohi T., Osaki S. and Sawaki K. (Eds) Recent Advances in Stochastic Operations Research II. World Scientific, Singapore, 243–251.

19. Zhao X. and Nakagawa T. (2013) Comparisons of periodic and random replacement policies. In: Frenkel I., Karagrigoriou A., Lisnianski A. and Kleyner A.V. (Eds) Applied Reliability Engineering and Risk Analysis, Probabilistic Models and Statistical Inference, Wiley, New York, 193–204.

2.3 *Papers*

(1) Maintenance Policies

1. Nakagawa T. and Osaki S. (1974) Optimum replacement policies with delay. J. of Applied Probability 11, 102–110.
2. Nakagawa T. and Osaki S. (1974) Optimum repair limit replacement policies. Operational Research Quarterly 25, 311–317.
3. Osaki S. and Nakagawa T. (1975) A note on age replacement. IEEE Tr. on Reliability R-24, 92–94.
4. Nakagawa T. (1976) On scheduling the delivery of spare units. IEEE Tr. on Reliability R-25, 35–37,
5. Nakagawa T. (1977) Optimum preventive maintenance and repair limit policies maximizing the expected earning rate. RAIRO Operations Research 11, 103–108.
6. Mine H. and Nakagawa T. (1977) Interval reliability and optimum preventive maintenance policy. IEEE Tr. on Reliability R-26, 131–133.
7. Nakagawa T. and Osaki S. (1977) Discrete time age replacement policies. Operational Research Quarterly 28, 881–885.
8. Mine H. and Nakagawa T. (1978) Age replacement model with mixed failure times. IEEE Tr. on Reliability R-27, 173.
9. Mine H. and Nakagawa T. (1978) A summary of optimum preventive maintenance policies maximizing interval reliability. J. of Operations Research Soc. of Japan 21, 205–216.
10. Nakagawa T. and Yasui K. (1978) Approximate calculation of block replacement with Weibull failure times. IEEE Tr. on Reliability R-27, 268–269.
11. Nakagawa T. (1979) Replacement problem of a parallel system in random environment. J. of Applied Probability 16, 203–205.
12. Nakagawa T. (1979) The decision to replace a unit early or late in an age replacement problem. Microelectronics and Reliability 19, 265–267.

13. Nakagawa T. (1979) Optimum policies when preventive maintenance is imperfect. IEEE Tr. on Reliability R-28, 331–332.
14. Nakagawa T. (1979) A summary of block replacement policies. RAIRO Operations Research 13, 351–361.
15. Nakagawa T. (1979) Further results of replacement problem of a parallel system in random environment. J. of Applied Probability 16, 923–926.
16. Nakagawa T. (1979) Imperfect preventive-maintenance. IEEE Tr. on Reliability R-28, 402.
17. Nakagawa T. and Yasui K. (1979) Approximate calculation of inspection policy with Weibull failure times. IEEE Tr. on Reliability R-28,403–404.
18. Nakagawa T. (1979) Optimum replacement policies for a used unit. J. of Operations Research Soc. of Japan 22, 338–346.
19. Nakagawa T. (1980) Optimum inspection policies for a standby unit. J. of Operations Research Soc. of Japan 23, 13–26.
20. Nakagawa T. and Yasui K. (1980) Approximate calculation of optimal inspection times. J. of Operations Research Soc. of Japan 31, 851–853.
21. Nakagawa T. (1980) Replacement models with inspection and preventive maintenance. Microelectronics and Reliability 20, 427–433.
22. Nakagawa T. (1980) A summary of imperfect preventive maintenance policies with minimal repair. RAIRO Operations Research 14, 249–255.
23. Nakagawa T. (1980) Mean time to failure with preventive maintenance. IEEE Tr. on Reliability R-29, 341.
24. Nakagawa T. (1980) Replacement policies for a unit with random and wearout failures. IEEE Tr. on Reliability R-29, 342–344.
25. Nakagawa T. and Yasui K. (1981) Calculation of age-replacement with Weibull failure times. IEEE Tr. on Reliability R-30, 163–164.
26. Nakagawa T. (1981) Modified periodic replacement with minimal repair at failure. IEEE Tr. on Reliability R-30, 165–168.
27. Nakagawa T. (1981) A summary of periodic replacement with minimal repair at failure. J. of Operations Research Soc. of Japan 24, 213–227.
28. Nakagawa T. (1981) Generalized models for determining optimal number of minimal repairs before replacement. J. of Operations Research Soc. of Japan 24, 325–337.
29. Nakagawa T. and Yasui K. (1982) Bounds of age replacement time. Microelectronics and Reliability 22, 603–609.
30. Nakagawa T. (1982) A modified block replacement with two variables. IEEE Tr. on Reliability R-31, 398–400.
31. Nakagawa T. and Kowada M. (1983) Analysis of a system with min-

imal repair and its application to replacement policy. European J. of Operations Research 12, 176–182.

32. Nakagawa T. (1983) Optimal number of failures before replacement time. IEEE Tr. on Reliability R-32, 115–116.

33. Nakagawa T. (1983) Combined replacement models. RAIRO Operations Research 17, 193–203.

34. Nakagawa T., Nishi K. and Sawa Y. (1983) Modified periodic preventive maintenance policies. Microelectronics and Reliability 23, 945–951.

35. Shima E. and Nakagawa T. (1984) Optimum inspection policy for a protective device. Reliability Engineering 7, 123–132.

36. Nakagawa T. (1984) Periodic inspection policy with preventive maintenance. Naval Research Logistics Quarterly 31, 33–40.

37. Nakagawa T. (1984) A summary of discrete replacement policies. European J. of Operations Research 17, 382–392.

38. Nakagawa T. (1984) Optimal policy of continuous and discrete replacement with minimal repair at failure. Naval Research Logistics Quarterly 31, 543–550.

39. Nakagawa T. (1985) Continuous and discrete age-replacement policies. J. of Operations Research Soc. 36, 147–154.

40. Nakagawa T. (1986) Periodic and sequential preventive maintenance policies. J. of Applied Probability 23, 536–542.

41. Nakagawa T. (1986) Modified discrete preventive maintenance policies. Naval Research Logistics Quarterly 33, 703–715.

42. Nakagawa T. (1987) Modified, discrete replacement models. IEEE Tr. on Reliability R-36, 243–245.

43. Nakagawa T. and Yasui K. (1987) Optimum policies for a system with imperfect maintenance. IEEE Tr. on Reliability R-36, 631–633.

44. Nakagawa T. (1988) Sequential imperfect preventive maintenance policies. IEEE Tr. on Reliability 37, 295–298.

45. Nakagawa T. (1989) A replacement policy maximizing MTTF of a system with spare units. IEEE Tr. on Reliability 38, 210–211.

46. Nakagawa T. (1989) A summary of replacement models with changing failure distributions. RAIRO Operations Research 23, 343–353.

47. Nakagawa T. and Yasui K. (1991) Periodic-replacement models with threshold levels. IEEE Tr. on Reliability 24, 46–49.

48. Sheu S.H., Kuo C.M. and Nakagawa T. (1993) Extended optimal age replacement policy with minimal repair. RAIRO Operations Research 27, 337–351.

49. Nakagawa T. and Murthy D.N.P. (1993) Optimal replacement policies

for a two-unit system with failure interactions. RAIRO Operations Research 27, 427–438.

50. Sheu S.H., Griffith W.S and Nakagawa T. (1995) Extended optimal replacement model with random minimal repair costs. European J. of Operations Research 85, 636–649.

51. Nakagawa T., Yasui K. and Sandoh H. (2004) Note on optimal partition problems in reliability models. J. of Quality in Maintenance Engineering 10, 282–287.

52. Nakagawa T. and Mizutani S. (2009) A summary of maintenance policies for a finite interval. Reliability Engineering & System Safety 94, 89–96.

53. Nakagawa T. and Mizutani S. (2009) Optimum problems in backward times of reliability models. IIE Transactions 41, 1–7.

54. Chen M., Nakamura S. and Nakagawa T. (2010) Replacement and preventive maintenance models with random working times. IEICE Trans. Fundamentals E93-A, 500–507.

55. Yun W.Y. and Nakagawa T. (2010) Replacement and inspection policies for products with random life cycle. Reliability Engineering & System Safety 95, 161–165.

56. Chen M., Mizutani S. and Nakagawa T. (2010) Random and age replacement policies. International J. of Reliability, Quality and Safety Engineering 17, 27–39.

57. Chen M., Mizutani S. and Nakagawa T. (2010) Optimal backward and backup policies in reliability theory. J. of Operations Research Soc. of Japan 53, 101–118.

58. Nakagawa T., Mizutani S. and Chen M. (2010) A summary of periodic and random inspection policies. Reliability Engineering & System Safety 95, 906–911.

59. Nakagawa T., Zhao X. and Yun W.Y. (2011) Optimal age replacement and inspection policies with random failure and replacement times. International J. of Reliability, Quality and Safety Engineering 18, 1–12.

60. Nakagawa T. and Zhao X. (2012) Optimization problems of a parallel system with a random number of units. IEEE Tr. on Reliability 61, 543–548.

61. Zhao X. and Nakagawa T. (2012) Optimization problems of replacement first or last in reliability theory. European J. of Operational Research 223, 141–149.

62. Nakagawa T. and Zhao X. (2013) Comparisons of replacement policies

with constant and random times. J. of Operations Research Soc. of Japan 56, 1–14.

(2) Reliability Analysis

1. Osaki S. and Nakagawa T. (1971) On a two-unit standby redundant system with standby failure. Operations Research 19, 510–523.
2. Nakagawa T. and Goel A.L. (1973) A note on availability for a finite interval. IEEE Tr. on Reliability R-22, 271–272.
3. Nakagawa T. and Osaki S. (1974) Stochastic behavior of a two-unit standby redundant system. INFOR 12, 66–70.
4. Nakagawa T. (1974) The expected number of visits to state k before a total system failure of a complex system with repair maintenance. Operations Research 22, 108–116.
5. Nakagawa T. and Osaki S. (1974) Stochastic behavior of a two-dissimilar-unit standby redundant system with repair maintenance. Microelectronics and Reliability 13, 143–148.
6. Nakagawa T. and Osaki S. (1974) Optimum preventive maintenance policies for a 2-unit redundant system. IEEE Tr. on Reliability R-23, 86–91.
7. Nakagawa T. and Osaki S. (1974) Off time distributions in an alternating renewal process with reliability applications. Microelectronics and Reliability 13, 181–184.
8. Nakagawa T. and Osaki S. (1974) Optimum preventive maintenance policies maximizing the mean time to the first system failure for a two-unit standby redundant system. Optimization Theory and Applications 14, 115–129.
9. Nakagawa T. and Osaki S. (1974) Combining drift and catastrophic failure modes. IEEE Tr. on Reliability R-23, 278–279.
10. Nakagawa T. and Osaki S. (1975) On a terminating renewal process with reliability applications. IEEE Tr. on Reliability R-24, 88–90.
11. Nakagawa T. and Osaki S. (1975) Stochastic behavior of 2 unit standby redundant systems with imperfect switchover. IEEE Tr. on Reliability R-24, 143–146.
12. Nakagawa T., Goel A.L. and Osaki S. (1975) Stochastic behavior of an intermittently used system. RAIRO Operations Research 2, 101–112.
13. Nakagawa T. and Osaki S. (1975) Stochastic behavior of a two-unit priority standby redundant system with repair. Microelectronics and Reliability 14, 309–313.

14. Nakagawa T. and Osaki S. (1975) The busy period of a repairman for redundant repairable systems. RAIRO Operations Research 3, 69–73.
15. Nakagawa T. and Osaki S. (1975) Stochastic behavior of two-unit parallel redundant systems with repair maintenance. Microelectronics and Reliability 14, 457–461.
16. Nakagawa T. and Osaki S. (1975) Applications of the sojourn-time problem to reliability. IEEE Tr. on Reliability R-24, 301–302.
17. Nakagawa T. and Osaki S. (1975) Stochastic behavior of a 2-unit parallel fuel charging system. IEEE Tr. on Reliability R-24, 302–304.
18. Nakagawa T. and Osaki S. (1976) Reliability analysis of a one-unit system with unrepairable spare units and its optimization applications. Operational Research Quarterly 27, 101–110.
19. Suzuki Y., Nakagawa T. and Sawa Y. (1976) Reliability analysis of intermittently used systems when failures are detected only during a usage period. Microelectronics and Reliability 15, 35–38.
20. Nakagawa T. and Osaki S. (1976) Analysis of a repairable system which operates at discrete times. IEEE Tr. on Reliability R-25, 110–112.
21. Nakagawa T. and Osaki S. (1976) Joint distribution of uptime and downtime for some repairable systems. J. of Operations Research Soc. of Japan 19, 209–216.
22. Osaki S. and Nakagawa T. (1976) Bibliography for reliability and availability of stochastic systems. IEEE Tr. on Reliability R-25, 284–287.
23. Nakagawa T. and Osaki S. (1976) A summary of optimum preventive maintenance policies for a two-unit standby redundant system. ZOR Operations Research 20, 171–187.
24. Nakagawa T. and Osaki S. (1976) Markov renewal processes with some non-regeneration points and their applications to reliability theory. Microelectronics and Reliability 15, 633–636.
25. Mine H. and Nakagawa T. (1976) Stochastic behavior of two-unit redundant systems which operate at discrete times. Microelectronics and Reliability 15, 551–554.
26. Nakagawa T. (1977) A 2-unit repairable redundant system with switching failure. IEEE Tr. on Reliability R-26, 128–130.
27. Nakagawa T. and Yasui K. (1977) Approximate calculation of system availability. IEEE Tr. on Reliability R-26, 133–134.
28. Nakagawa T. (1977) Optimum preventive maintenance policies for repairable systems. IEEE Tr. on Reliability R-26, 168–173.
29. Nakagawa T. (1978) Reliability analysis of standby repairable sys-

tems when an emergency occurs. Microelectronics and Reliability 17, 461–464.

30. Nakagawa T. (1984) Optimum number of units for a parallel system. J. of Applied Probability 21, 431–436.

31. Nakagawa T. (1985) Optimization problems in k-out-of-n systems. IEEE Tr. on Reliability R-34, 248–250.

32. Yasui K., Nakagawa T. and Osaki S.(1988) A summary of optimum replacement policies for a parallel redundant system. Microelectronics and Reliability 28, 635–641.

33. Teramoto K., Nakagawa T. and Motoori M. (1990) Optimal inspection policy for a parallel redundant system. Microelectronics and Reliability 30, 151–155.

34. Nakagawa T. and Qian C.H. (2002) Note on reliability of series-parallel and parallel-series systems. J. of Quality in Maintenance Engineering 8, 274–280.

35. Nakagawa T. and Yasui K. (2003) Note on reliability of a system complexity. Mathematical Computing and Modelling 38, 1365–1371.

36. Nakagawa T. and Yasui K. (2003) Note on reliability of a system complexity. J. of Quality in Maintenance Engineering 9, 83–91.

37. Nakagawa T. and Yasui K. (2005) Note on optimal redundant policies for reliability models. J. of Quality in Maintenance Engineering 11, 82–96.

(3) Damage Models

1. Nakagawa T. and Osaki S. (1974) Some aspects of damage models. Microelectronics and Reliability 13, 253–257.

2. Nakagawa T. (1975) On cumulative damage with annealing. IEEE Tr. on Reliability R-24, 90–91.

3. Nakagawa T. (1976) On a cumulative damage model with N different components. IEEE Tr. on Reliability R-25, 112–114.

4. Nakagawa T. (1976) On a replacement problem of a cumulative damage model. Operational Research Quarterly 27, 895–900.

5. Nakagawa T. and Kijima M. (1989) Replacement policies for a cumulative damage model with minimal repair at failure. IEEE Tr. on Reliability 38, 581–584.

6. Kijima M. and Nakagawa T. (1992) Replacement policies of a shock model with imperfect preventive maintenance. European J. of Operations Research 12, 176–182.

7. Satow T., Yasui K. and Nakagawa T. (1996) Optimal garbage collection policies for a database in a computer system. RAIRO Operations Research 30, 359–372.

8. Satow T., Yasui K. and Nakagawa T. (1996) Optimal garbage collection policies for a database with random threshold level. Electronics and Communications in Japan 79, 31–40.

9. Satow T. and Nakagawa T. (1997) Three replacement models with two kinds of damage. Microelectronics and Reliability 37, 909–913.

10. Satow T. and Nakagawa T. (1997) Replacement policies for a shock model with two kinds of damage. Lecture Notes in Economics and Mathematical System 445, 188–195.

11. Satow T. and Nakagawa T. (1997) Optimal replacement policy for a cumulative damage model with deteriorated inspection. International J. of Reliability, Quality and Safety Engineering 4, 387–393.

12. Qian C.H., Nakamura S. and Nakagawa T. (1999) Cumulative damage model with two kinds of shocks and its application to the backup policy. J. of Operational Research Soc. of Japan 42, 501–511.

13. Satow T., Teramoto K. and Nakagawa T. (2000) Optimal replacement policy for a cumulative damage model with time deterioration. Mathematical and Computer Modelling 31, 313–319.

14. Qian C.H., Nakamura S. and Nakagawa T. (2000) Replacement policies for cumulative damage model with maintenance cost. Scientiae Mathematicae 3, 117–126.

15. Qian C.H., Ito K. and Nakagawa T. (2005) Optimal preventive maintenance policies for a shock model with given damage level. J. of Quality in Maintenance Engineering 11, 216–227.

16. Zhao X., Qian C.H. and Nakagawa T. (2009) Study on preventive software rejuvenation policy for two kinds of bugs. J. of System Science and Information 7, 103–110.

17. Zhao X. and Nakagawa T. (2010) Optimal replacement policies for damage models with the limit number of shocks. International J. of Reliability and Quality Performance 2, 13–20.

18. Zhao X., Nakamura S. and Nakagawa T. (2011) Two generational garbage collection models with major collection time. IEICE Transactions Fundamentals E94-A, 1558–1566.

19. Zhao X., Zhang H., Qian C.H., Nakagawa T. and Nakamura S. (2012) Replacement models for combining additive independent damages. International J. of Performability Engineering 8, 91–100.

20. Zhao X., Nakagawa T. and Qian C.H. (2012) Optimal imperfect preven-

tive maintenance policies for a used system. International J. of Systems Science 43, 1632–1641

21. Zhao X., Nakamura S. and Nakagawa T. (2012) Optimal tenuring and major collection times for a generational garbage collector. Asia-Pacific J. of Operational Research 29, 1240018 (17 pages).

18. Nakamura S., Zhao X. and Nakagawa T. (2013) Stochastic modeling of database backup policy for a computer system. J. of Software Engineering and Applications 6, 53–58.

19. Zhao X., Qian C.H. and Nakagawa T. (2013) Optimal policies for cumulative damage models with maintenance last and first. Reliability Engineering & System Safety 110, 50–59.

20. Zhao X., Nakamura S. and Nakagawa T. (2013) Optimal maintenance policies for cumulative damage models with random working times. J. of Quality in Maintenance Engineering 19, 25–37

(4) Reliability Applications

1. Nakagawa T. (1976) On scheduling the delivery of spare units. IEEE Tr. on Reliability R-25, 35–37.

2. Nakagawa T. and Osaki S. (1978) Optimum ordering policies with lead time for an operating unit. RAIRO Operations Research 12, 383–393.

3. Ito K. and Nakagawa T. (1992) Optimal inspection policies for a system in storage. Computers & Mathematics with Applications 24, 87–90.

4. Ito K., Nakagawa T. and Nishi K. (1995) Extended optimal inspection policies for a system in storage. Mathematical Computing and Modelling 22, 83–87.

5. Ito K. and Nakagawa T. (1995) An optimal inspection policy for a storage system with three types of hazard rate functions. J. of Operations Research Soc. of Japan 38, 423–441.

6. Ito K. and Nakagawa T. (1995) An optimal inspection policy for a storage system with high reliability. Microelectronics and Reliability 35, 875–886.

7. Ito K., Teramoto K. and Nakagawa T. (1997) Optimal times of burn-in tests for multilevel assembly times. Lecture Notes in Economics and Mathematical Systems 445, 236–245.

8. Ito K. and Nakagawa T. (2000) Optimal inspection policies for a storage system with degradation at periodic tests. Mathematical Computer Modelling 31, 191–195.

9. Sandoh H. and Nakagawa T. (2003) How much should we reweigh ? J. of Operations Research Soc. 54, 318–321.
10. Ito K. and Nakagawa T. (2003) Optimal self-diagnosis policy for FADEC of gas turbins engines. Mathematical and Computing Modeling 38, 1243–1248.
11. Ito K. and Nakagawa T. (2004) Comparison of cyclic and delayed maintenances for a phased array radar. J. of Operations Research Soc. of Japan 47, 51–61.
12. Sandoh H., Igaki N. and Nakagawa T. (2004) Inspection policy for a scale considering accidental detections. J. of Quality in Maintenance Engineering 10, 148–153.
13. Nakagawa T. and Ito K. (2008) Optimal maintenance policies for a system with multiechelon risks. IEEE Tr. on System, Man, and Cybernetics 38, 461–469.
14. Qian C.H., Chen J. and Nakagawa T. (2009) Comparison of two information structures with noise and its application to Bays decision analysis. Quality Technology & Quantitative Management 6, 1–10.
15. Chen M. and Nakagawa T. (2012) Optimal scheduling of random works with reliability applications. Asia-Pacific J. of Operational Research 29, 1250027 (14 pages).
16. Chen M. and Nakagawa T. (2013) Optimal redundant systems for works with random processing time. Reliability Engineering & System Safety 116, 99–104.

(5) Computer Analysis

1. Nakagawa T. (1982) Reliability analysis of a computer system with hidden failure. Policy and Information 6, 43–49.
2. Nakagawa T., Nishi K. and Yasui K. (1984) Optimum preventive maintenance policies for a computer system with resrart. IEEE Tr. on Reliability R-33, 272–276.
3. Yasui K. and Nakagawa T. (1989) A simple fault-diagnosis system with periodic testing. IEEE Tr. on Reliability 38, 571–572.
4. Nakagawa T. and Yasui K. (1989) Optimal testing-policies for intermittent faults. IEEE Tr. on Reliability 38, 577–580.
5. Nakagawa T., Motoori M. and Yasui K. (1990) Optimal testing policy for a computer system with intermittent faults. Reliability Engineering and System Safety 27, 213–218.

6. Nakagawa T., Yasui K. and Motooti M. (1990) Optimal fault margin in a computer system. Microelectronics and Reliability 30, 1117–1121.

7. Yasui K., Nakagawa T. and Koike S. (1992) Reliability consideration on error control policies for a data communication system. Computers & Mathematics with Applications 24, 51–55.

8. Hayashi I., Ishii N., Sandoh H. and Nakagawa T. (1995) Optimal charging times of a battery for memory backup. Mathematical Computing and Modelling 22, 71–75.

9. Koike S., Nakagawa T. and Yasui K. (1995) Optimal block length for basic mode data transmission control procedure. Mathematical Computing and Modelling 22, 161–171.

10. Yasui K. and Nakagawa T. (1995) Reliability consideration of a selective-repeat ARQ policy for a data communication system. Microelectronics and Reliability 35, 41–44.

11. Sawada K., Sandoh H. and Nakagawa T. (1998) A study on ARQ policies for data transmission based on Kullback-Leibler information. International J. of Reliability, Quality and Safety Engineering 5, 5–13.

12. Sandoh H., Hirakoshi H. and Nakagawa T. (1998) A new modified discrete preventive maintenance policy and its application to hard disk management. J. of Quality in Maintenance Engineering 4, 284–290.

13. Yasui K., Nakagawa T. and Imaizumi M. (1998) Reliability evaluations of hybrid ARQ policies for a data communication system. International J. of Reliability, Quality and Safety Engineering 5, 15–28.

14. Imaizumi M., Yasui K. and Nakagawa T. (1998) Reliability analysis of microprocessor systems with watchdog processors. J. of Quality in Maintenance Engineering 4, 263–272.

15. Imaizumi M., Yasui K. and Nakagawa T. (2000) An optimal number of microprocessor units with watchdog processor. Mathematical and Computer Modeling 31, 183–190.

16. Qian C.H., Pan Y. and Nakagawa T. (2002) Optimal policies for a database system with two backup schemes. RAIRO Operations Research 36, 227–235.

17. Imaizumi M., Yasui K. and Nakagawa T. (2003) Optimal reset number of microprocessor system with network processing. Computer and Mathematics with Applications 46, 1047–1054.

18. Imaizumi M., Yasui K. and Nakagawa T. (2003) Reliability of a job execution process using signature. Mathematical and Computer Modelling 38, 1219–1223.

19. Kimura M., Yasui K., Nakagawa T. and Ishii N. (2003) Optimal check-